Processing, Properties and Applications of Biopolymer Composites

Processing, Properties and Applications of Biopolymer Composites

Edited by Lauren Dixon

www.statesacademicpress.com

States Academic Press,
109 South 5th Street,
Brooklyn, NY 11249, USA

Visit us on the World Wide Web at:
www.statesacademicpress.com

ISBN: 978-1-63989-720-9

Trademark Notice: Registered trademark of products or corporate names are used only for explanation and identification without intent to infringe.

Cataloging-in-Publication Data

Processing, properties and applications of biopolymer composites / edited by Lauren Dixon.
        p. cm.
Includes bibliographical references and index.
ISBN 978-1-63989-720-9
1. Biopolymers. 2. Polymeric composites. I. Dixon, Lauren.
QP801.B69 P76 2023
572.33--dc23

Table of Contents

Preface

In my initial years as a student, I used to run to the library at every possible instance to grab a book and learn something new. Books were my primary source of knowledge and I would not have come such a long way without all that I learnt from them. Thus, when I was approached to edit this book; I became understandably nostalgic. It was an absolute honor to be considered worthy of guiding the current generation as well as those to come. I put all my knowledge and hard work into making this book most beneficial for its readers.

Biopolymers refer to the type of polymers created by living organisms. Biopolymer composites are biodegradable composites that are manufactured by reinforcing natural fibers derived from plant and animals sources with natural or synthetic biopolymers. Natural fiber in discontinuous phase is added to the biopolymer matrix in continuous phase to improve the tensile strength and stiffness of the manufactured composite. The goal of making this type of composite is to create a product with strong mechanical and durability properties, which are provided through natural fiber and biopolymers. Natural fiber-reinforced biopolymer composites are quickly developing as potential alternative to synthetic-based materials in food packaging, sporting goods, automotive, aerospace, biomedical, furniture, and electronic industries. This book aims to shed light on the processing, properties, and applications of biopolymer composites. It consists of contributions made by international experts. This book, with its detailed analyses and data, will prove immensely beneficial to professionals and students involved in the study and application of biopolymers at various levels.

I wish to thank my publisher for supporting me at every step. I would also like to thank all the authors who have contributed their researches in this book. I hope this book will be a valuable contribution to the progress of the field.

Editor

Fabrication of Porous Materials from Natural/Synthetic Biopolymers and their Composites

Udeni Gunathilake T.M. Sampath [1], Yern Chee Ching [1,*], Cheng Hock Chuah [2], Johari J. Sabariah [2] and Pai-Chen Lin [3]

[1]   Department of Mechanical Engineering, Faculty of Engineering, University of Malaya, 50603 Kuala Lumpur, Malaysia; sampath@gmail.com
[2]   Department of Chemistry, Faculty of Science, University of Malaya, 50603 Kuala Lumpur, Malaysia; chchuah@um.edu.my (C.H.C.); sabariah@gmail.my (J.J.S.)
[3]   Department of Mechanical Engineering, National Chung Cheng University, 621 Chiayi Country, Taiwan; impcl@ccu.edu.tw
*   Correspondence: chingyc@um.edu.my

Academic Editor: Rafael Luque Alvarez de Sotomayor

Abstract: Biopolymers and their applications have been widely studied in recent years. Replacing the oil based polymer materials with biopolymers in a sustainable manner might give not only a competitive advantage but, in addition, they possess unique properties which cannot be emulated by conventional polymers. This review covers the fabrication of porous materials from natural biopolymers (cellulose, chitosan, collagen), synthetic biopolymers (poly(lactic acid), poly(lactic-*co*-glycolic acid)) and their composite materials. Properties of biopolymers strongly depend on the polymer structure and are of great importance when fabricating the polymer into intended applications. Biopolymers find a large spectrum of application in the medical field. Other fields such as packaging, technical, environmental, agricultural and food are also gaining importance. The introduction of porosity into a biomaterial broadens the scope of applications. There are many techniques used to fabricate porous polymers. Fabrication methods, including the basic and conventional techniques to the more recent ones, are reviewed. Advantages and limitations of each method are discussed in detail. Special emphasis is placed on the pore characteristics of biomaterials used for various applications. This review can aid in furthering our understanding of the fabrication methods and about controlling the porosity and microarchitecture of porous biopolymer materials.

Keywords: natural biopolymers; synthetic biopolymers; fabrication; biocomposites; porosity; sustainable

1. Introduction

Biopolymers and their derivatives are abundant, varied, and important for living beings, they exhibit special properties and have greater importance for miscellaneous applications. These properties and the possibility of formation of these substances using renewable resources make biopolymers a popular initiative in industrial applications. In recent years, there has been an increasing interest in the use of biodegradable materials for packaging, medicine, agriculture, and other areas [1,2].

Biopolymers consist of monomeric units covalently bonded to form macromolecules. There are primarily two classes of biopolymers, namely, natural biopolymers and synthetic biopolymers. Natural biopolymers are obtained from living organisms and the synthetic biopolymers represent the macromolecules synthesized with biomolecules. Natural biopolymers are further divided into polysaccharides, proteins, polynucleotides, polyisoprenes, and polyesters. Synthetic biopolymers can be classified according to the way of preparation such as, biopolymers synthesized by addition and condensation polymerization reaction are listed separately [3]. Biocomposite materials are materials made from two or more constituent biomaterials that result in significant properties than those of the characteristics of individual components. Biodegradability and other properties of biopolymers strongly depend on the polymer structure. The properties of a polymer can be categorized into three broad classes: (1) intrinsic properties, which are inherent to the polymer itself; (2) processing properties, which refer to the behavior of material during forming; and (3) product properties in principle determined by combinations of intrinsic and processing properties. The practitioner needs more detailed information about processing properties such as viscosity, melt strength, melt flow index at the various stages of production [4,5].

Many of the applications of biopolymers can be found in the medical field, such as drug delivery systems, surgical implant devices, wound closure and healing products due to possessing of certain properties like, biocompatibility, biodegradation to non-toxic end products, high bio-activity, low antigenicity, ability to support cell growth and proliferation with appropriate mechanical properties, processability to complex shapes with appropriate porosity, as well as maintaining mechanical strength. Due to the film forming and barrier properties, biopolymers are widely used for applications including food containers, soil retention sheeting, agriculture film, waste bags and use as packaging material in general. They are also popular in areas such as automotive development, hazardous waste removal, paper industries and development of new building materials [6].

The introduction of porosity into a biomaterial broadens the scope of applications. Porous materials made from biopolymers having properties such as biocompatibility and biodegradability are of special interest for medical, cosmetic, pharmaceutical, and other applications. Three-dimensional scaffolds for tissue engineering, "green" packaging, delivery matrices and eco-friendly insulating materials are only a few examples of these applications [7].

Total porosity is one of the important structural parameters of a matrix to be used as a porous biomaterial. Total porosity is defined as the ratio of the total pore volume, to the overall (or bulk) volume. However, for certain applications, total porosity alone does not have a direct impact on its features. Pore size and pore interconnectivity are more important. Usually, high total porosity is accomplished by poor mechanical properties. Inside a porous biopolymer, there may be closed (isolated) pores and open (connected) pores. Pore interconnectivity is important for the accessiblity of gas, liquid, and particulate suspensions. Pore interconnectivity is defined as the ratio of the pore volume accessible from the matrix surrounding by a sphere of known diameter, to the total pore volume.

Apart from the total porosity and the pore interconnectivity, pore size and pore size distribution are also important for most kinds of applications. Pore size <10 μm, defined as micropores and pore size >50 μm considered as macropores.

Figure 1 shows the porous features of a porous biomaterial. Micropores usually found in the struts of the porous biomaterials and the connected macropores are often irregularly distributed along the pathway. Therefore, there are two distinct apertures, namely throat and stomach. Pore morphology has a greater impact on tissue engineering applications as cell/tissue ingrowth behavior depend on porous structure. If the throat size is very small, cells and/or tissues are not able to penetrate or grow into the pores. Various methods have been developed for the fabrication of biomaterials. However, some processing methods do not guarantee high pore interconnectivity and large throat, even though a high total porosity is ensured [8].

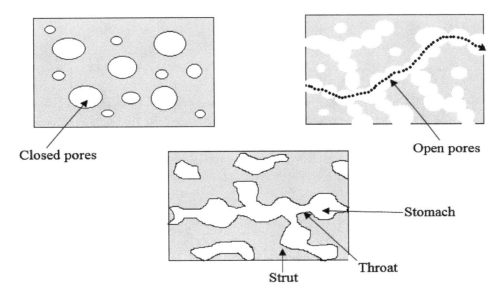

Figure 1. Porous features of porous biomaterial.

Porous scaffolds can be manufactured using biopolymers with specific surface-area-to-volume ratio, crystallinity, pore size, and porosity. The three-dimensional porous scaffolds with enhanced porosities having a homogeneous interconnected pore network are required for tissue engineering applications. Ideal pore sizes vary for different cells and tissues. Porous controlled-release systems should contain pores that are sufficiently large to enable diffusion of the drug [9].

There are many techniques used to manufacture porous polymers. Those can be divided into two categories, designed manufacturing techniques and non-designed manufacturing techniques. Non-designed manufacturing techniques include freeze-drying or emulsion freezing, melt molding, phase separation, solvent casting or particulate leaching, gas foaming or high-pressure processing, electrospinning and combination of these techniques. Designed manufacturing technique includes 3D printing, rapid prototyping of solid free-form technologies. A modern method for creating porous structures using biodegradable fibers by electrospinning is the latest development in this field [9].

This paper aims to present a review regarding different fabrication methods used for the porous fabrication of biopolymers such as chitosan, cellulose, collagen, PLA, poly(lactic-*co*-glycolic acid) and their composite materials. Fabrication methods, including the basic and conventional techniques to the more recent ones, are tabulated. Despite significant development in fabrication methods, no single technique exceeds all the others, so this review will summarize the advantages and limitations of each method in detail.

2. Fabrication of Natural Biopolymers

2.1. Chitosan

Chitosan is produced by deacetylation of chitin, which is present in the exoskeleton of crustaceans (such as crabs and shrimps) and the bone plates of cuttlefish and squids. Chitin is recognized as second most abundant biopolymer in nature. Also, it is a major constituent of the cell wall of fungi. Due to the biological and mechanical properties of chitosan, it has been used to produce powder, hydrogels, membranes, fibers, porous scaffolds and beads that have been investigated with various biological and medical applications. High adaptability of chitosan for a vast range of applications is due to a high degree of chemical reactive amino groups present in D-glucosamine residues. When compared with the higher deacetylate chitosan scaffolds, lower deacetylate chitosan scaffolds possess smaller pore sizes, higher mechanical strength, moderate swelling properties and greater cellular activities [10]. The structure of chitosan is shown in Figure 2.

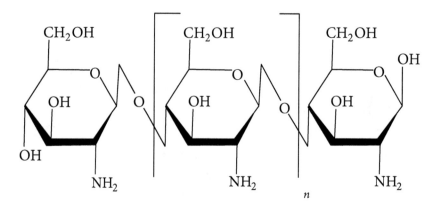

Figure 2. Chemical structure of chitosan [poly-(β-1/4)-2-amino-2-deoxy-D-glucopyranose].

Chitosan porous membranes can be produced to be used as scaffolds by thermally induced phase separation method, in which temperature is decreased to freezing conditions, to induce the phase separation of the polymer. The pore size of the scaffold varied according to the temperature and water content. Low temperature and high water content will result smaller pore sizes. Hydrated porous chitosan membranes have a high surface area and volume when compared with the nonporous chitosan membranes, but their elasticity and resistant to the fraction is smaller than the nonporous membranes. Resistant and elasticity can be improved by the addition of glutaraldehyde, polyethylenglycol, heparin, or collagen, but this can make chitosan insoluble in acid solutions and to form closed pore structures [11,12].

Nwe, Furuike and Tamura [10] reported that porosity of the chitosan scaffolds produced by fungal mycelia was greater than that of the chitosan obtained from crab shells and squid bone plates. Chitosan from shrimp shells possessed the lowest porosity. During the scaffold construction process, when chitosan dissolved with acetic acid the molecular alignment disappeared and with freeze-drying step the formation of hydrogen bonds between polymer chains caused interconnecting of a network of pores. It was also observed that polygonal pores resulted when low molecular weight chitosan is used and elongated pores ensued from chitosan with high molecular weight. Porosity, pore size, and distribution pattern of pores highly affected the mechanical properties, water absorption, and water vapor permeability of the scaffold.

3D micro-porous chitosan scaffolds were produced by dissolving chitosan in acetic acid, followed by stirring and finally adding the solution into a NaOH solution. Here the foaming is achieved by mechanical stirring without adding a chemical foaming agent. It is mentioned that pore diameter of chitosan scaffold decreases with increasing the stirring rate of the sample due to the high shear force of the homogenizer's cutting head. Also, the pore diameter decreased with decreasing the concentration of chitosan in the solution. Concentrations of chitosan solutions less than 1%, lowered the sample's viscosity for foaming and on the other hand concentrations greater than 3%, caused sample's viscosity too high and stick to the homogenizer's cutting heads which resulted in difficulties in mixing [13].

Porous chitosan matrices can be used for the removal of heavy metals from contaminated water. Chitosan and chitosan modified with glutaraldehyde microparticles were prepared for Pb(II) biosorption and compared the properties such as morphology of particles, solubility and pore characteristics. It is reported that glutaraldehyde crosslinked chitosan produces more interconnected pores than that of chitosan microparticles. Porosity increased with the proportion of glutaraldehyde used for the crosslinking. The proportion of the crosslinking agent was found to have no effect on particle size. However, the surface morphology was found to be changed with the crosslinking agent [14].

Formation of a nonporous skin layer is a barrier to the cell growth in tissue engineering scaffolds. Chitosan hydrogels were fabricated with dense gas CO_2 and porous structure on top, bottom surface and in cross sections were obtained without formation of a nonporous skin layer. Due to the high porosity, the crosslinked chitosan hydrogel fabricated under dense gas conditions showed higher equilibrium swelling ratio than the samples fabricated under atmospheric conditions. Hydrogels exhibited equilibrium swelling ratios of 17.2 ± 0.8 and 10.3 ± 0.4 at the dense gas condition and atmospheric condition, respectively [15].

Low viscosity and high diffusivity of supercritical CO_2 offers fabrication of highly porous tissue engineering scaffolds. Chitosan hydrogel was fabricated using supercritical CO_2 for the implantation of osteoblast cells. Hydrogel fabricated under 250 bar, 45 °C, 2 h at 5 g/min CO_2 flow rate yielded 87.03% porosity which was similar to lyophilization (88.68%) operated at 55 °C for 48 h. Even though the porosity is similar for both conditions, the fabrication using supercritical CO_2 found to be more time and energy efficient [16].

Macroporous chitosan scaffolds were developed to load either with bone morphogenetic proteins (BMP-2) or insulin-like growth factor (IGF-1) to study the bone healing property in vivo. Porosity was developed by mechanical stirring the chitosan solution using homogenizer and transferring the bubbled solution into sodium hydroxide solution to conduct liquid hardening process. Chitosan scaffold with pore sizes from 70 to 900 μm was obtained using liquid hardening method. The absorption efficiency of IGF-1 and BMP-2 was found to be $90\% \pm 2\%$ and $87\% \pm 2\%$, respectively. In vivo studies revealed that chitosan scaffolds loaded with IGF-1 showed significant osteoblastic differentiation than BMP-2-loaded chitosan scaffolds [17].

2.2. Cellulose

Cellulose is the most abundant organic polymer on earth. It is an important structural component of the primary cell wall of plant cells and tissues [18]. The building block of the cellulose polymer is monosaccharide glucose molecules. Polymer consists of repeated glucose units attached together by β-1,4 glycosidic linkages as shown in Figure 3. β-1,4 glycosidic bond is formed by covalent bonding of oxygen to the C1 of one glucose ring and the C4 of the connecting ring. Three hydroxyl groups containing in the repeating unit and their ability to make hydrogen bonds between cellulose chains responsible for the physical properties of cellulose [19–21].

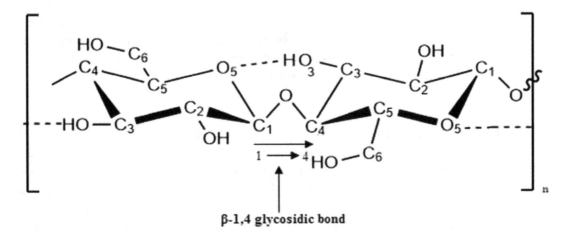

β-1,4 glycosidic bond

Figure 3. β-1,4 glycosidic bond of a cellulose unit.

Cellulose can be synthesized by fungi and some species of bacteria (*Acetobacter xylinum*). Bacterial cellulose is similar to plant cellulose in chemical structure, but the lack of contaminant molecules (lignin and hemicelluloses). Hence, it does not require intensive purification methods. Due to the significance of mechanical strength and biocompatibility, cellulose is widely used in tissue engineering applications [22].

Cellulose is used as a raw material in industries such as veterinary, foods, fibers, textile, wood, paper, cosmetic and pharmaceuticals. Derivatives of cellulose also play a major role in the applications of fibers, textiles, coatings, thermoplastic films, pharmaceutical technologies and as food additives [21].

Development of highly porous structure in cellulose is important because of their potential uses in biomedical applications, filtration, controlled flow of fluids, aircraft, automotive, building and packaging industries [23].

Foams are closed pore structures with the presence of cavities that are not interconnected. These foam structures used to enhance lightness, impact strength, softness, and thermal insulating properties in automotive, aircraft, building and packaging industries. The porous structure can be produced by introducing a blowing agent into the polymer solution. Supercritical CO_2 is widely used in medical applications as CO_2 can be completely removed from the product. In supercritical CO_2 process, rapid depressurization rates cause homogeneous pore distributions with closed pores. Decreasing the depressurization rates will produce wide and large pore size distributions, and more interconnected pores [23,24].

Direct addition of antimicrobial agents to initial food formulations may decrease its concentrations on the food surface due to its diffusion into inner parts of the food. Also, it may cause neutralization of the agent due to interaction with the constituents of food. Controlled release of antimicrobial agent can be achieved by loading it into the food packaging material. Baldino, et al. [25] produced cellulose acetate antimicrobial membrane using supercritical CO_2 as pore forming agent. Cellulose acetate dissolved with the antimicrobial agent put into the membrane-preparing vessel and filled with supercritical CO_2. The vessel was then flushed with CO_2 and depressurized for about 30 min. They observed that the mean pore size decreased with increasing the operative pressure. It was also observed that the pore size decreased with decreasing the operative temperature.

Freeze-drying process and introducing of pore forming chemicals were successfully tested for the creation of porous structure in cellulose materials for building insulation applications. The introduction of various substances, which producing CO_2 such as brewer's yeast, baking powder and $NaHCO_3$ have been tested for foam formation in cellulose matrices. Porous structures in cellulose insulating materials were also obtained by the sublimation of water during the freeze-drying process [19].

Three-dimensional macroporous scaffold from bacterial cellulose was developed for culture of breast cancer cells and patterned pores were created by using an infrared laser. Different pore sizes were able to fabricate by adjusting the distance between specimens and laser focus. In fact, that the cancer cells were larger than the tissue cells, satisfactory biocompatibility was obtained with the macroporous scaffold produced by bacterial cellulose [26].

2.3. Collagen

Collagen is the most abundant structural protein in the vertebrate body. It is a major component of connective tissues, skin, bone, cartilage, and tendons. Moreover, collagen is the most abundant protein type of the extracellular matrix of connecting tissues, which provides structural integrity and conferred the mechanical and biochemical properties. At present, 28 types of collagen have been identified, and among these, the dominant collagen present in extracellular matrix, in tissues such as skin, tendon and bone is type I collagen. Type II collagen found in cartilage and type III occurs in adult skin [27].

Collagen protein has a complex hierarchy of structural order in primary, secondary, tertiary and quaternary structures. In primary structure, every third amino acid is a glycine, with strict repeating as shown in Figure 4a. About 35% of the non-glycine positions in the repeating unit are consist of proline, can be mostly found in x-position and 4-hydroxyproline, predominant in y-position. In secondary structure, glycine and hydroxyproline units lead to form a helical macromolecule. In tertiary structure, three helical units twist to form a right-handed triple-helical collagen molecule as shown in Figure 4b. In quaternary structure, triple-helical collagen molecules stagger into fibrils, which then arranged into fibers or even larger fiber bundles as illustrated in Figure 4c [28].

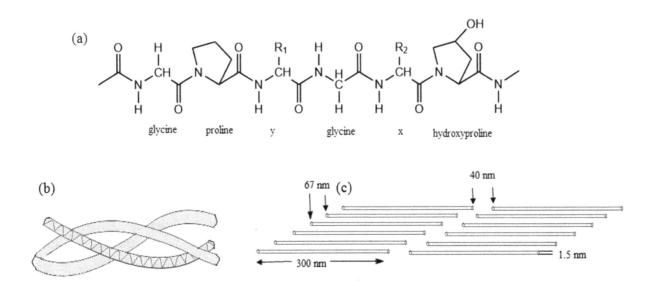

Figure 4. Chemical structure of collagen type I (**a**) Primary amino acid sequence; (**b**) secondary left handed helix and tertiary right handed triple-helix; (**c**) staggered quaternary structure.

Collagen is important as a biomaterial for various applications due to its specific properties such as abundance, biocompatibility, high porosity, easy processing, a facility for combination with other materials, low antigenicity, hydrophilicity and absorbability in the body. Collagen can be processed into various forms such as sheets, sponges, tubes, fleeces, powders, injectable solutions and dispersions, which then can be used as systems for drug delivery, scaffolding materials which promoting cell migration, wound healing, and tissue regeneration. Various methods have been developed for the modification of processing characteristics of collagen, including porosity development to suit with a wide range of applications [29].

Oh, et al. [30] developed porous collagen scaffold with using ice particulates as templates. Templates were prepared by dispersing the water droplets on a cooled copper plate wrapped with perfluoroalkoxy film. Ice templates were cooled to $-5\,°C$ and introduced the collagen solution followed by gradual freezing to $-5\,°C$. Finally, ice crystals were removed by freeze-drying and micropatterned pores were obtained. Two types of pores were obtained; one is the negative replica of ice templates and other is from ice crystals developed from freeze-drying. Pore sizes were able to control by freezing temperatures. Decreasing the freezing temperature produced scaffolds with smaller pores.

Collagen scaffold with controlled insulin release properties was developed with freeze-drying method for cartilage tissue engineering. Insulin was microencapsulated with poly(lactic-co-glycolic acid) beads and introduced to the collagen aqueous solution. Prepared ice particulates were added to the collagen bead mixture solution and freeze-dried to obtain the required pore size of the scaffold. Scaffold with interconnected pore structure was obtained with pore sizes equivalent to ice particulates. Drug release studies revealed that the scaffold exhibited a zero order release kinetics of insulin up to a period of 4 weeks [31].

Highly porous hydrogel with enhanced salt and pH resistance properties was prepared using the hydrolyzed collagen as the backbone of the hydrogel. Acrylic acid and 2-acrylamido-2-methylpropanesulfonic acid were polymerized and crosslinked to hydrolyze the collagen backbone. The porous structure was achieved by partially neutralizing the grafted polymer after gel formation. The pores formed in the gel due to the water evaporation as a result of neutralization heat [32].

3. Fabrication of Synthetic Biopolymers

3.1. Poly(lactic acid)

Among the biobased materials, poly(lactic acid) (PLA) is one of the most promising biomaterials and has remarkable properties, which make it suitable for different applications. It is cheaper and commercially available with wide range of grades. Since the basic monomer unit (lactic acid) synthesized by the fermentation of renewable resources (carbohydrates), PLA complies the concept of sustainable development and is classified as an eco-friendly material [33].

Figure 5 shows the three main routes for the synthesis of PLA. In the first step, lactic acid is condensation-polymerized to form low molecular weight prepolymer and employed with chain coupling agent to increase chain length and to form high molecular weight PLA. The second step involves azeotropic dehydrative condensation, a one-step process to form high molecular weight PLA from the monomer unit. The third and the main process is a ring-opening process of lactide into high molecular weight PLA which is patented by Cargill (US) in 1992 [34].

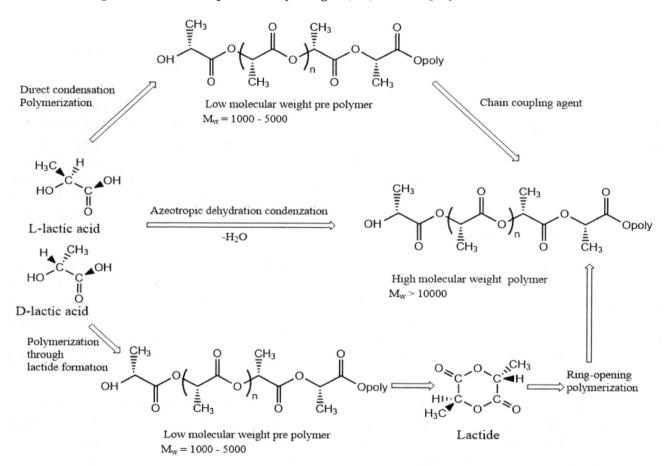

Figure 5. Three main routes for obtaining high molecular weight poly(lactic acid) (PLA).

PLA has been widely used for tissue engineering applications due to its biocompatibility, biodegradability and ease of fabrication into porous structures. Conde, et al. [35] developed a poly(L-lactic acid) scaffold using solvent-casting/particulate leaching to investigate the influence of pore size on the proliferation and differentiation of dental pulp stem cells. The scaffolds were prepared in pulp chambers of 1 mm thick tooth slices and porosity was developed using salt crystals of two sizes (150–250 μm or 251–450 μm) as porogen. It was observed that scaffolds with different pore sizes allowed the proliferation and differentiation of dental pulp stem cells into odontoblast-like cells.

PLA tubes were fabricated by atomizing the polymer solution over a rotating shaft. Polymer solution was prepared using a mixture of ethanol and dichloromethane. Due to the low boiling point of dichloromethane, it evaporated before ethanol. Also, dichloromethane is a good solvent for PLA than ethanol. Then ethanol would become more concentrated and reached to the critical concentration stage where PLA became to precipitation. At last, ethanol evaporation took place and created PLA tubes with internal pore structure with a rough surface [36].

Room temperature ionic liquid based on 1-butyl-3-methylimidazolium bearing hydrophilic anion Cl was used for the preparation of three-dimensional (3D) porous PLA scaffold for tissue regeneration. Hydrophobic PLA and relatively hydrophilic room temperature ionic liquid containing anion Cl was dissolved in dichloromethane and formed a homogeneous mixture. It later formed a continuous bicomponent network by the phase separation process. After the ethanol washing step, complete removal of room temperature ionic liquid created the pore structure in the biopolymer scaffold. Open porous PLA network with pore sizes greater than 100 μm and porosities of about 86%–94% was obtained by using this method [37].

Gelatin particles were used as a porogen for the fabrication of poly(L-lactic acid) (PLLA) scaffolds for chondrocyte regeneration. Gelatin particles (280–450 μm) sieved from raw gelatin was boned by incubation in saturated water vapor and immerged into PLLA/1,4-dioxane solution. After freeze-drying, gelatin immerged PLLA/1,4-dioxane solution was treated with water to remove the gelatin particles to obtain the porous scaffold. Scaffolds with interconnected pore structure and pre-designed pore sizes (280–450 μm) with the porosity >94% were developed by using this method [38].

Solid state foaming of biomaterials can be achieved by using gasses such as N_2 and CO_2. However, the main drawback of this method is, it formed mostly closed pores, which are not suitable for applications such as tissue engineering. A method was developed to fabricate PLA using solid-state foaming and ultrasound for tissue engineering applications. CO_2 blowing was carried out at room temperature with gas pressures at 3–5 MPa. Ultrasound was applied to break the pore walls of the solid state foams (closed pores). It was observed that ultrasound can successfully apply to improve the interconnectivity of pores of PLA created by CO_2 blowing method [39].

The porous scaffold of PLA was developed by applying solid state extrusion combined with porogen (NaCl) leaching method. Poor mechanical properties exist in porogen leaching method to porous scaffold was overcome by utilizing solid state extrusion process. The introduction of biocompatible poly(ethylene glycol) (PEG) as a plasticizer caused the development of PLA ductile and enabled the formation of interconnected pores. Highly interconnected porous architecture with high connectivity exceeding 97% and with enhanced porosity over 60% was obtained in PLA scaffold with the composition of NaCl higher than 75.00 wt % and PEG more than 1.25 wt % [40].

3.2. Poly(lactic-co-glycolic acid)

Poly(lactic-*co*-glycolic acid) (PLGA) is a copolymer synthesized by means of ring-opening *co*-polymerization of poly(lactic acid) (PLA) and poly(glycolic acid) (PGA). Since PLA contains an asymmetric α-carbon, it is typically described by two enantiomeric forms of poly(D-lactic acid) (PDLA) and poly(L-lactic acid) (PLLA). PLGA is known as poly(D,L-lactic-*co*-glycolic acid) when poly(D-lactic acid) (PDLA) and poly(L-lactic acid) (PLLA) are present in equal ratios. PLGA also can be prepared by different ratios of its monomeric units. Different types of PLGA can be obtained by the different ratios of monomer units. These are identified according to the ratio of two types of monomers. For instance, PLGA 75:25 refers to a copolymer consists of 75% lactic acid and 25% glycolic acid [41]. Figure 6 shows the structure of PLA and its constituent monomers, lactic and glycolic acid.

Unlike the two types of monomers, PLGA dissolved in a wide range of common solvents such as tetrahydrofuran, chlorinated solvents acetone or ethyl acetate. PLGA degrades by hydrolysis in aqueous environments and produces lactic acid and glycolic acid as byproducts. Degradation rates depend on the molecular weight of the polymer, the ratio of glycolic acid to lactic acid, stereochemistry (depends on the D and L-lactic acid monomers), and end group functionalization (polymers end capped with esters degrades slowly than the presence of free carboxylic acid groups at the end). Since the T_g is above 37 °C, PLGA shows a glassy behavior in nature. It is reported T_g decreases with decreasing the amount of lactic acid present in the copolymer [42].

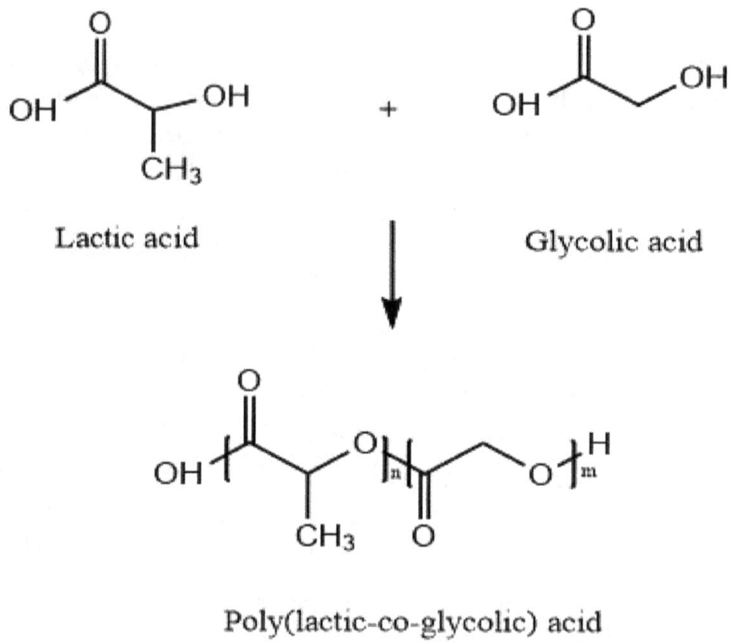

Figure 6. Poly(lactic-*co*-glycolic acid) (PLGA) and its constituent monomers, lactic and glycolic acid.

Due to biocompatibility, tailored biodegradation, potential to modify surface properties and ease of fabrication into porous structures, PLGA is being considered and investigated for a wide range of biomedical applications [42]. They are mainly used in the pharmaceutical industry to develop drug delivery systems, as a sutures for wound closure and as scaffolding materials for tissue engineering [43].

Krebs, et al. [44] developed an injectable, PLGA scaffold with in situ pore formation via phase inversion. Porogen and a small amount of water were introduced into the PLGA polymer solution for the in situ pore formation, as with previous studies, a porous structure was not given alone with the polymer solution. When water-insoluble PLGA contact with the aqueous solution, it precipitated by phase inversion. Due to the porogen and the small amount of water, it created a microporous interconnected architecture on the surface and within the bulk.

PLGA foams were produced by the pressure quench method using supercritical CO_2 as the blowing agent. Due to the prolonged exposure of the polymer to supercritical CO_2, it decreased its glass transition temperature than the processing condition of the vessel and made a polymer/gas solution. After the rapid depressurization, the solubility of CO_2 decreased with the polymer and caused bubble nucleation due to the supersaturation. This resulted in interconnected PLGA foams with relative pore densities ranging from 0.107 to 0.232 and porosities as high as 89% [45].

Thermally induced phase separation method was used to fabricate PLGA tubular foam scaffold for tissue engineering applications. The polymer solution was filtered using 0.45 μm nylon filter and cast to a petri dish. Then frozen in liquid nitrogen, and the solvent was sublimated in a vacuum. Porosity >93% with macropores of ~100 μm average diameter and interconnected micropores of 10–50 μm diameter were obtained with this method. Composition and porosity can be tightly controlled by thermally induced phase separation method. This ensures the optimization of scaffold for particular tissue engineering application [46].

Porous PLGA microparticles were prepared by water-in-oil-in-water ($W_1/O/W_2$) multi-emulsion method, to study the pH related drug release properties. Tiotropium (drug) dissolved in sodium tetraborate aqueous solution with a porogen and pH-sensitive drug release activator (3-diethylaminopropyl-conjugated hyaluronate) (W_1 phase) and vigorously emulsified with PLGA dissolved in dichloromethane (O phase). The mixed solution was injected to polyvinyl alcohol and NaCl aqueous solution (W_2 phase). After that, emulsification was carried out in homo-mixer at 4000 rpm for 5 min and hardened by mild stirring for 1 h at 50 °C. Microparticles were collected by centrifugation. The average pore diameters of the microparticles with 0, 20 and 50 mg of pH-sensitive drug release activator at pH 7.4 were 11, 12, and 27 nm, respectively. However, the average pore diameters of the microparticles with 0, 20 and 50 mg of pH-sensitive drug release activator at pH 6 were 13, 23, and 120 nm, respectively. These results indicated that the incorporation of pH-sensitive drug release activator increased the average pore diameter and surface area of microparticles in acidic medium [47].

Chemical structures of biopolymers commonly used in the preparation of porous materials are shown in Figure 7. The pore characteristics of biomaterials formed by natural biopolymers and fabricated with different methods are listed in Table 1.

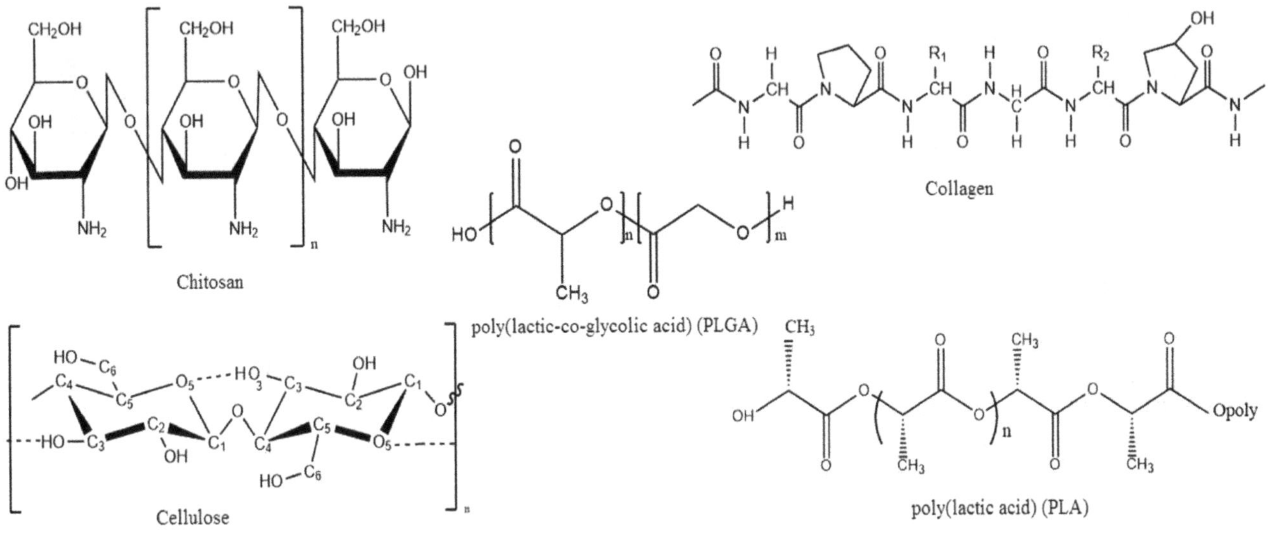

Figure 7. Chemical structures of biopolymers commonly used in the preparation of porous biomaterials.

Table 1. An overview of pore characteristics of biomaterials fabricated with different methods.

Biopolymer	Fabrication Method	Application	Pore Characteristics	Reference
Chitosan	Freeze-drying	Scaffold	Polygonal pores formed with low molecular weight chitosan and elongated pores formed with high molecular weight chitosan. Average pore sizes of scaffold were approximately 60–90 μm.	[10]
Chitosan	Liquid hardening	Scaffold	Pore diameter decreased with increasing the stirring rate and decreasing the concentration of chitosan. Average pore sizes of 200–500 μm and 80% porosity could be obtained by varying the concentration of chitosan and the stirring rate.	[13]
Chitosan	Dense gas CO_2	Scaffold	The porous structure obtained without formation of a nonporous skin layer. The average pore size in the scaffold produced at 60 bar and 4 °C was 30 to 40 μm using glutaraldehyde and genipin as crosslinker, respectively.	[15]
Chitosan	Supercritical CO_2	Scaffold	Under optimum condition (CO_2 pressure of 250 bar, 45 °C, 5 g/min CO_2 flow rate for 2 h) that yielded 87.03% porosity. The pore sizes were in the range of 20–100 μm.	[16]
Chitosan	Liquid hardening	Scaffold	Pore sizes from 70 to 900 μm were obtained when transferring the stirred chitosan solution to sodium hydroxide solution. Macroporous chitosan scaffold with porosity 85% ± 2% was obtained.	[17]
Cellulose	Supercritical CO_2	Polymeric foam	Pore size increased as decreasing the pressure and pore morphology varied with the depressurization rate.	[22]
Cellulose	Supercritical CO_2	Antimicrobial food packaging	Mean pore size decreased with increasing the operative pressure and decreasing the operative temperature. Porosity increased with decreasing the pressure and with increasing the temperature.	[24]
Cellulose	Infrared laser	Scaffold	Pore size varied with adjusting the distance between specimen and laser focus. Patterned macropores with smooth surface and diameter larger than 100 μm was introduced to the scaffold with this method.	[25]
Collagen	Freeze-drying (using ice particulates as templates)	Scaffold	Two types of pores formed; one from the negative replica of ice templates and other from ice crystals developed by freeze-drying. Pore size decreased with decreasing the freezing temperature. The micropatterned pores of the scaffold can be controlled by designing a desirable micropattern for the ice template.	[30]
Collagen	Freeze-drying (mixing with ice particulates)	Drug delivery system	Interconnected pore structure obtained with pore size equivalent to ice particulates (150–250 μm). All the scaffolds had large controlled pore structure.	[31]
Poly(L-lactic acid)	Solvent-casting and particulate leaching	Scaffold	Two ranges of pore size formed using two particle sizes of NaCl as porogen: 150–250 μm and 251–425 μm. Pore structures were formed after Poly(L-lactic acid) dissolved in chloroform was dropped over the salt and leached with distilled water.	[35]

Table 1. *Cont.*

Biopolymer	Fabrication Method	Application	Pore Characteristics	Reference
PLA	Phase-separation	Scaffold	Open porous PLA network formed with pore sizes greater than 100 μm and porosities of about 86%–94%.	[37]
Poly(L-lactic acid)	Solvent casting and particulate leaching	Scaffold	Interconnected pore structure formed with pre-designed pore sizes (280–450 μm) and porosity >94%.	[38]
PLA	CO_2 blowing with the application of ultrasound	Scaffold	Interconnectivity of pores improved by ultrasound (by breaking the pore walls of closed pores). The diameters of the closed pores were from 30 to 70 μm. After the ultrasound treatment, pore sizes changed to 30–90 μm due to the formation of interconnected pores.	[39]
PLA	Solid state extrusion combined with porogen (NaCl) leaching method	Scaffold	Interconnected porous architecture formed with high connectivity exceeding 97% and with enhanced porosity over 60%. Smaller pore sizes (9 μm) were resulted due to the fragmentation of bulky NaCl during the processing.	[40]
PLGA	Phase inversion	Scaffold	Microporous interconnected architecture formed on the surface and within the bulk. The total porosities were 32.19% ± 11.4% and 72.24% ± 4.0% for the control (nonporous) and porous scaffolds, respectively.	[44]
PLGA	Supercritical CO_2	Scaffold	Highly interconnected pores formed with relative pore densities ranging from 0.107 to 0.232 and porosities as high as 89%. The pore sizes were within the range from 30 to 100 μm.	[45]
PLGA	Thermally induced phase separation	Scaffold	Macropores with average diameter ~100 μm and interconnected micropores of 10–50 μm diameter formed with porosity > 93%. Tubular pores consited of radially oriented.	[46]
PLGA	Multi-emulsion method	Drug delivery system	Incorporation of pH-sensitive drug release activator increased the average pore diameter and surface area of microparticles in acidic medium. The average pore diameters of the microparticles at pH 7.4 were within the range of 11, 12, and 27 nm, respectively. It decreased at pH 6.0 to 13, 23, and 120 nm.	[47]

4. Biocomposite Materials

Composite material is a material made from two or more constituent materials that result in significant properties than those of the characteristics of individual components. Composites can be produced with the view of tailoring physical, chemical, or mechanical properties, to fulfil the requirements of different applications such as automotive, packaging, aeronautic, naval, and so on [48,49].

In view of their potential for high performance, composite biomaterials have been studied and tested for various kind of applications as shown in Table 2. To be considered as biocomposite material, each constituent of the composite must be biocompatible. Extreme modifications of the properties of biomaterials can be improved by incorporation of nanosized filler. This properties improvement depends on both nanofiller geometry and on the surface (interface) area. There are three types of nanofillers namely, spherical, layered and acicular, based on their aspect ratio and geometry. A wide range of nanobiocomposites has been produced by introducing nanosized fillers and tested to overcome the conventional drawbacks of biopolymer [50].

To be successful for the applications, any biomaterial requires a wide range of study about its fabrication and properties. Fabrication of porous structures in biocomposites with various material combinations has become an increased research interest due to their wide range of applications in areas such as tissue engineering, nanocomposites, drug delivery systems, packaging and the automotive industry [51].

4.1. Fabrication of Porous Biocomposite Materials

4.1.1. Chitosan/PLA Composite Materials

PLA is biocompatible and undergoes scission in the body to form lactic acid, which is a natural intermediate in the process of metabolic conversion of carbohydrates. These characteristics make this polymer compatible for use in biomedical applications. However, PLA has several drawbacks such as acidic degradation product, past biodegradation, and hydrophobicity. Chitosan is another biodegradable and biocompatible material, hydrophilic and alkaline in nature, which is widely used in biomedical applications. Acidic byproduct produced by PLA can be readily neutralized due to the alkaline nature of chitosan when using these two polymers as biocomposite material [52].

Li, Ding and Zhou [52] obtained the porosity of Chitosan/PLA scaffold by introducing NaCl to the mixture of chitosan and PLA. The mixture was then melted at 160 °C and molded, after that, NaCl grains were excluded by dipping the composite material in distilled water. Porosity was able to regulate by controlling the weight fraction of NaCl.

The blending of chitosan with PLA is difficult as chitosan decomposes before melting, due to high glass transition temperature of chitosan. Also, it is difficult to find a cosolvent to dissolve these two substances as chitosan dissolves in very few dilute acids and PLA only dissolves in few organic solvents. Therefore, methods were developed to grafting lactic acid onto the amino groups in chitosan by vacuum freeze-drying and vacuum reaction without a catalyst. It was observed that the porosity decreased with increasing lactic acid/chitosan feed ratio. Also, a higher porosity with large pores was formed with chitosan than with the PLA in the grafted copolymer [53].

By mixing of chitosan with PLA, formed a composite material with improved intensity and elasticity, to be used as nerve conduits. The chitosan and PLA solutions were mixed with best constituent ratios and sprayed over a rotating glass rod mandrel with compressed nitrogen gas. Infrared rays were used for dehydration and solidification. The detached coatings were sterilized and used as conduits. The conduits were biodegradable and provided many micropores to enhance the permeability. The pore size of chitosan/PLA conduits were less than 1 μm which allowed the permeability only for nutrient, other molecules and not for other cells [54].

Electrospinning is a method of production of polymer fibers in the diameter range from few nanometers to several micrometers. Also, electrospinning is used to form nanofibers with large surface-area-to-volume ratio and high porosity with small pore size. PLA/chitosan composite nanofibers were produced using coaxial electrospinning method and tested the potential of application as porous antibacterial material using antibacterial test. Results showed that antibacterial effect depended on the fabrication core feed rate [55].

4.1.2. PLA/Cellulose Nanocomposites

Due to the biocompatibility, PLA is widely used in biomedical applications. The degradation product, lactic acid is also biocompatible due to the possibility of incorporation into the carbohydrate metabolism. However, due to the poor mechanical properties of PLA, various types of fillers such as carbon fibers, carbon nanotubes, hydroxyapatite and cellulose microcrystals have been introduced for the reinforcement of the polymer [56].

Cellulose nanocrystals, also known as nanowhiskers, have been introduced as a filler for the polymers due to exceptional mechanical properties, high aspect ratio, and large specific surface area. Cellulose nanowhisker can be obtained from plant cellulose as well as bacterial cellulose. They are exceptionally hydrophilic and shows high tensile strength properties [56].

Porous bacterial cellulose nanowhisker/PLA composites were fabricated using solvent casting and freeze-drying method to evaluate its possibility to use as tissue engineering scaffold and drug carriers. Bacterial cellulose nanowhisker suspension in water was introduced to PLA dissolved in 1,4-dioxane. After that, the mixture was frozen in liquid nitrogen and then freeze-dried to remove most of the 1,4-dioxane and water to obtain the porous composite. It was observed that the porosity increased from 79% to 92% by adding 5 wt % of bacterial cellulose nanowhisker to PLA [57].

Producing of polymer foams using supercritical CO_2 was successfully tested for various kinds of applications such as large volume packaging applications and biomedical implants. Cell size and cell density of the foams of PLA produced using supercritical carbon dioxide were affected by the addition of cellulose nanofibers. Cellulose nanofibers were processed for surface modification, prior to mixing with the polymer in order to enhance the binding affinity. A mixture of PLA and cellulose nanofiber was prepared by hot pressing and foams were obtained by pressure quenching of supercritical CO_2. Results showed that the foam cell size decreased and cell density increased with increasing the cellulose nanofiber concentration [57].

The role of cellulose nanofibers in the supercritical foaming process of PLA was investigated with different loadings of the nanofibers. Nanocomposite sheets were obtained using a casting–kneading–hot pressing procedure. Supercritical CO_2 was used as a plasticizer and blowing agent. The presence of cellulose nanofibers speeded up the nucleating process, but lowered the cell growth rate and delayed the coalescence. Also, the addition of cellulose nanofibers reduced the cell size and increased the cell density of the foam [58].

PLA/cellulose nanocrystal nanocomposite fibers produced by electrospinning method was investigated for control release of nonionic compounds with varying the concentration of cellulose nanocrystals (0, 1, and 10 wt %). The incorporation of cellulose nanocrystals increased the crystallinity of PLA, accelerated the hydrolytic degradation and decreased the hydrophobicity of the polymer. At the same time, the mean pore size of the composite material was increased with increasing the cellulose nanocrystals. Due to the bigger pore size and high water uptake with 10% cellulose nanocrystals incorporated PLA, favored higher diffusion, and hence, increased the nonionic compound release percentage [59].

4.1.3. Cellulose/Chitosan Composite Materials

Cellulose is polysaccharide made from glucose subunits and chitosan is made from amino polysaccharide subunits. The chemical structure of chitosan backbone is very similar to that of cellulose. Therefore, studies have been done to test the miscibility of chitosan with cellulose and introduction of amino groups to cellulose to improve the physical, chemical, mechanical and biological properties of developed composites [55].

Spray dried chitosan mixed with cellulose using N-methylmorpholine-N-oxide and deodorizing properties, metal ion sorption properties were investigated with varying concentrations of both polymers. A mixture of two polymers extruded from a syringe to form droplets and lyophilized until drying. Lyophilization process and freezing created open pores with a high degree of interconnectivity. Pore diameter decreased with increasing the chitosan concentration. This was due to the increase in interactions among chitosan, cellulose, and water, hence, decrease in ice crystal growth. Increasing the concentration of chitosan contributed to increased sorption of deodorizing agent and metal ions due to high interactions of these particles with chitosan than cellulose [60].

In contrast to the cellulose derived from plants, bacterial cellulose is free from contaminants such as natural fiber like lignin and hemicellulose. Low molecular weight chitosan was introduced to bacterial cellulose produced by *Acetobacto xylinum* and properties were modified with varying concentration of chitosan with different molecular weights. Chitosan was directly introduced to the culture medium and the product formed with bacterial cellulose was separated as sheets. It was observed that the pore size and surface area of the dried films were increased with the addition of low molecular weight chitosan. Also, the mechanical properties and water absorption capacity increased with the introduction of chitosan, but water vapor transmission rates, average crystallinity index and anti-microbial ability remained unchanged [61].

Chitosan/cellulose hydrogel beads were prepared using ionic liquid and investigated the dye adsorbent properties with varying dye concentrations, the number of beads and initial pH. The pore diameters of the beads were within the range of 10 to 20 nm. The maximum adsorption capacity of hydrogel beads was found to be 40 mg/g, for congo red dye removal from aqueous solutions, which was more efficient when compared with other economical methods [62].

Bacterial cellulose and chitosan composite material for potential biomedical application was produced by immersion of wet bacterial cellulose pellicle in chitosan solution followed by freeze-drying method. A very well interconnected porous network structure with larger aspect surface, which suitable for cell adhesion and proliferation, was obtained with using this method. Cell adhesion studies revealed that the biocompatibility increased with the addition of chitosan to the bacterial cellulose [63].

4.1.4. Chitosan/PLGA Composite Materials

Due to the hydrophobic characteristics of PLGA, it is difficult to apply alone as a biopolymer in many applications like tissue engineering. To overcome these difficulties, composite materials have been made with PLGA by mixing with a hydrophilic biopolymer. Due to the presence of β-(1-4)-linked between D-glucosamine and N-acetyl-D-glucosamine and the ability of D-glucosamine to immobilization of ligands and glycoproteins through covalent bonding, chitosan has been widely used to mix with PLGA to make biocomposite materials, especially related to tissue engineering applications [64].

Kim, Yang, Chun, Chae, Jang and Shim [64] prepared a chitosan/poly(D,L-lactic-*co*-glycolic acid) composite fibrous scaffold by the coelectrospinning process and compared the properties with chitosan and PLGA scaffolds. Fibrous scaffold with high surface-to-volume ratio, high porosity, and variable pore size distribution was obtained by electrospinning method. The weak mechanical properties of chitosan were improved by electrospinning with PLGA; on the other hand, chitosan provided the hydrophilic property to PLGA. In the electrospinning process, the spinning parameters, solution viscosity, polymer concentration, applied voltage, and flow rate greatly influenced the porosity and pore size distribution of the composite material.

Chitosan/PLGA nanocomposite scaffold was produced via electrospinning and unidirectional freeze-drying techniques for tissue engineering applications. The porosity of composite material was found to be more than 96% and it decreased with increasing the chitosan concentration. In addition, the scaffold exhibited high surface area-to-volume ratio due to the incorporation of PLGA nanofiber [65].

Porous chitosan/poly(D,L-lactic-co-glycolic acid) nanocomposite scaffold was prepared using electrospinning and freeze-drying method and properties were studied with varying processing parameters, such as electrospinning time and the concentration of chitosan solution. The water absorption capacity and porosity of the nanocomposite scaffold decreased with an increase in the chitosan solution concentration and electrospinning time. In addition, compressive strength and the modulus of compressibility of the nanocomposite material improved due to the introduction of PLGA nanofibers [66].

Ajalloueian, et al. [67] developed novel nanofibers using a blend of chitosan and PLGA through emulsion electrospinning method for biomedical applications. Polyvinyl alcohol was introduced to the electrospinning solution as an emulsifier. After that, the polyvinyl solution was removed by extraction with ethanol solution. Porosity and hydrophilicity of the composite membrane increased due to the removal of polyvinyl alcohol from the mats. The improved hydrophilicity was expected to lead to higher cell affinity of the composite scaffold.

4.1.5. Chitosan/Collagen Composite Material

Collagen is a one of promising biomaterial used in tissue engineering applications due to its superior biocompatibility and biodegradability. The main difficulty with collagen as a tissue engineering scaffold is rapid degradation when exposure to body fluids and cell culture media. Chitosan has been widely used in biomedical applications due to many advantageous such as antibacterial activity, wound healing property, and accelerating tissue regeneration. In addition, chitosan can function as a bridge to improve the mechanical strength of collagen scaffolds due to the presence of a large number of amino groups attached to its backbone [68].

Chitosan/collagen scaffolds were fabricated using freezing and lyophilizing methods with varying concentration of both biomaterials. Results showed that the pore size of the composite scaffold was comparatively smaller than chitosan scaffolds. Porosity decreased with increasing the concentration of chitosan but remained unchanged with the concentration of collagen. Moreover, the addition of collagen decreased the mean pore size of the scaffold, which is expected to improve the binding capacity of fibroblasts [68].

Chitosan/collagen scaffolds were formed using glutaraldehyde (GA) as a crosslinking agent and freeze-drying method for skin tissue engineering applications. It is reported that the mean pore size of crosslinked scaffold is higher than the pore size of the uncrosslinked scaffold. Rehydration and relyophilization process accompanied with GA cross-linking treatment further increased the pore sizes due to the fusion of smaller pores and by inducing the combination of collagen fibers [69].

Porous collagen membranes were placed in a chitosan solution and crosslinked with glutaraldehyde vapor to obtain chitosan-coated collagen membrane. Chitosan solution was added on collagen membrane without applying any pressure to allow the chitosan to penetrate into the pores of the collagen membranes. Results showed that the pore size and the porosity decreased in coated collagen scaffolds when comparing with the non-coated collagen scaffolds [70].

Table 2. An overview of pore characteristics of biocomposite materials fabricated with different methods.

Biocomposite Material	Fabrication Method	Application	Pore Characteristics	Reference
Chitosan/PLA	Melt molding and particulate (NaCl) leaching	Scaffold	The pore sizes were larger than 100 μm and all the pores including inner pores were interconnected. Porosity increased with the weight fraction of NaCl.	[52]
Chitosan/PLA	Freeze drying	Scaffold	Scaffold with interconnected porous structures and pore size around 100–500 μm was obtained. The pore size of the scaffolds decreased with increasing lactic acid/chitosan feed ratio. The chitosan scaffold had a porosity of 62.3% and pore size of 500 μm, and the lactic acid/chitosan scaffold (4:1, wt/wt) had a porosity of 34.37% and pore size of 100 μm.	[53]
Chitosan/PLA	Mold casting/infrared dehydration	Scaffold	Well-distributed 0.2 μm pores on the surface of the conduit was formed.	[54]
PLA/nanocellulose	Electrospinning	Release of nonionic compounds	There was no significant difference in the mean pore size between the nonwoven fabrics electrospun from PLA containing 0% and 1% cellulose nanocrystals. The mean pore size increased twice as big with PLA containing 10% cellulose nanocrystals. The mean pore sizes of the PLA nonwoven fabrics with 0%, 1% and 10% of cellulose nanocrystals were 0.48 ± 0.04 μm, 0.51 ± 0.08 μm and 0.94 ± 0.14 μm, respectively.	[59]
Cellulose/chitosan	Freeze drying	Sorption of trimethylamine and metal ions	The mean pore diameter was within the range of 100–300 μm. The pore diameters decreased with increasing chitosan concentration.	[60]
Cellulose/chitosan	Freeze drying	Dye adsorption	The beads were nanoporous with pore sizes from 10 nm to 20 nm.	[62]
Bacterial cellulose nanofiber/chitosan	Freeze drying	Scaffold	After the bacterial cellulose was treated by chitosan, porous structure remained but pore sizes became larger. Nanofibrous bacterial cellulose and bacterial cellulose/chitosan composite had well interconnected pore network structure.	[63]
Chitosan/PLGA	Electrospinning	Scaffold	In the electrospinning process, the spinning parameters, solution viscosity, polymer concentration, applied voltage, and flow rate highly influenced the porosity and pore size distribution of the composite material.	[64]
Chitosan/PLGA nanocomposite	Electrospinning and unidirectional freeze-drying	Scaffold	The porosity was found to be more than 96% and it decreased with increasing the chitosan concentration.	[65]

Table 2. *Cont.*

Biocomposite Material	Fabrication Method	Application	Pore Characteristics	Reference
Chitosan/PLGA nanocomposite	Electrospinning and freeze drying	Scaffold	The porosity of chitosan/PLGA nanocomposite scaffolds decreased with increasing the chitosan solution concentration and electrospinning time.	[66]
Chitosan/collagen	Freeze drying	Scaffold	The chitosan scaffold showed the pore sizes between 500 and 700 μm while the chitosan/collagen composite scaffold showed a smaller pore sizes of 100–400 μm. The addition of collagen decreased the pore size of the composite scaffold. All samples composed of different proportions of chitosan and collagen showed porosities higher than 90%. The addition of collagen did not change the porosity.	[68]
Chitosan/collagen	Freeze drying	Scaffold	The mean pore size of the scaffold increased from 100 μm to >200 μm by crosslinking with glutaraldehyde. Elongated pores were formed with high concentration of glutaraldehyde. Refreeze-drying induced the fusion of some smaller pores to generate larger ones.	[69]
Chitosan/collagen	Freeze drying	Scaffold	At the highest chitosan/collagen ratio (75/25), the gels showed a sponge-like structure with larger pores than the gels containing lower chitosan content for both crosslinked and uncrosslinked scaffolds.	[71]

Most common method to create chitosan/collagen composite scaffold is freeze-drying process. Chitosan/collagen matrices often need an additional crosslinking step to increase the mechanical strength, stability and to prevent from degradation. Glutaraldehyde is the most commonly used crosslinking agent for both chitosan and collagen due to its high effectiveness. In most cases, cells cannot be incorporated into the matrix during the fabrication stage due to the high cytotoxicity of glutaraldehyde, and therefore it must be implanted into the exterior of scaffolds post-fabrication. Chitosan/collagen hydrogels were prepared using glyoxal, a dialdehyde with relatively low toxicity as a crosslinking agent. Gels were freeze-dried and morphology was studied with varying concentration of both biopolymers. Cell studies revealed that glyoxal was cytocompatible at concentrations less than 1 mM for the time of exposure up to 15 h. Scanning electron micrographs showed that gels containing a high content of chitosan exhibited larger pores when compare with the gels containing lower chitosan concentrations. Crosslinked gels had even larger pores and a plate-like structure, while un-crosslinked gels showed a more uniform appearance [71].

5. Fabrication Methods

Fabrication methods for biobased porous materials more related to the choice of material. This can be classified into three main types. First, natural polymers, such as collagen and chitosan, are heat sensitive, so freeze-drying is mainly used to produce porosity, although electrospinning is also possible. Secondly, synthetic polymers, such as PLGA and PLA, often known as thermoplastics, so they can be fabricated by a wide variety of techniques. Bioceramics, such as hydroxyapatite tricalcium phosphate usually introduce as additives into polymeric matrices, since, pure ceramic matrices suffer from low hardness. Freeze-drying can be used to fabricate pure ceramic biomaterials, but this needs the use of sintering as a post-processing step which leads to additional porosity within the matrix walls [72].

Porous fabrication techniques can be categorized into two categories, designed manufacturing techniques and non-designed manufacturing techniques. Designed manufacturing technique includes 3D printing, rapid prototyping of solid free-form technologies. Non-designed manufacturing techniques include freeze-drying or emulsion freezing, solvent casting or particulate leaching, gas foaming, phase separation, electrospinning and combination of these techniques.

5.1. Solvent Casting and Particulate Leaching

Solvent casting and particulate leaching is one of easy and cheapest way of porous scaffold fabrication. Figure 8 shows the detailed process of solvent casting-particulate leaching fabrication technique. The polymer is first dissolved in an organic solvent. Particles, mainly water soluble salts (e.g., sodium chloride, sodium citrate) with specific dimension are then added to the solution. After that, the mixture is poured into a mold of the desired shape. Next, the solvent is removed either by lyophilization or evaporation and allowed salt particles leached into the polymer matrix. Finally, the mold is dipped in a water bath for sufficient time to dissolve the salt particles, leached inside the polymer matrix. Porosity and pore size can be easily controlled by the amount and the size of salt particles added to the matrix. Though, difficulty confronted in the removal of leached salt particles from the matrix limits the thickness of the matrix to 0.5–2 mm [73].

Interconnected porous chitosan scaffold was prepared using sodium acetate particulate leaching method. Sodium acetate was mixed with chitosan solution and injected into the mold. Then the mold was freeze dried and lyophilized to evaporate the solvent. After that, washed with series of ethanol solution (100%, 90%, 80%, 70% and 50% v/v) sequentially for 2 h each and salt-leached in distilled water for 48 h. Finally, freeze-dried at −70 °C for 24 h and lyophilized for 24 h. It was observed that the porosity and pore interconnectivity increased with the ratio of sodium acetate. Moreover, with 90% sodium acetate ratio, many minute pores (7–30 μm) were formed between the main pores (200–500 μm) [74].

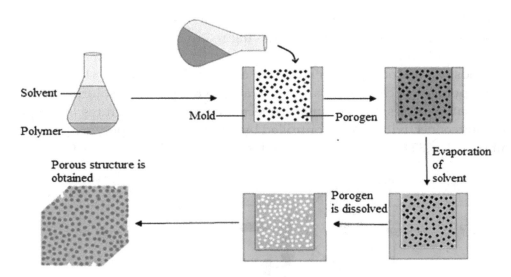

Figure 8. The schematic diagram of solvent casting and particulate leaching technique.

5.2. Thermally Induced Phase Separation

Thermally induced phase separation is a simple and versatile method for the preparation of microporous membranes. This method involves the dissolution of a polymer in a particular solvent having a high boiling point and low molecular weight at elevated temperature to form a homogeneous solution. Then the hot polymer solution is cast onto a mold followed by a cooling step. When a homogeneous solution at high temperature is cooled down, it induced solidification and phase separation into a polymer-rich phase and a polymer poor phase. After the solvent is removed by extraction or freeze-drying, a microporous structure is formed. This method is applicable for wide range of polymers, including those having poor solubility. The thermally induced phase separation process can be used to generate macro and microporous structure with an overall porosity as high as 90% [75,76].

Since this method has fewer influencing factors such as diluent, cooling rate, polymer concentration, and additives, it is easy for controlling membrane structures. As diluent is closely related to phase separation, different diluent cause different pore structures [77]. The advantages of this method are simplicity of the process, high reproducibility, low defects rate, high porosity and narrow pore size distribution [78].

Nano-hydroxyapatite/poly(L-lactic acid) composite scaffold was developed for bone tissue engineering and the morphologies, mechanical properties and protein adsorption capacities of the composite scaffolds was investigated. The porosity more than 90% was easily achieved and the pore sizes were able to adjust by varying the phase separation parameters [79].

PLLA/PLA scaffolds were prepared via thermally induced phase separation starting from ternary systems where dioxane as the solvent and water as the non-solvent. The porosity was within the range from 87% to 92%. Average pore size, pore distribution, pore interconnectivity and mechanical properties depended on the combination of the operating conditions such as solvent/non-solvent ratio, polymer concentration, remixing temperature and time [80].

5.3. Gas Foaming

Gas foaming is being used to fabricate the polymers with high porosity without using any organic solvent [81]. This technique uses high-pressure CO_2 for saturation of the polymer in an isolated chamber for a certain period of time. It needs high-pressure CO_2 (800 psi) to saturate the polymer with gas [82]. When the polymer is saturated with CO_2 at high-pressure, intermolecular interactions between CO_2 and the polymer molecules become higher and causes a reduction of glass transition temperature of the polymer. Rapid depressurization causes thermodynamic instability and leads to form nucleated gas cells creating pores inside the polymer matrix. This technique is suitable for amorphous and semicrystalline polymers having relatively low T_g or T_m and high affinity for CO_2 [83]. Instead of carbon dioxide, nitrogen gas can also be used for this method. The disadvantage is that it yields mostly a nonporous skin layer and closed pore structure [84]. This can be overcome by introducing a porogen such as salt particles (NaCl) to the polymer solution before gas foaming. Leaching of this salt particles, formed interconnected open pore structures in the polymer matrix. Porosity, pore interconnectivity can be controlled by altering the salt/polymer ratio and the particle size of the salt particles [85,86]. Figure 9 displays the schematic diagram of a CO_2 gas foaming device.

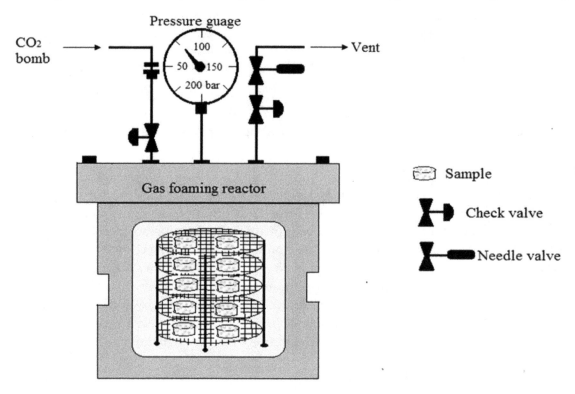

Figure 9. The schematic diagram of CO_2 gas foaming device.

The effect of high-pressure CO_2 on the characteristics of elastin-based hybrid hydrogel was investigated. Compared to fabrication at atmospheric pressure condition, fabrication at high-pressure CO_2 eliminated the skin-like layer formation on top of the hydrogel and formed larger pores with an average pore size of 78 $\pm$ 17 μm. However, the swelling ratio of the hydrogels fabricated at high-pressure CO_2 decreased due to a higher degree of cross-linking. In addition, dense gas CO_2 substantially increased the compressive and tensile modulus of fabricated hydrogels [87].

5.4. Emulsion Freeze-Drying

In this method polymer is dissolved in its solvent and water is added. Then polymer solvent solution and water homogenized to form an emulsion. Before the separation of two phases, the emulsion is rapidly cooled to lock in the liquid state structure. Finally, solvent and water are removed by freeze-drying [88]. Figure 10 shows the schematic diagram of emulsion freez-drying process. This technique can be used to obtain porosity level above 90% and to control the pore size for targeted application. Porosity and pore structure can be controlled by polymer concentration, solvent, and water phase percentage, and freeze-drying parameters [89]. This technique is advantageous due to lack of leaching step but, the addition of organic solvent is a concern for the tissue engineering applications [90].

Hydroxyapatite/poly(hydroxybutyrate-*co*-valerate) composite scaffold was fabricated through the emulsion freezing/freeze-drying process and effect of polymer solution concentration, solvent and water phase on the morphology of composite scaffold was investigated. It was observed that at the same volume fraction of the water phase, the porosity of scaffolds decreased with increasing the polymer concentration. When the volume fraction of the water phase was increased, the porosity was found to be increased. It was reported that the produced scaffolds were highly porous with interconnected porous structures. Scaffolds exhibited pore sizes ranging from several microns to around 300 μm [91].

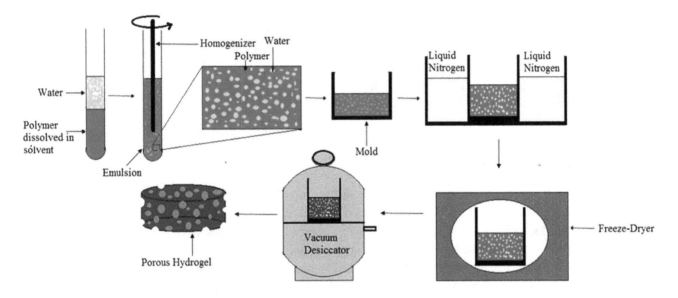

Figure 10. The schematic diagram of emulsion freeze-drying process.

5.5. Rapid Prototyping Technique

Rapid prototyping, generally known as solid freeform fabrication technique and one of the most promising techniques for designing and producing scaffolds with 100% interconnected pores, fully computer controlled architecture with high porosities [92,93]. The inherent limitations such as long fabrication periods, incomplete removal of residual chemicals or volatile porogenic elements, labor intensive processes, poor repeatability, insufficient interconnectivity of pores and thin wall structures, irregularly shaped pores, of the conventional methods have led to use the rapid prototyping techniques to customize design and fabricate 3D porous scaffolds [94,95]. All current rapid prototyping techniques are based on the use of computer-aided design information that is converted to a stereo lithography type file format. Rapid prototype machine software processes this file to produce a solid model by a variety of processes. Starting from the bottom, the first layer of the physical model is created. The next layer is glued or bonded to the previous layer. This process is continued until the whole model is completed. Any supports are removed from the finished surface model and cleaned [96]. Figure 11 displays the schematic diagram of rapid prototyping technique. Main advantages of rapid prototyping process are rapid processing time, customization and efficiency [97]. Limitations of this techniques are high machine cost, high processing temperatures limiting the ability to process temperature-sensitive polymers, and need of multidisciplinary collaboration [98].

Hybrid poly(L-lactide)/chitosan scaffolds were developed using the rapid freeze prototyping technique. It was found that the mechanical properties of the scaffolds depend on the ratio of chitosan microspheres to poly(L-lactide) and cryogenic temperature used in the rapid freeze prototyping fabrication process. The results showed that scaffolds with greater porosity and enhanced pore size distribution compared to dispensing-based rapid prototyping technique [99].

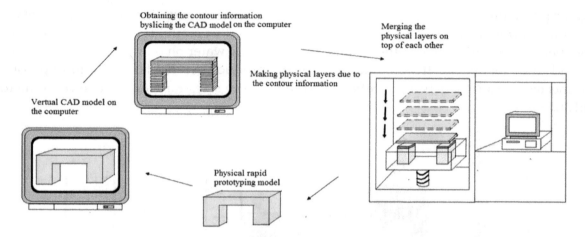

Figure 11. The schematic diagram of rapid prototyping technique.

5.6. 3D Printing

3D printing, also known as additive manufacturing, and inkjet printing liquid binder is used to make a three-dimensional object from digital model data. This technique involves printing liquid binder to bind the loose powder to create 3D objects. Since biomaterials widely exist as solid or liquid form, most of them can be directly utilized in this technology. The first step of 3D printing is modeling of virtual model from computer-aided design (CAD) or animation modeling software. The machine uses these data as a guideline to print [100]. Then, a thin layer of powder is deposited onto the building platform. In the printing step, the machine reads the design from digital model data and a printer head selectively lay down liquid binder solution onto a powder bed to form the 2D pattern. This process is repeated layer by layer until the material/binder layering is completed and the final 3D model has been printed. Final objects are extracted from the powder bed by removing or dissolving the unbound powder [101]. Pore size and the spacing can be controlled by the pattern used. The advantage to 3D printing is the control of pore size and distribution. Both rapid prototyping (RP) and 3D printing technologies build models layer by layer using computer-aided design. But, there are still some differences such as 3D printers usually make smaller parts, 3D printing costs less, less material choices for 3D printers, 3D printers are less complex and easier to use than rapid prototyping machines. Figure 12 displays the schematic diagram of rapid prototyping technique.

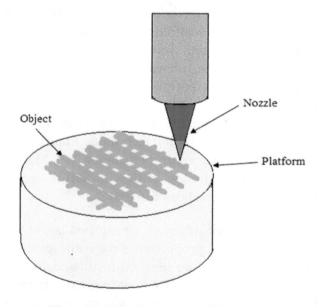

Figure 12. The schematic diagram of 3D printing technique.

Cylindrical scaffolds of five different designs were fabricated using a unique blend of starch-based polymer powders (cornstarch, dextran and gelatin) by 3D printing process. It was observed that the scaffold porosity corresponded to the designed porosities. A unique microporosity resulted due to the voids formed between granules or particles of the bulk material. Microporosity of the scaffolds of all designs were found to be within the range of 0.335–0.590. A highly interconnected porous network with suitable mechanical properties was fabricated using 3D printing process [102].

Indirect 3D printing protocol was employed to overcome the limitations of the direct technique for the preparation of porous scaffolds. 3D structures were fabricated by inkjet printing liquid binder droplets onto particulate matter. In indirect 3D printing protocol, molds are printed and the final materials are cast into the mold cavity. Scanning electron micrographs showed that well interconnected, highly open, uniform pore architecture (~100–150 μm), which is essential for uniform cell seeding, proliferation, growth, and migration in three dimensions [103].

5.7. Electrospinning Technique

A combination of two techniques namely electrospray and spinning is applied in electrospinning technique to form loosely connected 3D porous mats with high porosity and high surface area. A high electric field is applied to a fluid or melt which may extrude from a metallic syringe needle and acts as one of the electrodes as shown in Figure 13. When the electrostatic forces overcome the liquid surface tension forces the droplet comes to the end of the needle and deformed [104]. Then a fine, charged jet of the polymer solution is ejected from the tip of the needle to the counter electrode leading to the formation of continuous fibers. The fiber diameter and porosity of the matrix depend on the parameters such as voltage, polymer flow rate, the distance between the needle and the plate, and polymer concentration in the solution [105]. A wide range of polymers can be used in this technique such as synthetic polymers, natural polymers or a blend of both. The most important thing in electrospinning is that it can be used with various polymers, both in solution and in melt form. In melt electrospinning, it does not require the dissolution of the polymer in organic solvent, therefore it is environmentally safe and no mass loss due to solvent evaporation [106]. Electrospinning can be used to encapsulate drugs to the fibers. Also, this technique is the most economical way of producing nanofibers [107]. The disadvantage of this technique is limited control of pore size. The pore size of the matrix depends on the fiber diameter. Fibers with small diameter lead to form smaller average pore sizes. To overcome this limitation, a dual electrospinning setup has been developed with the additional stream of polymer which acts as a sacrificial fiber to increase the void space in the matrix [108,109].

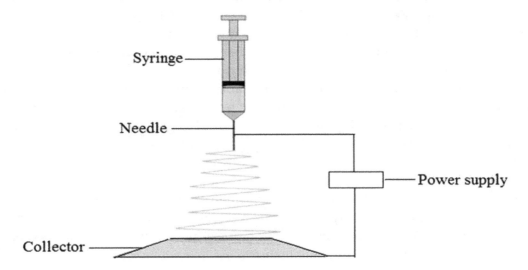

Figure 13. The schematic diagram of electrospinning technique.

A porous scaffold consisting of biopolymer nanofibers is one of the most promising candidates for tissue engineering applications. Immiscible biopolymers of gelatin and polycaprolactone were first electrospun to form composite fiber of gelatin/polycaprolactone. A leaching method was carried out to generate porous nanofibers to selectively remove the water soluble component of gelatin. After the leaching treatment grooves, ridges, and elliptical pores were appeared on the surface as well as inside of the resultant individual nanofibers [110].

Electrospun poly(ε-caprolactone)/chitosan nanofibers were prepared to study the effects of chitosan concentration on the bovine serum albumin (BSA) protein release behavior. Poly(ε-caprolactone) (PCL) and chitosan nanofibers with different ratios of chitosan were electrospun with using formic acid/acetic acid solvent system. Based on the scanning electron micrograph images, with increasing the chitosan content in nanofiber exhibited a higher fiber diameter and pore size. In addition, compared to PCL/chitosan nanofibers, PCL/chitosan/BSA nanofibers showed higher fiber diameter and larger pore size. Results showed that the chitosan ratio affected significantly in the protein release profile from the PCL/chitosan/BSA blend [111].

6. Conclusions

Each biopolymer molecule has a unique primary structure. Material-specific properties are due to the unique primary structure that they possess. In this review, we discussed two classes of biopolymers, namely, natural biopolymers (chitosan, cellulose, collagen) and synthetic biopolymers (PLA, PLGA) and their composite materials (chitosan/PLA, PLA/Cellulose, cellulose/chitosan, chitosan/PLGA, chitosan/collagen). Chitosan is a widely used biopolymer with attractive properties. It has been widely used to fabricate porous membranes, hydrogels, scaffolds and microparticles using a variety of simple fabrication techniques. Development of highly porous structure in cellulose is important because of their potential uses in tissue engineering, building insulating, antimicrobial food packaging applications, and, automotive, packaging and aircraft industries. Collagen can be easily fabricated into various forms such as sheets, sponges, tubes, fleeces, powders, injectable solutions, and dispersions. PLA has been widely using to fabricate porous scaffolds for tissue engineering applications. PLGA, a synthetic biopolymer which dissolved in a wide range of common solvents. Fabrication methods such as phase inversion, supercritical CO_2, and thermally induced phase separation are widely used to fabricate porous PLGA scaffolds. Porosity, pore size and pore connectivity of fabricated biomaterials vary depending on the biomaterial, processing method and process conditions. All these factors can be controlled and varied depending on which application *is* needed. The properties of composite materials are superior to the properties of the individual components which it is constructed. Extreme modifications of the properties of biomaterials can be achieved by producing nanobiocomposites. Porous fabrication methods are two types; designed manufacturing techniques and non-designed manufacturing techniques. Natural polymers are heat sensitive and normally fabricated by methods with no heat generation. Synthetic biopolymers are often known as thermoplastics, so they can be fabricated by a wide range of techniques. Each fabrication method has its own advantageous and disadvantageous. A modern method for creating porous structures using biodegradable fibers by electrospinning is the latest development in this field. This method also has disadvantages like limitations of controlling the pore size. 3D printing, rapid prototyping, solvent casting and particulate leaching techniques are most suitable for controlling porosity and pore diameter as compared to other fabrication methods.

Biopolymer offers developers the tremendous flexibility to design porous matrices to broaden its applications in different areas day by day. In view of broadening the scope of biobased porous polymer applications, it is vital to take full advantageous of unique structure and properties of biopolymers to develop novel materials with pioneering new features.

Acknowledgments: The authors would like to acknowledge the financial support from the Ministry of Education Malaysia: FP030-2013A and FP053-2015A; University Malaya research grant: RP011A-13AET, RU022A-2014, RG031-15AET, PG160-2016A and RU018I-2016 for the success of this project.

Author Contributions: Udeni Gunathilake T.M. Sampath wrote the paper; Udeni Gunathilake T.M. Sampath and Yern Chee Ching initiated and contributed to the scope of the manuscript; Yern Chee Ching, Johari J. Sabariah and Pai-Chen Lin planned the review of the literature and revised the manuscript; Yern Chee Ching and Cheng Hock Chuah critically reviewed the manuscript.

References

1. Vroman, I.; Tighzert, L. Biodegradable polymers. *Materials* **2009**, *2*, 307–344. [CrossRef]

2. Lim, K.M.; Ching, Y.C.; Gan, S.N. Effect of palm oil bio-based plasticizer on the morphological, thermal and mechanical properties of poly(vinyl chloride). *Polymers* **2015**, *7*, 2031–2043. [CrossRef]

3. Othman, S.H. Bio-nanocomposite materials for food packaging applications: Types of biopolymer and nano-sized filler. *Agric. Agric. Sci. Procedia* **2014**, *2*, 296–303. [CrossRef]

4. Niaounakis, M. *Biopolymers: Applications and Trends*; William Andrew Publishing: Oxford, UK, 2015; pp. 91–138.

5. Ershad, A.; Yong, K.C.; Ching, Y.C.; Chuah, C.H.; Liou, N.S. Effect of single and double stage chemically treated kenaf fibers on mechanical properties of polyvinyl alcohol film. *Bioresource* **2015**, *10*, 822–838.

6. Tan, B.K.; Ching, Y.C.; Poh, S.C.; Luqman, C.A.; Gan, S.N. Review of natural fiber reinforced poly(vinyl alcohol) based composites: Application and Opportunity. *Polymers* **2015**, *7*, 2205–2222. [CrossRef]

7. Gavillon, R.; Budtova, T. Aerocellulose: New highly porous cellulose prepared from cellulose-NaOH aqueous solutions. *Biomacromolecules* **2007**, *9*, 269–277. [CrossRef] [PubMed]

8. Miao, X.; Sun, D. Graded/gradient porous biomaterials. *Materials* **2009**, *3*, 26–47. [CrossRef]

9. Dhandayuthapani, B.; Yoshida, Y.; Maekawa, T.; Kumar, D.S. Polymeric scaffolds in tissue engineering application: A review. *Int. J. Polym. Sci.* **2011**, *2011*, 290602. [CrossRef]

10. Nwe, N.; Furuike, T.; Tamura, H. The mechanical and biological properties of chitosan scaffolds for tissue regeneration templates are significantly enhanced by chitosan from gongronella butleri. *Materials* **2009**, *2*, 374–398. [CrossRef]

11. Rodriguez-Vazquez, M.; Vega-Ruiz, B.; Ramos-Zuniga, R.; Saldana-Koppel, D.A.; Quinones-Olvera, L.F. Chitosan and its potential use as a scaffold for tissue engineering in regenerative medicine. *BioMed Res. Int.* **2015**, *2015*, 821279. [CrossRef] [PubMed]

12. Croisier, F.; Jerome, C. Chitosan-based biomaterials for tissue engineering. *Eur. Polym. J.* **2013**, *49*, 780–792. [CrossRef]

13. Hsieh, W.-C.; Chang, C.-P.; Lin, S.-M. Morphology and characterization of 3D micro-porous structured chitosan scaffolds for tissue engineering. *Colloids Surf. B Biointerfaces* **2007**, *57*, 250–255. [CrossRef] [PubMed]

14. Simonescu, C.M.; Marin, I.; Tardei, C.; Marinescu, V.; Oprea, O.; Capatina, C. Chitosan and chitosan modified with glutaraldehyde microparticles for Pb(II) biosorption I. Microparticles preparation and characterization. *Rev. Chim.* **2014**, *65*, 627–632.

15. Ji, C.; Annabi, N.; Khademhosseini, A.; Dehghani, F. Fabrication of porous chitosan scaffolds for soft tissue engineering using dense gas CO_2. *Acta Biomater.* **2011**, *7*, 1653–1664. [CrossRef] [PubMed]

16. Ozdemir, E.; Sendemir-Urkmez, A.; Yesil-Celiktas, O. Supercritical CO_2 processing of a chitosan-based scaffold: Can implantation of osteoblastic cells be enhanced? *J. Supercrit. Fluids* **2013**, *75*, 120–127. [CrossRef]

17. Sivashankari, P.; Prabaharan, M. Prospects of chitosan-based scaffolds for growth factor release in tissue engineering. *Int. J. Biol. Macromol.* **2016**, *93 Pt B*, 1382–1389. [CrossRef] [PubMed]

18. Ching, Y.C.; Ershad, A.; Luqman, C.A.; Choo, K.W.; Yong, C.K.; Sabariah, J.J.; Chuah, C.H.; Liou, N.S. Rheological properties of cellulose nanocrystal-embedded polymer composites: A review. *Cellulose* **2016**, *23*, 1011–1030. [CrossRef]

19. Ching, Y.C.; Ng, T.S. Effect of preparation conditions on cellulose from oil palm empty fruit bunch fiber. *Bioresource* **2014**, *9*, 6373–6385. [CrossRef]

20. Rubentheren, V.; Thomas, W.; Ching, Y.C.; Praveena, N. Effects of heat treatment on chitosan nanocomposite film reinforced with nanocrystalline cellulose and tannic acid. *Carbohydr. Polym.* **2016**, *140*, 202–208. [CrossRef] [PubMed]

21. Goh, K.Y.; Ching, Y.C.; Chuah, C.H.; Luqman, C.A.; Liou, N.S. Individualization of microfibrillated celluloses from oil palm empty fruit bunch: Comparative studies between acid hydrolysis and ammonium persulfate oxidation. *Cellulose* **2016**, *23*, 379–390. [CrossRef]

22. Mhd Haniffa, M.A.C.; Ching, Y.C.; Luqman, C.A.; Poh, S.C.; Chuah, C.H. Review of bionanocomposite coating films and their applications. *Polymers* **2016**, *8*, 246. [CrossRef]

23. Medina-Gonzalez, Y.; Camy, S.; Condoret, J.-S. Cellulosic materials as biopolymers and supercritical CO_2 as a green process: Chemistry and applications. *Int. J. Sustain. Eng.* **2012**, *5*, 47–65. [CrossRef]

24. White, L.J.; Hutter, V.; Tai, H.; Howdle, S.M.; Shakesheff, K.M. The effect of processing variables on morphological and mechanical properties of supercritical CO_2 foamed scaffolds for tissue engineering. *Acta Biomater.* **2012**, *8*, 61–71. [CrossRef] [PubMed]

25. Baldino, L.; Cardea, S.; Reverchon, E. Production of antimicrobial membranes loaded with potassium sorbate using a supercritical phase separation process. *Innov. Food Sci. Emerg. Technol.* **2016**, *34*, 77–85. [CrossRef]

26. Xiong, G.; Luo, H.; Gu, F.; Zhang, J.; Hu, D.; Wan, Y. A novel in vitro three-dimensional macroporous scaffolds from bacterial cellulose for culture of breast cancer cells. *J. Biomater. Nanobiotechnol.* **2013**, *4*, 316–326. [CrossRef]

27. Chang, S.-W.; Shefelbine, S.J.; Buehler, M.J. Structural and mechanical differences between collagen homo-and heterotrimers: Relevance for the molecular origin of brittle bone disease. *Biophys. J.* **2012**, *102*, 640–648. [CrossRef] [PubMed]

28. Friess, W. Collagen-biomaterial for drug delivery. *Eur. J. Pharm. Biopharm.* **1998**, *45*, 113–136. [CrossRef]

29. Ferreira, A.M.; Gentile, P.; Chiono, V.; Ciardelli, G. Collagen for bone tissue regeneration. *Acta Biomater.* **2012**, *8*, 3191–3200. [CrossRef] [PubMed]

30. Oh, H.H.; Ko, Y.G.; Lu, H.; Kawazoe, N.; Chen, G. Preparation of porous collagen scaffolds with micropatterned structures. *Adv. Mater.* **2012**, *24*, 4311–4316. [CrossRef] [PubMed]

31. Nanda, H.S.; Chen, S.; Zhang, Q.; Kawazoe, N.; Chen, G. Collagen scaffolds with controlled insulin release and controlled pore structure for cartilage tissue engineering. *BioMed Res. Int.* **2014**, *2014*, 623805. [CrossRef] [PubMed]

32. Pourjavadi, A.; Kurdtabar, M.; Ghasemzadeh, H. Salt-and ph-resisting collagen-based highly porous hydrogel. *Polym. J.* **2008**, *40*, 94–103. [CrossRef]

33. Yee, Y.Y.; Ching, Y.C.; Shaifulazuar, R.; Awanis, H.; Ramesh, S. Preparation and characterization of poly(lactic acid) based composite reinforced with oil palm empty fruit bunch film and nanosilica. *Bioresource* **2016**, *11*, 2269–2286. [CrossRef]

34. Ching, Y.C.; Rosiyah, Y.Y.; Li, G. Preparation and characterization of nano particle reinforced polyactides composite. *J. Nano Res.* **2013**, *25*, 128–136.

35. Conde, C.M.; Demarco, F.F.; Casagrande, L.; Alcazar, J.C.; Nör, J.E.; Tarquinio, S.B.C. Influence of poly-L-lactic acid scaffold's pore size on the proliferation and differentiation of dental pulp stem cells. *Braz. Dent. J.* **2015**, *26*, 93–98. [CrossRef] [PubMed]

36. Chavalitkul, J.; Likittanaprasong, N.; Seansala, P.; Suttiruengwong, S.; Seadan, M. Application of ultrasonic atomization for biopolymer particles and tube fabrication. *Energy Procedia* **2014**, *56*, 458–465. [CrossRef]

37. Lee, H.-Y.; Jin, G.-Z.; Shin, U.S.; Kim, J.-H.; Kim, H.-W. Novel porous scaffolds of poly(lactic acid) produced by phase-separation using room temperature ionic liquid and the assessments of biocompatibility. *J. Mater. Sci. Mater. Med.* **2012**, *23*, 1271–1279. [CrossRef] [PubMed]

38. Gong, Y.; Ma, Z.; Zhou, Q.; Li, J.; Gao, C.; Shen, J. Poly(lactic acid) scaffold fabricated by gelatin particle leaching has good biocompatibility for chondrogenesis. *J. Biomater. Sci. Polym. Ed.* **2008**, *19*, 207–221. [CrossRef] [PubMed]

39. Wang, X.; Li, W.; Kumar, V. A method for solvent-free fabrication of porous polymer using solid-state foaming and ultrasound for tissue engineering applications. *Biomaterials* **2006**, *27*, 1924–1929. [CrossRef] [PubMed]

40. Yin, H.-M.; Qian, J.; Zhang, J.; Lin, Z.-F.; Li, J.-S.; Xu, J.-Z.; Li, Z.-M. Engineering porous poly(lactic acid) scaffolds with high mechanical performance via a solid state extrusion/porogen leaching approach. *Polymers* **2016**, *8*, 213. [CrossRef]

41. Makadia, H.K.; Siegel, S.J. Poly lactic-*co*-glycolic acid (PLGA) as biodegradable controlled drug delivery carrier. *Polymers* **2011**, *3*, 1377–1397. [CrossRef] [PubMed]

42. Gentile, P.; Chiono, V.; Carmagnola, I.; Hatton, P.V. An overview of poly(lactic-*co*-glycolic) acid (PLGA)-based biomaterials for bone tissue engineering. *Int. J. Mol. Sci.* **2014**, *15*, 3640–3659. [CrossRef] [PubMed]

43. Azimi, B.; Nourpanah, P.; Rabiee, M.; Arbab, S. Poly(lactide-*co*-glycolide) fiber: An overview. *J. Eng. Fabr. Fibers (JEFF)* **2014**, *9*, 47–66.

44. Krebs, M.D.; Sutter, K.A.; Lin, A.S.; Guldberg, R.E.; Alsberg, E. Injectable poly(lactic-*co*-glycolic) acid scaffolds with in situ pore formation for tissue engineering. *Acta Biomater.* **2009**, *5*, 2847–2859. [CrossRef] [PubMed]

45. Singh, L.; Kumar, V.; Ratner, B.D. Generation of porous microcellular 85/15 poly(DL-lactide-*co*-glycolide) foams for biomedical applications. *Biomaterials* **2004**, *25*, 2611–2617. [CrossRef] [PubMed]

46. Day, R.M.; Boccaccini, A.R.; Maquet, V.; Shurey, S.; Forbes, A.; Gabe, S.M.; Jérôme, R. In vivo characterisation of a novel bioresorbable poly(lactide-*co*-glycolide) tubular foam scaffold for tissue engineering applications. *J. Mater. Sci. Mater. Med.* **2004**, *15*, 729–734. [CrossRef] [PubMed]

47. Kim, S.; Kwag, D.S.; Lee, D.J.; Lee, E.S. Acidic pH-stimulated tiotropium release from porous poly(lactic-*co*-glycolic acid) microparticles containing 3-diethylaminopropyl-conjugated hyaluronate. *Macromol. Res.* **2016**, *24*, 176–181. [CrossRef]

48. Rubentheren, V.; Ward, T.A.; Chee, C.Y.; Tang, C.K. Processing and analysis of chitosan nanocomposites reinforced with chitin whiskers and tannic acid as a crosslinker. *Carbohydr. Polym.* **2015**, *115*, 379–387. [CrossRef] [PubMed]

49. Ching, Y.C.; Rahman, A.; Ching, K.Y.; Sukiman, N.L.; Cheng, H.C. Preparation and characterization of polyvinyl alcohol-based composite reinforced with nanocellulose and nanosilica. *BioResources* **2015**, *10*, 3364–3377. [CrossRef]

50. Yong, K.C.; Ching, Y.C.; Afzan, M.; Lim, Z.K.; Chong, K.E. Mechanical and thermal properties of chemical treated oil palm empty fruit bunches fiber reinforced polyvinyl alcohol composite. *J. Biobased Mater. Bioenergy* **2015**, *9*, 231–235.

51. Kumar, A.; Negi, Y.S.; Choudhary, V.; Bhardwaj, N.K. Microstructural and mechanical properties of porous biocomposite scaffolds based on polyvinyl alcohol, nano-hydroxyapatite and cellulose nanocrystals. *Cellulose* **2014**, *21*, 3409–3426. [CrossRef]

52. Li, L.; Ding, S.; Zhou, C. Preparation and degradation of PLA/chitosan composite materials. *J. Appl. Polym. Sci.* **2004**, *91*, 274–277. [CrossRef]

53. Zhang, Z.; Cui, H. Biodegradability and biocompatibility study of poly(chitosan-g-lactic acid) scaffolds. *Molecules* **2012**, *17*, 3243–3258. [CrossRef] [PubMed]

54. Xie, F.; Li, Q.F.; Gu, B.; Liu, K.; Shen, G.X. In vitro and in vivo evaluation of a biodegradable chitosan-PLA composite peripheral nerve guide conduit material. *Microsurgery* **2008**, *28*, 471–479. [CrossRef] [PubMed]

55. Choo, K.; Ching, Y.C.; Chuah, C.H.; Sabariah, J.; Liou, N.S. Preparation and characterization of polyvinyl alcohol-chitosan composite films reinforced with cellulose nanofiber. *Materials* **2016**, *9*, 644. [CrossRef]

56. Ng, T.S.; Ching, Y.C.; Awanis, N.; Ishenny, N.; Rahman, M.R. Effect of bleaching condition on thermal properties and UV-transmittance of PVA/cellulose biocomposites. *Mater. Res. Innov.* **2014**, *18*, 400–404. [CrossRef]

57. Cho, S.Y.; Park, H.H.; Yun, Y.S.; Jin, H.-J. Influence of cellulose nanofibers on the morphology and physical properties of poly(lactic acid) foaming by supercritical carbon dioxide. *Macromol. Res.* **2013**, *21*, 529–533. [CrossRef]

58. Dlouha, J.; Suryanegara, L.; Yano, H. The role of cellulose nanofibres in supercritical foaming of polylactic acid and their effect on the foam morphology. *Soft Matter* **2012**, *8*, 8704–8713. [CrossRef]

59. Xiang, C.; Taylor, A.G.; Hinestroza, J.P.; Frey, M.W. Controlled release of nonionic compounds from poly(lactic acid)/cellulose nanocrystal nanocomposite fibers. *J. Appl. Polym. Sci.* **2013**, *127*, 79–86. [CrossRef]

60. Twu, Y.-K.; Huang, H.-I.; Chang, S.-Y.; Wang, S.-L. Preparation and sorption activity of chitosan/cellulose blend beads. *Carbohydr. Polym.* **2003**, *54*, 425–430. [CrossRef]

61. Rubentheren, V.; Ward, T.A.; Chee, C.Y.; Nair, P. Physical and chemical reinforcement of chitosan film using nanocrystalline cellulose and tannic acid. *Cellulose* **2015**, *22*, 2529–2541. [CrossRef]

62. Li, M.; Wang, Z.; Li, B. Adsorption behaviour of congo red by cellulose/chitosan hydrogel beads regenerated from ionic liquid. *Desalination Water Treat.* **2016**, *57*, 16970–16980. [CrossRef]

63. Kim, J.; Cai, Z.; Lee, H.S.; Choi, G.S.; Lee, D.H.; Jo, C. Preparation and characterization of a bacterial cellulose/chitosan composite for potential biomedical application. *J. Polym. Res.* **2011**, *18*, 739–744. [CrossRef]

64. Kim, S.J.; Yang, D.H.; Chun, H.J.; Chae, G.T.; Jang, J.W.; Shim, Y.B. Evaluations of chitosan/poly(D,L-lactic-*co*-glycolic acid) composite fibrous scaffold for tissue engineering applications. *Macromol. Res.* **2013**, *21*, 931–939. [CrossRef]

65. Yuanyuan, Z.; Song, L. Preparation of chitosan/poly(lactic-*co* glycolic acid)(PLGA) nanocoposite for tissue engineering scaffold. *Optoelectron. Adv. Mater.-Rapid Commun.* **2012**, *6*, 516–519.

66. Cui, Z.; Zhao, H.; Peng, Y.; Han, J.; Turng, L.-S.; Shen, C. Fabrication and characterization of highly porous chitosan/poly(DL lactic-*co*-glycolic acid) nanocomposite scaffolds using electrospinning and freeze drying. *J. Biobased Mater. Bioenergy* **2014**, *8*, 281–291. [CrossRef]

67. Ajalloueian, F.; Tavanai, H.; Hilborn, J.; Donzel-Gargand, O.; Leifer, K.; Wickham, A.; Arpanaei, A. Emulsion electrospinning as an approach to fabricate PLGA/chitosan nanofibers for biomedical applications. *BioMed Res. Int.* **2014**, *2014*, 475280. [CrossRef] [PubMed]

68. Peng, L.; Cheng, X.R.; Wang, J.W.; Xu, D.X.; Wang, G. Preparation and evaluation of porous chitosan/collagen scaffolds for periodontal tissue engineering. *J. Bioact. Compat. Polym.* **2006**, *21*, 207–220. [CrossRef]

69. Ma, L.; Gao, C.; Mao, Z.; Zhou, J.; Shen, J.; Hu, X.; Han, C. Collagen/chitosan porous scaffolds with improved biostability for skin tissue engineering. *Biomaterials* **2003**, *24*, 4833–4841. [CrossRef]

70. Mighri, N.; Mao, J.; Mighri, F.; Ajji, A.; Rouabhia, M. Chitosan-coated collagen membranes promote chondrocyte adhesion, growth, and interleukin-6 secretion. *Materials* **2015**, *8*, 7673–7689. [CrossRef]

71. Wang, L.; Stegemann, J.P. Glyoxal crosslinking of cell-seeded chitosan/collagen hydrogels for bone regeneration. *Acta Biomater.* **2011**, *7*, 2410–2417. [CrossRef] [PubMed]

72. Ashworth, J.; Best, S.; Cameron, R. Quantitative architectural description of tissue engineering scaffolds. *Mater. Technol.* **2014**, *29*, 281–295. [CrossRef]

73. Mallick, S.; Tripathi, S.; Srivastava, P. Advancement in scaffolds for bone tissue engineering: A review. *IOSR J. Pharm. Biol. Sci.* **2015**, *10*, 37–54.

74. Lim, J.I.; Lee, Y.-K.; Shin, J.-S.; Lim, K.-J. Preparation of interconnected porous chitosan scaffolds by sodium acetate particulate leaching. *J. Biomater. Sci. Polym. Ed.* **2011**, *22*, 1319–1329. [CrossRef] [PubMed]

75. Lee, J.S.; Lee, H.K.; Kim, J.Y.; Hyon, S.H.; Kim, S.C. Thermally induced phase separation in poly(lactic acid)/dialkyl phthalate systems. *J. Appl. Polym. Sci.* **2003**, *88*, 2224–2232. [CrossRef]

76. Li, D.; Krantz, W.B.; Greenberg, A.R.; Sani, R.L. Membrane formation via thermally induced phase separation (TIPS): Model development and validation. *J. Membr. Sci.* **2006**, *279*, 50–60. [CrossRef]

77. Yang, H.C.; Wu, Q.Y.; Liang, H.Q.; Wan, L.S.; Xu, Z.K. Thermally induced phase separation of poly(vinylidene fluoride)/diluent systems: Optical microscope and infrared spectroscopy studies. *J. Polym. Sci. B Polym. Phys.* **2013**, *51*, 1438–1447. [CrossRef]

78. Kim, J.F.; Kim, J.H.; Lee, Y.M.; Drioli, E. Thermally induced phase separation and electrospinning methods for emerging membrane applications: A review. *AIChE J.* **2016**, *62*, 461–490. [CrossRef]

79. Wei, G.; Ma, P.X. Structure and properties of nano-hydroxyapatite/polymer composite scaffolds for bone tissue engineering. *Biomaterials* **2004**, *25*, 4749–4757. [CrossRef] [PubMed]

80. La Carrubba, V.; Pavia, F.C.; Brucato, V.; Piccarolo, S. PLLA/PLA scaffolds prepared via thermally induced phase separation (TIPS): Tuning of properties and biodegradability. *Int. J. Mater. Form.* **2008**, *1*, 619–622. [CrossRef]

81. Liu, X.; Ma, P.X. Polymeric scaffolds for bone tissue engineering. *Ann. Biomed. Eng.* **2004**, *32*, 477–486. [CrossRef] [PubMed]

82. Sachlos, E.; Czernuszka, J. Making tissue engineering scaffolds work. Review: The application of solid freeform fabrication technology to the production of tissue engineering scaffolds. *Eur. Cell Mater.* **2003**, *5*, 39–40.

83. Poursamar, S.A.; Hatami, J.; Lehner, A.N.; da Silva, C.L.; Ferreira, F.C.; Antunes, A.P.M. Gelatin porous scaffolds fabricated using a modified gas foaming technique: Characterisation and cytotoxicity assessment. *Mater. Sci. Eng. C* **2015**, *48*, 63–70. [CrossRef] [PubMed]

84. Bhamidipati, M.; Scurto, A.M.; Detamore, M.S. The future of carbon dioxide for polymer processing in tissue engineering. *Tissue Eng. B Rev.* **2013**, *19*, 221–232.

85. Mooney, D.J.; Baldwin, D.F.; Suh, N.P.; Vacanti, J.P.; Langer, R. Novel approach to fabricate porous sponges of poly(D,L-lactic-*co*-glycolic acid) without the use of organic solvents. *Biomaterials* **1996**, *17*, 1417–1422. [CrossRef]

86. Bak, T.-Y.; Kook, M.-S.; Jung, S.-C.; Kim, B.-H. Biological effect of gas plasma treatment on CO_2 gas foaming/salt leaching fabricated porous polycaprolactone scaffolds in bone tissue engineering. *J. Nanomater.* **2014**, *2014*, 657542. [CrossRef]

87. Annabi, N.; Mithieux, S.M.; Weiss, A.S.; Dehghani, F. Cross-linked open-pore elastic hydrogels based on tropoelastin, elastin and high pressure CO_2. *Biomaterials* **2010**, *31*, 1655–1665. [CrossRef] [PubMed]

88. Whang, K.; Thomas, C.; Healy, K.; Nuber, G. A novel method to fabricate bioabsorbable scaffolds. *Polymer* **1995**, *36*, 837–842. [CrossRef]

89. Sultana, N.; Wang, M. Fabrication of tissue engineering scaffolds. In *Integrated Biomaterials in Tissue Engineering*; Scrivener Publishing: Beverly, MA, USA, 2012; pp. 63–89.

90. Mikos, A.G.; Temenoff, J.S. Formation of highly porous biodegradable scaffolds for tissue engineering. *Electron. J. Biotechnol.* **2000**, *3*, 23–24. [CrossRef]

91. Sultana, N.; Wang, M. Fabrication of HA/PHBV composite scaffolds through the emulsion freezing/freeze-drying process and characterisation of the scaffolds. *J. Mater. Sci. Mater. Med.* **2008**, *19*, 2555–2561. [CrossRef] [PubMed]

92. Torabi, K.; Farjood, E.; Hamedani, S. Rapid prototyping technologies and their applications in prosthodontics, a review of literature. *J. Dent. (Shiraz)* **2014**, *16*, 1–9.

93. Sobral, J.M.; Caridade, S.G.; Sousa, R.A.; Mano, J.F.; Reis, R.L. Three-dimensional plotted scaffolds with controlled pore size gradients: Effect of scaffold geometry on mechanical performance and cell seeding efficiency. *Acta Biomater.* **2011**, *7*, 1009–1018. [CrossRef] [PubMed]

94. Hoque, M.E.; Chuan, Y.L.; Pashby, I. Extrusion based rapid prototyping technique: An advanced platform for tissue engineering scaffold fabrication. *Biopolymers* **2012**, *97*, 83–93. [CrossRef] [PubMed]

95. Alvarez, K.; Nakajima, H. Metallic scaffolds for bone regeneration. *Materials* **2009**, *2*, 790–832. [CrossRef]

96. Stanek, M.; Manas, D.; Manas, M.; Navratil, J.; Kyas, K.; Senkerik, V.; Skrobak, A. Comparison of different rapid prototyping methods. *Intern. J. Math. Comput. Simul.* **2012**, *6*, 550–557.

97. Peltola, S.M.; Melchels, F.P.; Grijpma, D.W.; Kellomäki, M. A review of rapid prototyping techniques for tissue engineering purposes. *Ann. Med.* **2008**, *40*, 268–280. [CrossRef] [PubMed]

98. Abdelaal, O.A.; Darwish, S.M. Fabrication of tissue engineering scaffolds using rapid prototyping techniques. *Int. J. Mech. Aerosp. Ind. Mechatron. Manuf. Eng.* **2011**, *5*, 2317–2325.

99. Zhu, N.; Li, M.; Cooper, D.; Chen, X. Development of novel hybrid poly(L-lactide)/chitosan scaffolds using the rapid freeze prototyping technique. *Biofabrication* **2011**, *3*, 034105. [CrossRef] [PubMed]

100. Li, X.; Cui, R.; Sun, L.; Aifantis, K.E.; Fan, Y.; Feng, Q.; Cui, F.; Watari, F. 3D-printed biopolymers for tissue engineering application. *Int. J. Polym. Sci.* **2014**, *2014*, 829145. [CrossRef]

101. Cox, S.C.; Thornby, J.A.; Gibbons, G.J.; Williams, M.A.; Mallick, K.K. 3D printing of porous hydroxyapatite scaffolds intended for use in bone tissue engineering applications. *Mater. Sci. Eng. C* **2015**, *47*, 237–247. [CrossRef] [PubMed]

102. Lam, C.X.F.; Mo, X.; Teoh, S.-H.; Hutmacher, D. Scaffold development using 3d printing with a starch-based polymer. *Mater. Sci. Eng. C* **2002**, *20*, 49–56. [CrossRef]

103. Lee, M.; Dunn, J.C.; Wu, B.M. Scaffold fabrication by indirect three-dimensional printing. *Biomaterials* **2005**, *26*, 4281–4289. [CrossRef] [PubMed]

104. Shi, X.; Zhou, W.; Ma, D.; Ma, Q.; Bridges, D.; Ma, Y.; Hu, A. Electrospinning of nanofibers and their applications for energy devices. *J. Nanomater.* **2015**, *2015*, 140716. [CrossRef]

105. Agarwal, S.; Wendorff, J.H.; Greiner, A. Use of electrospinning technique for biomedical applications. *Polymer* **2008**, *49*, 5603–5621. [CrossRef]

106. Sundararaghavan, H.G.; Metter, R.B.; Burdick, J.A. Electrospun fibrous scaffolds with multiscale and photopatterned porosity. *Macromol. Biosci.* **2010**, *10*, 265–270. [CrossRef] [PubMed]

107. Bhardwaj, N.; Kundu, S.C. Electrospinning: A fascinating fiber fabrication technique. *Biotechnol. Adv.* **2010**, *28*, 325–347. [CrossRef] [PubMed]

108. Dahlin, R.L.; Kasper, F.K.; Mikos, A.G. Polymeric nanofibers in tissue engineering. *Tissue Eng. B Rev.* **2011**, *17*, 349–364. [CrossRef] [PubMed]

109. Chew, S.; Wen, Y.; Dzenis, Y.; Leong, K.W. The role of electrospinning in the emerging field of nanomedicine. *Curr. Pharm. Des.* **2006**, *12*, 4751–4770. [CrossRef] [PubMed]
110. Zhang, Y.; Feng, Y.; Huang, Z.; Ramakrishna, S.; Lim, C.T. Fabrication of porous electrospun nanofibres. *Nanotechnology* **2006**, *17*, 901–908. [CrossRef]
111. Roozbahani, F.; Sultana, N.; Almasi, D.; Naghizadeh, F. Effects of chitosan concentration on the protein release behaviour of electrospun poly(ε-caprolactone)/chitosan nanofibers. *J. Nanomater.* **2015**, *2015*, 11. [CrossRef]

2

Eco-Friendly and Biodegradable Biopolymer Chitosan/Y$_2$O$_3$ Composite Materials in Flexible Organic Thin-Film Transistors

Bo-Wei Du [ID], Shao-Ying Hu, Ranjodh Singh, Tsung-Tso Tsai, Ching-Chang Lin * and Fu-Hsiang Ko *

Department of Materials Science and Engineering, National Chiao Tung University, 1001 University Road, Hsinchu 30010, Taiwan; dbw6522@gmail.com (B.-W.D.); Gj94ekup8896@gmail.com (S.-Y.H.); chemrjd@gmail.com (R.S.); kenny2870@gmail.com (T.-T.T.)
* Correspondence: kyo@nctu.edu.tw (C.-C.L.); fhko@mail.nctu.edu.tw (F.-H.K.)

Abstract: The waste from semiconductor manufacturing processes causes serious pollution to the environment. In this work, a non-toxic material was developed under room temperature conditions for the fabrication of green electronics. Flexible organic thin-film transistors (OTFTs) on plastic substrates are increasingly in demand due to their high visible transmission and small size for use as displays and wearable devices. This work investigates and analyzes the structured formation of aqueous solutions of the non-toxic and biodegradable biopolymer, chitosan, blended with high-k-value, non-toxic, and biocompatible Y$_2$O$_3$ nanoparticles. Chitosan thin films blended with Y$_2$O$_3$ nanoparticles were adopted as the gate dielectric thin film in OTFTs, and an improvement in the dielectric properties and pinholes was observed. Meanwhile, the on/off current ratio was increased by 100 times, and a low leakage current was observed. In general, the blended chitosan/Y$_2$O$_3$ thin films used as the gate dielectric of OTFTs are non-toxic, environmentally friendly, and operate at low voltages. These OTFTs can be used on surfaces with different curvature radii because of their flexibility.

Keywords: non-toxic; organic thin-film transistors (OTFTs); chitosan; Y$_2$O$_3$; flexible

1. Introduction

For two decades, the global developments of the electronics industry have focused on flexible electronic devices, such as curved full-color displays [1,2] integrated sensors [3], flexible solar cells, and the amazing achievement of E-paper [4,5]. Using a solution-based process achieves many advantages that are cost-effective and simple to fabricate, and produces mechanically flexible thin-film transistors compared to conventional semiconductor technologies, which depend on vacuum-based thin film fabrication [6,7]. In the past decade, eco-friendly, biocompatible, and green materials have been the subject of many economic and scientific projects [8–10], and have caused less damage to the environment. Oxide thin films under low annealing temperatures have been fabricated by using an inexpensive "water-inducement" technique [11] that combines a high-k-value YO$_X$ dielectric material with an eco-friendly water-inducement process [12]. Because of their superior performance, organic thin-film transistors (OTFTs) can be used to replace conventional thin-film transistors (TFTs). Chitin is a natural amino polysaccharide and is the largest nitrogenous natural organic compound on the planet after protein and the cellulose polysaccharides found in nature [13]. It has many outstanding characteristics, such as biocompatibility, non-toxicity, biodegradability, antimicrobial activity, and excellent mechanical strength, which make it suitable for use in the biomedical field [14,15]. Chitosan can be transformed into chitin and has the same excellent characteristics, such as biocompatibility, biodegradability, non-toxicity, antimicrobial activity, and an outstanding film-forming ability [16–18]. Therefore, in previous studies, chitosan has also been used as dielectric layer in organic transistors [19,20]. Yttrium (III) oxide (Y_2O_3) is a type of rare earth oxide that is non-toxic, thermodynamically stabile, stabile at high temperature (T_m = 2430 °C), and has a high dielectric constant (ε = 15~18), light transparency, and a linear transmittance in the infrared spectra. Y_2O_3 is commonly known as a high-k dielectric material that can replace SiO_2, because it has a high-k dielectrics value and a phase with cubic symmetry. Its lattice constant, a = 10.6 Å, is two times as large as the lattice constant of Si (a = 5.43 Å). Y_2O_3 can be deposited by different deposition techniques, including pulsed laser deposition (PLD) [21], sputtering metal-organic chemical vapor deposition (MOCVD) [22], and electron beam evaporation [23].

Therefore, due to its high-k-value, non-toxicity, and biocompatibility, Y_2O_3 nanoparticles were blended into the chitosan solution in this study to improve their dielectric properties and pinholes. To achieve good electrical performance with the chitosan-based metal-insulator-metal (MIM) structure, various concentrations of Y_2O_3 nanoparticles were blended into the chitosan to decrease the leakage current and improve the depth of the pinholes. Chitosan thin films have an electric-double-layer effect that gives OTFTs the property of low-voltage operation. Furthermore, the thin films of chitosan blended with Y_2O_3 nanoparticles were used as the dielectric material in OTFTs, and the performance of these OTFTs was enhanced.

2. Experimental

Yttrium (III) oxide (Y_2O_3) was provided by Alfa-Aesar (Heysham, UK). Chitosan, poly(3-hexylthiophene) (P3HT) and acetic acid were provided by Sigma-Aldrich (St. Louis, MO, USA). All other reagents and anhydrous solvents were obtained from local suppliers and used without further purification, unless otherwise noted.

The Y_2O_3/chitosan thin film as the dielectric gate of the flexible OTFT, and P-type organic semiconductor, poly(3-hexylthiophene) (P3HT), as the semiconductor layer on polyimide substrate were demonstrated in this study. The basic process flow for the fabrication of this flexible device is shown in Figure 1. Moreover, the performance of the flexible OTFT during bending tests with different curvature radii was also observed. In detail, a 5-nm Cr metal layer was deposited as an adhesive layer on the polyimide film by thermal deposition and then a 30-nm Au layer was deposited on the adhesive layer as the bottom gate. The adhesive layer was used to stabilize the Au layer and guarantee that the device would be stable under bending tests. The polyimide substrate was first cleaned before placing it into thermal coater chamber, after which a Cr and Au metal layer was deposited. Before the dielectric layer was deposited, we covered the adhesive tape on the bottom Au electrode, for the bottom gate can only be exposed in the final step.

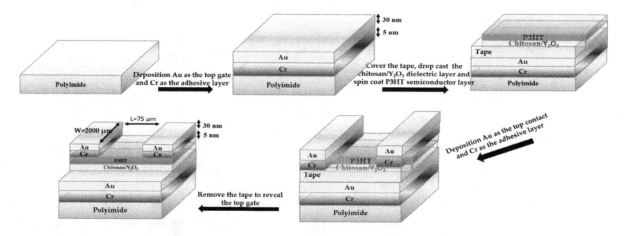

Figure 1. The fabrication process of the bottom-gate top-contact flexible organic thin-film transistor.

Chitosan (deacetylated $\geq$75%) was dissolved in aqueous acetic acid (0.5 wt %) and heated with a hot plate at 50 °C for 24 h, following which its solutions of various concentrations were filtered with a 25-mm syringe filter containing a 0.45-μm polyvinylidene difluoride (PVDF) membrane. The Y_2O_3 nanoparticles were blended with deionized water at a ratio of 0.5 wt % and the hybrid solution was obtained by mixing chitosan aqueous solution with Y_2O_3 nanoparticles aqueous solution, with a specific volume ratio CS:Y_2O_3 (20:1). In other words, the weight percentage of Y_2O_3 in the blended solutions was 0.023 wt %. The drop casting was used to form the 0.023 wt % Y_2O_3/chitosan film on the bottom-gate as the dielectric gate, which was then dried in an oven at room temperature for 24 h.

After depositing the dielectric gate, a P3HT channel layer was deposited on the Y_2O_3/chitosan dielectric film by spin-coating at 1200 rpm for 30 s and then 1500 rpm for 30 s. The polyimide substrate was placed in an oven at 60 °C to remove residual solvent in the P3HT active channel layer. Finally, a 5-nm Cr and a 30-nm Au layer were deposited with a mask to form the source and drain top contacts. Cr metal was used as an adhesive layer between the P3HT channel layer and Au contacts as before. In the meantime, the bottom Au electrode was revealed by carefully removing the tape. Therefore, as the above procedure was finished, the bottom-gate top-contact flexible organic thin-film transistor was successfully designed.

3. Results and Discussion

3.1. Materials and Films Characterization

Chitosan (deacetylated $\geq$75%) was dissolved in aqueous acetic acid (0.5 wt %) and heated by using a hot plate at 55 °C for 24 h. The impurities in the chitosan solutions of various concentrations were filtered using a 25-mm syringe filter with a 0.45-μm PVDF membrane. The chitosan solution with various concentrations was transferred by spin-coating onto separate single silicon substrates that were already coated with aluminum metal as an electrode by spin-coating. Then, we removed the water in the chitosan thin film by heating on a hot plate at 80 °C for 1 h. Finally, we deposited the aluminum metal as the top electrode on the chitosan, which formed a so-called MIM structure. We found that the lower leakage current was 6.827×10^{-10} A at an applied voltage of 2 V in the MIM based on a 1.0 wt % chitosan thin film, and observed that the 1.0 wt % chitosan film had the lowest leakage current. This result was attributed to the size and the number of the pinholes in the surface of the chitosan, as shown in Figure 2. The thin film with 0.5 wt % chitosan had many small pinholes (10–20 nm), as shown in Figure 2a, and the thin film with 1.5 wt % chitosan had some large pinholes (80–100 nm), as shown in Figure 2c. The thin film with 1.0 wt % chitosan had the optimized hole size (30–50 nm) and number of holes, as shown in Figure 2b; this is why the 1.0 wt % sample had the lowest leakage current.

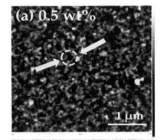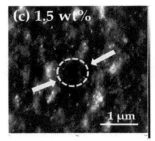

Figure 2. (**a**) 0.5 wt % chitosan shows many small holes (**b**) 1.0 wt % chitosan shows a few middle-sized holes and (**c**) 1.5 wt % chitosan shows many large holes.

The high-k-value Y_2O_3 nanoparticles were blended into a 1 wt % chitosan water solution with various weight percentages of Y_2O_3 (from 0.012 wt % to 0.016 wt %, 0.023 wt % and 0.045 wt %). While the weight percentage increased, the leakage current decreased. The blended 0.045 wt % Y_2O_3

in 1 wt % chitosan thin film had the lowest leakage current of 1.81×10^{-11} A, as shown in Figure 3a. The pure chitosan thin films were almost transparent, the weight percentage of the Y_2O_3 increased, as shown in Figure 3b. In the meantime, we discovered a decrease in the relative depth (from 1.025 to 0.356 nm) when the concentration of the Y_2O_3 increased, as shown in Figure 4a–d. However, the relative depth and roughness were increased when the weight percentage of Y_2O_3 reached 0.045 wt %, as shown in Figure 4e. The purpose of blending the Y_2O_3 nanoparticles into the chitosan thin film was not to only reduce the leakage current, but also to improve the pinholes at the surface.

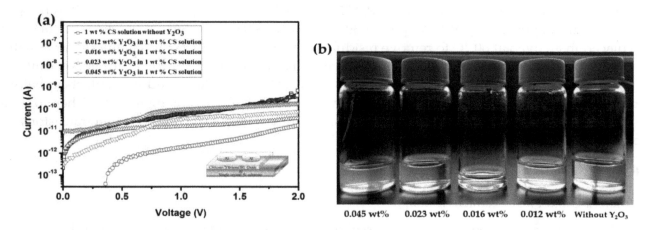

Figure 3. (**a**) Leakage current of the thin films measured with various concentrations of Y_2O_3 solutions (without Y_2O_3, 0.012 wt %, 0.016 wt %, 0.023 wt %, 0.045 wt %) in 1 wt % chitosan solution (**b**) Pictures showing the various weight percentages of Y_2O_3 in the mixed solution.

The cross-sectional images of the blended Y_2O_3/chitosan thin films are shown in Figure 5, and the thicknesses of the blended thin films were approximately 120 nm–200 nm. We also analyzed the distribution of the Y_2O_3 nanoparticles in the surface of the blended thin films by energy-dispersive X-ray spectroscopy (EDX, JEOL, Freising, Germany). We discovered that the 0.012 wt %, 0.016 wt %, and 0.023 wt % blended thin films showed a uniform dispersion of the Y_2O_3 nanoparticles, as shown in Figure 6a–c. The Y_2O_3 nanoparticles attracted each other and clusters formed when the weight percentage of Y_2O_3 reached 0.045 wt % (Figure 6d). In Figure 4e, we measured the relatively large profile depth of 1.051 nm, and we attributed this phenomenon to the clustering of the Y_2O_3 nanoparticles. It was observed that the 0.023 wt % thin film had smoothest surface, and its pinholes were much improved.

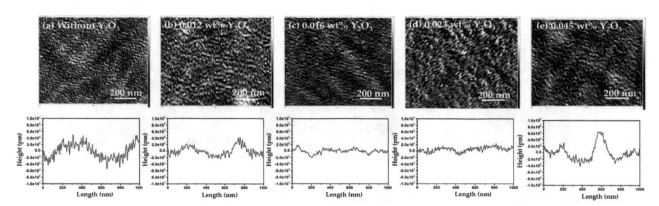

Figure 4. AFM morphology and profile depth of the pinholes in the surface of the Y_2O_3/chitosan thin film with various Y_2O_3 concentrations: (**a**) Without Y_2O_3; (**b**) 0.012 wt %; (**c**) 0.016 wt %; (**d**) 0.023 wt %; and (**e**) 0.045 wt %.

The FTIR spectra analysis (PerkinElmer, Waltham, MA, USA) of the blended Y_2O_3/chitosan thin films was used to obtain information on the chemical bonding, as shown in Figure 7a. The pure chitosan thin film showed broad absorption peaks at 3000–5000 cm^{-1} that were attributed to NH_2 asymmetric stretching and the hydrogen-bonded OH. The peak at approximately 2879 cm^{-1} was attributed to the CH_3 asymmetric stretching vibrations, and the absorption peak at 1541 cm^{-1} was attributed to the asymmetric bending modes [24–26] of NH_3^+. Figure 7a also shows the FTIR spectra analysis of all the blended Y_2O_3/chitosan thin films, which are listed in Table 1. The NH_3^+ was assigned to the bending frequency at 1541 cm^{-1} for pure chitosan, which shifted to a higher frequency at 1559 cm^{-1} and 1580 cm^{-1} for Y_2O_3 = 0.012 wt % and 0.016 wt %, respectively, then shifted to a higher frequency at 1598 cm^{-1} for Y_2O_3 = 0.023 wt % and shifted to a lower frequency as 1578 cm^{-1} for Y_2O_3 = 0.045 wt %. The shifts were due to the H-bonds between the oxygen of yttrium (III) oxide and the amine of chitosan, as shown in Figure 7b. The AFM morphology (Veeco, Plainview, NY, USA) also proved that a dense structure and fewer pinholes were formed by blending Y_2O_3 nanoparticles into the thin film, as shown in Figure 4.

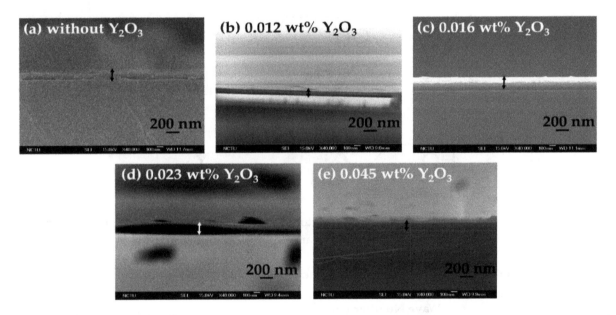

Figure 5. Cross-section images of the blended Y_2O_3/chitosan thin films: (**a**) Without Y_2O_3; (**b**) 0.012 wt %; (**c**) 0.016 wt %; (**d**) 0.023 wt %; and (**e**) 0.045 wt %.

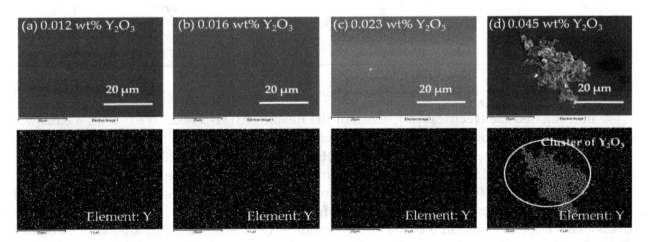

Figure 6. EDX images showing the distribution of yttrium: (**a**) 0.012 wt %; (**b**) 0.016 wt %; (**c**) 0.023 wt %; and (**d**) 0.045 wt %.

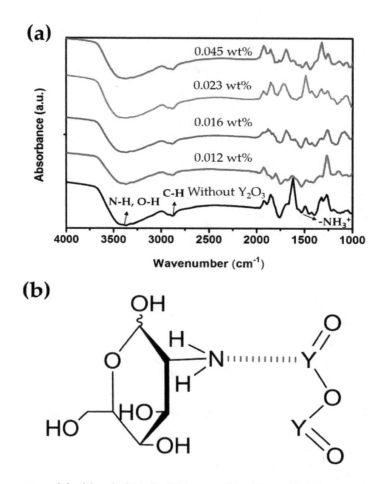

Figure 7. (a) FTIR spectra of the blended Y_2O_3/chitosan thin films; **(b)** Schematic diagram of hydrogen bonding between the Y_2O_3 nanoparticles and the amine group of chitosan.

Table 1. FTIR spectra analysis of the blended Y_2O_3/chitosan thin film.

Y_2O_3 wt %	Wave Number (cm^{-1})	CH Stretching	NH_3^+ Bending
	O–H and N–H Stretching Broad Absorption Peaks		
n/a	3368	2879	1541
0.012	3368	2880	1559
0.016	3368	2880	1580
0.023	3367	2880	1598
0.045	3367	2879	1578

3.2. Electric Characteristics of the Flexible Organic Thin Film Transistor

We investigated the electrical properties of the flexible P3HT-based OTFTs with a 0.023 wt % blended Y_2O_3/chitosan dielectric gate on the polyimide substrate. Figure 8 shows the transfer plots for concave and convex bending of the OTFTs with concave bending radii, R, of 3.5 cm and 2.8 cm and convex bending radii, R, of 3.5 cm and 2.8 cm. In this figure, the electrical characterization of the OTFTs on polyimide substrates without bending is similar to that of the OTFTs on silicon wafers. After concave bending, the I_{off} value decreased from 7.421×10^{-10} A to 5.740×10^{-11} A due to extrusion, causing a decrease in the size of the pinholes. On the other hand, after convex bending, the I_{off} value increased to 1.722×10^{-9} A. No matter the direction of bending, the I_{on} value would decrease. In Figure 8, we compared the electrical characterization of the OTFTs with different bending radii (3.5 cm and 2.8 cm), and the comparison chart is listed in Table 2. The output characterization (I_{DS}-V_{DS}) of this device, as shown in Figure S1.

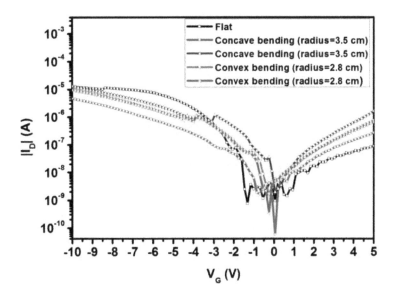

Figure 8. Comparison of the electrical characterization of the P3HT-based organic thin-film transistors (OTFTs) with a 0.023 wt % blended Y_2O_3/chitosan dielectric gate for bending tests at different bending radii and without bending.

Table 2. Electrical characterization of the P3HT-based OTFTs with a 0.023 wt % blended Y_2O_3/chitosan dielectric gate on the polyimide substrate for bending tests.

Condition	V_{th} (V)	I_{on} (A)	I_{off} (A)	I_{off}/I_{on} Ratio	Mobility (cm^2/Vs)
Flat	−2.0	1.268×10^{-5}	7.421×10^{-10}	10^5	2.50×10^{-2}
Concave bending 3.50 (cm)	−2.5	4.592×10^{-6}	5.740×10^{-11}	10^5	3.33×10^{-2}
Concave bending 2.85 (cm)	−2.7	1.298×10^{-5}	1.913×10^{-9}	10^4	2.70×10^{-2}
Flat	−2.1	1.294×10^{-5}	4.571×10^{-10}	10^5	1.87×10^{-4}
Convex bending 3.50 (cm)	−3.0	9.899×10^{-6}	1.722×10^{-9}	10^3	8.53×10^{-2}
Convex bending 2.85 (cm)	−3.3	9.480×10^{-6}	3.296×10^{-10}	10^3	4.80×10^{-2}
Flat	−1.5	1.059×10^{-5}	2.539×10^{-9}	10^4	3.33×10^{-2}

4. Conclusions

A solution-based processed and low-voltage operating P3HT-based OTFT with a Y_2O_3/chitosan gate dielectric layer was demonstrated in this study. To improve the electrical performance of the chitosan-based MIM, various concentrations of Y_2O_3 nanoparticles were blended into the chitosan, which achieved a decreased leakage current and improved the depth of the pinholes. Furthermore, the P3HT-based OTFT with a 0.023 wt % blended Y_2O_3/chitosan gate dielectric layer was manufactured on polyimide for bending tests. The electrical performance of the flexible device incurred no obvious changes except for a slight increase in the leakage current and off current. The non-toxic and eco-friendly biopolymer chitosan, blended with Y_2O_3 nanoparticles, was successfully used in flexible OTFTs as the gate dielectric, enabling the OTFT to operate under low voltages, and producing a I_{on}/I_{off} ratio of 10^5 at a gate voltage of −10 V and a drain voltage of −1 V.

Supplementary Materials:
Figure S1: Output characterization of P3HT-based OTFTs with 0.023 wt % blended Y_2O_3/chitosan dielectric gate.

Acknowledgments: The authors would like to thank the Ministry of Science and Technology of Taiwan for supporting this work under Contract MOST 104-2113-M-009-008-MY3 and MOST 105-2622-M-009-004-CC3.

Author Contributions: Fu-Hsiang Ko, Ranjohn Singh and Ching-chang Lin designed the experiments. Shao-Ying Hu and Tsung-Tso Tsai carried out the experimental work. Bo-Wei Du wrote the manuscript. All authors contributed to the discussions of this work.

References

1. Forrest, S.R. The path to ubiquitous and low-cost organic electronic appliances on plastic. *Nature* **2004**, *428*, 911–918. [CrossRef] [PubMed]

2. Li, G.; Zhu, R.; Yang, Y. Polymer solar cells. *Nat. Photonics* **2012**, *6*, 153–161. [CrossRef]

3. Vichare, N.M.; Pecht, M.G. Prognostics and health management of electronics. *IEEE Trans. Compon. Packag. Technol.* **2006**, *29*, 222–229. [CrossRef]

4. Pagliaro, M.; Ciriminna, R.; Palmisano, G. Flexible solar cells. *ChemSusChem* **2008**, *1*, 880–891. [CrossRef] [PubMed]

5. Jang, J. Displays develop a new flexibility. *Mater. Today* **2006**, *9*, 46–52. [CrossRef]

6. Fortunato, E.M.; Barquinha, P.M.; Pimentel, A.C.; Goncalves, A.M.; Marques, A.J.; Martins, R.F.; Pereira, L.M. Wide-bandgap high-mobility ZnO thin-film transistors produced at room temperature Wide-bandgap high-mobility ZnO thin-film transistors produced at room temperature. *Appl. Phys. Lett.* **2004**, *85*, 2541–2543. [CrossRef]

7. Yang, Z.; Hao, J.; Yuan, S.; Lin, S.; Yau, H.M.; Dai, J.; Lau, S.P. Field-Effect Transistors Based on Amorphous Black Phosphorus Ultrathin Films by Pulsed Laser Deposition. *Adv. Mater.* **2015**, *27*, 3748–3754. [CrossRef] [PubMed]

8. Arthur, T.; Harjani, J.R.; Phan, L.; Jessop, P.G.; Hodson, P.V. Effects-driven chemical design: The acute toxicity of CO_2-triggered switchable surfactants to rainbow trout can be predicted from octanol-water partition coefficients. *Green Chem.* **2012**, *14*, 357–362. [CrossRef]

9. Hwang, S.W.; Park, G.; Edwards, C.; Corbin, E.A.; Kang, S.K.; Cheng, H.; Lee, J.E. Dissolution chemistry and biocompatibility of single-crystalline silicon nanomembranes and associated materials for transient electronics. *ACS Nano* **2014**, *8*, 5843–5851. [CrossRef] [PubMed]

10. Agharkar, M.; Kochrekar, S.; Hidouri, S.; Azeez, M.A. Trends in green reduction of graphene oxides, issues and challenges: A review. *Mater. Res. Bull.* **2014**, *59*, 323–328. [CrossRef]

11. Liu, G.; Liu, A.; Zhu, H.; Shin, B.; Fortunato, E.; Martins, R.; Shan, F. Low-Temperature, Nontoxic Water-Induced Metal-Oxide Thin Films and Their Application in Thin-Film Transistors. *Adv. Funct. Mater.* **2015**, *25*, 2564–2572. [CrossRef]

12. Cao, X.; Shen, F.; Zhang, M.; Bie, J.; Liu, X.; Luo, Y.; Sun, C. Facile synthesis of chitosan-capped ZnS quantum dots as an eco-friendly fluorescence sensor for rapid determination of bisphenol A in water and plastic samples. *RSC Adv.* **2014**, *4*, 16597–16606. [CrossRef]

13. Pillai, C.K.S.; Paul, W.; Sharma, C.P. Chitin and chitosan polymers: Chemistry, solubility and fiber formation. *Prog. Polym. Sci.* **2009**, *34*, 641–678. [CrossRef]

14. Chung, Y.C.; Tsai, C.F.; Li, C.F. Preparation and characterization of water-soluble chitosan produced by Maillard reaction. *Fish. Sci.* **2006**, *72*, 1096–1103. [CrossRef]

15. Rinaudo, M. Chitin and chitosan: Properties and applications. *Prog. Polym. Sci.* **2006**, *31*, 603–632. [CrossRef]

16. Couto, D.S.; Hong, Z.; Mano, J.F. Development of bioactive and biodegradable chitosan-based injectable systems containing bioactive glass nanoparticles. *Acta Biomater.* **2009**, *5*, 115–123. [CrossRef] [PubMed]

17. Jayakumar, R.; Menon, D.; Manzoor, K.; Nair, S.V.; Tamura, H. Biomedical applications of chitin and chitosan based nanomaterials-A short review. *Carbohydr. Polym.* **2010**, *82*, 227–232. [CrossRef]

18. Wan, Y.; Wu, H.; Cao, X.; Dalai, S. Compressive mechanical properties and biodegradability of porous poly (caprolactone)/chitosan scaffolds. *Polym. Degrad. Stab.* **2008**, *93*, 1736–1741. [CrossRef]

19. Morgado, J.; Pereira, A.T.; Bragança, A.M.; Ferreira, Q.; Fernandes, S.C.M.; Freire, C.S.R.; Alcácer, L. Self-standing chitosan films as dielectrics in organic thin-film transistors. *Express Polym. Lett.* **2013**, *7*, 960–965. [CrossRef]

20. Liu, Y.H.; Zhu, L.Q.; Feng, P.; Shi, Y.; Wan, Q. Freestanding Artificial Synapses Based on Laterally Proton-Coupled Transistors on Chitosan Membranes. *Adv. Mater.* **2015**, *27*, 5599–5604. [CrossRef] [PubMed]

21. Yi, S.S.; Bae, J.S.; Moon, B.K.; Jeong, J.H.; Park, J.C.; Kim, I.W. Enhanced luminescence of pulsed-laser-deposited Y_2O_3: Eu^{3+} thin-film phosphors by Li doping. *Appl. Phys. Lett.* **2002**, *81*, 3344–3346. [CrossRef]

22. Durand, C.; Dubourdieu, C.; Vallée, C.; Loup, V.; Bonvalot, M.; Joubert, O.; Renault, O. Microstructure and electrical characterizations of yttrium oxide and yttrium silicate thin films deposited by pulsed liquid-injection plasma-enhanced metal-organic chemical vapor deposition. *J. Appl. Phys.* **2004**, *96*, 1719–1729. [CrossRef]

23. Atanassov, G.; Thielsch, R.; Popov, D. Optical properties of TiO$_2$, Y$_2$O$_3$ and CeO$_2$ thin films deposited by electron beam evaporation. *Thin Solid Films* **1993**, *223*, 288–292. [CrossRef]

24. Ko, Y.G.; Lee, H.J.; Shin, S.S.; Choi, U.S. Dipolar-molecule complexed chitosan carboxylate, phosphate, and sulphate dispersed electrorheological suspensions. *Soft Matter* **2012**, *8*, 6273–6279. [CrossRef]

25. Li, Z.; Zhuang, X.P.; Liu, X.F.; Guan, Y.L.; Yao, K.D. Study on antibacterial *O*-carboxymethylated chitosan/cellulose blend film from LiCl/*N*, *N*-dimethylacetamide solution. *Polymer* **2002**, *43*, 1541–1547. [CrossRef]

26. Liu, X.F.; Guan, Y.L.; Yang, D.Z.; Li, Z.; Yao, K.D. Antibacterial action of chitosan and carboxymethylated chitosan. *J. Appl. Polym. Sci.* **2001**, *79*, 1324–1335.

3

Hybrid Cellulose-Glass Fiber Composites for Automotive Applications

Cindu Annandarajah [1] , **Amy Langhorst** [2] , **Alper Kiziltas** [2], **David Grewell** [3],
Deborah Mielewski [2] **and Reza Montazami** [4,*]

[1] Department of Agricultural and Biosystems Engineering, Iowa State University, Ames, IA 50011, USA;
cindu@iastate.edu

[2] Ford Motor Company, Research and Advanced Engineering, Dearborn, MI 48124, USA;
alangho1@ford.com (A.L); akizilt1@ford.com (A.K.); dmielews@ford.com (D.M.)

[3] Department of Industrial and Manufacturing Engineering, North Dakota State University, Fargo, ND 58102,
USA; david.grewell@ndsu.edu

[4] Department of Mechanical Engineering, Iowa State University, Ames, IA 50011, USA

[*] Correspondence: reza@iastate.edu

Abstract: In the recent years, automakers have been striving to improve the carbon footprint of their vehicles. Sustainable composites, consisting of natural fibers, and/or recycled polymers have been developed as a way to increase the "green content" and reduce the weight of a vehicle. In addition, recent studies have found that the introduction of synthetic fibers to a traditional fiber composite such as glass filled plastics, producing a composite with multiple fillers (hybrid fibers), can result in superior mechanical properties. The objective of this work was to investigate the effect of hybrid fibers on characterization and material properties of polyamide-6 (PA6)/polypropylene (PP) blends. Cellulose and glass fibers were used as fillers and the mechanical, water absorption, and morphological properties of composites were evaluated. The addition of hybrid fibers increased the stiffness (tensile and flexural modulus) of the composites. Glass fibers reduced composite water absorption while the addition of cellulose fibers resulted in higher composite stiffness. The mechanical properties of glass and cellulose filled PA6/PP composites were optimized at loading levels of 15 wt% glass and 10 wt% cellulose, respectively.

Keywords: composites; hybrid fibers; cellulose; glass fiber; automotive; compatibilizer

1. Introduction

Increasing global industrialization has resulted in environmental deterioration, including land and air pollution, leading to more global environmental awareness and promoting the investigation of environmentally friendly and sustainable materials. In addition, new legislation in large industrial markets, such as the European Union, has driven the automotive industry to prioritize global sustainability. Even though there is no federal law governing extended producer responsibility (EPR) in the United States, "product stewardship" practices call for shared responsibility among manufacturers and consumers to reduce product impact on the environment [1].

Currently, about 50% of the volume of materials in the cars are made of polymeric materials. The average usage of plastics in automotive in developed countries and globally averages are 120 kg and 105 kg, respectively, accounting for 10–12% of the total weight of a vehicle [2–4]. The Corporate Average Fuel Economy (CAFE) estimates that a reduction of 10% of an automobile's weight will reduce its fuel consumption by 6–8% [3]. In attempts to lightweight and decrease the carbon footprint of vehicles, automakers have expressed increased interest in bio-based materials, especially natural-fiber reinforced polymer composites including kenaf, hemp, sisal, jute, and flax [5,6]. Automobile manufacturers have incorporated these natural fibers as the reinforcing phase for polymer composites in door panels, seat backs, headliners, package trays, dashboards, and interior parts. Use of these materials not only increases the "green content" of each vehicle, but can also contribute to reduction in weight, cost, carbon footprint, and lead to less dependence on foreign and domestic petroleum-based fuels [2,5]. These fiber based composites have the advantage to meet diverse design requirements with a 30% weight reduction and a cost reduction of 20% [7].

Development of hybrid composite reinforced with more than two types of fibers in matrix provides a more favorable balance of material properties. In hybridization, the formulation is made so that the fibers are able to support the loads while the matrix adheres the fiber together for efficient load transfer in the composite [8]. The properties of a hybrid composite are dependent on several factors, such as the nature of matrix; the nature, length, and relative composition of the reinforcements, fiber–matrix interface, and hybrid design [9]. For instance, Kalaprasad et al. studied the hybrid composited of sisal/glass reinforced polyethylene (SGRP) and reported that the mechanical properties were increased with increase in volume fraction of glass fibers [10]. In another study, Langhorst et al. found that increased fiber loading level in agave fiber filled polypropylene composite enhanced the stiffness of the material [11]. Joseph et al. investigated the mechanical properties and water sorption behavior of phenol–formaldehyde (PF) hybrid composites reinforced with banana fiber and glass fiber. It is found that the tensile, impact, and the flexural properties of the banana fiber-PF composites have been increased by the hybridization of glass fibers [12]. In another study, Kahl et al. reported that the glass fiber content helped in significant increase of the tensile strength and compared to cellulose fiber composites at an overall 16% vol. ratio [13].

The automotive industry has been looking into adapting polymer blends, such as the polypropylene (PP) and Nylon 6 (PA6) blends, for "under-the-hood" applications where thermal stability is a key parameter. PP is commonly used for automobile parts because it is inexpensive, highly processable, and exhibits high water/chemical resistance. However, the relatively low modulus and poor heat resistance of PP makes it unsuitable for use in under-the-hood components [14]. PA6, on the other hand, exhibits high heat resistance, tensile strength, modulus, resistance to corrosive chemicals, and has attractive surface appearance, but readily absorbs moisture resulting in dimensional instability. Thus, blends of PP and PA6 yield intermediate properties that can be suitable for engine covers, air intake manifolds, cooling, and heating components, and cylinder head cover [15,16]. On the contrary, the opposing polarities of PP and PA6 causes phase separation in the blend and could result in poor mechanical properties [16–18]. In order to improve the mechanical properties and the morphology of PP/PA6 blends, PP grafted maleic-acid anhydride (PPgMA) has been used as a reactive compatibilizer [17,19]. Many studies have investigated the use of polymers blends of PP/PA6 along with compatibilizing agents [20–23]. In addition, there are also works done on the use of short glass fiber as a filler or reinforcement in these polymer blends [24–26]. However, to the authors' best recollection, there are no studies that have been published investigating the mechanical properties of natural fiber reinforced PA6/PP blends or even cellulose-glass fiber reinforced PP/ PA6 composites via injection molding, a technique that is used in the automotive industry. The main objective was to study the dispersion of cellulose and glass fibers in recycled PP/PA6 blends, and to examine the fiber hybridization effect on the mechanical, morphological, and water absorption properties of these polymer blends.

2. Materials and Methods

2.1. Procurement

Post-consumer recycled polyamide 6 (PA6) and polypropylene (PP) were supplied by Wellman Advanced Materials (Johnsonville, SC, USA). Maleic anhydride-*grafted*-polypropylene (PPgMA) with a grafting level of 0.5 wt% maleic-anhydride (Fusabond P613) was supplied by DuPont (Wilmington,

DE, USA). Cellulose fiber (~150 μm $\times$ 20 μm $\times$ ~2 μm) and glass fiber (~6–10 mm) were obtained from International Paper (Memphis, TN, USA) and PPG Industries Inc. (Pittsburgh, PA, USA), respectively. The supplied PP and PA6 had melting points of approximately 160 °C and 220 °C, respectively.

2.2. Extrusion and Injection Molding

Nine samples were produced, and their formulations are shown in Table 1. Each formulation consisted of a 70:30 wt% ratio of PA6: PP and, 6 wt% PPgMA. All composites contained 15–30 wt% glass fibers and 10–30 wt% cellulose fibers. Control groups consisting of an unfilled PA6/PP blend, a 30% glass filled PA6/PP blend, and a 30% cellulose filled PA6/PP blend were produced.

Table 1. Experimental design for the preparation of recycled polypropylene and Nylon 6 blends with hybrid fibers composites.

Formulation	Final Composition				
	PA6 (wt%)	PP (wt%)	PPgMA(wt%)	Glass (wt%)	Cellulose (wt%)
PA6/ PP/ PPgMA (Blend)	65.8	28.2	6	0	0
Blend + 30% Glass fiber	44.8	19.2	6	30	0
Blend + 30% Cellulose	44.8	19.2	6	0	30
Blend + 15% Glass fiber + 10% Cellulose	48.3	20.7	6	15	10
Blend + 15% Glass fiber + 15% Cellulose	44.8	19.2	6	15	15
Blend + 20% Glass fiber + 10% Cellulose	44.8	19.2	6	20	10
Blend + 20% Glass fiber + 15% Cellulose	41.3	17.7	6	20	15

Extrusion was completed using a twin-screw laboratory extruder (ThermoHaake Rheomex Model PTW25, Thermo Fisher Scientific, Waltham, MA, USA). Prior to extruding, the materials were dried for at least 12 hours at 60 °C in a conventional oven to reduce the moisture content to less than 1% in the fibers and the possibility of hydrolysis the nylon in the blend. The blend pellets and the hybrid fibers (cellulose and glass fiber) were separately starve-fed into the extruder using K-Tron gravimetric feeders (Coperion, Stuttgart, Germany), and the extruded samples were immediately immersed into a room-temperature water bath. Extruder temperatures ranged from 205 to 220 °C from the hopper to the die. The compounded materials were pelletized using a lab-scale chopper. Pellets were dried in a conventional oven for at least 12 hours at 60 °C to reduce moisture content less than 1% before being injection molded into test specimens using a Boy Machine model 80M injection molding machine (BOY, Exton, PA, USA). Molding occurred using barrel temperatures ranging from 175 to 185 °C, with a mold temperature of 26 °C.

3. Test Procedures

3.1. Mechanical Testing

Tensile, flexural, and impact tests were performed according to ASTM D638-10 (2010), ASTM D790-10 (2010), and ASTM D256-10 (2010), respectively. All samples were tested in an environmentally conditioned room maintained at 23 °C and 50% relative humidity. Tensile tests were performed on an Instron 3366 machine (Instron, Norwood, MA, USA) using a 5 kN load cell, 50 mm extensometer, and extension rate of 5 mm/min. Three-point bend flexural tests were performed using a 50 mm wide support span and were run at a crosshead speed of 1.25 mm/min. Notched Izod impact tests were performed using a TMI machine (model 43-02-03), (Testing Machines Inc., New Castle, DE, USA) fitted with a 2 lb pendulum. At least six tensile, five flexural, and ten impact test specimens were tested for each formulation.

The ultimate tensile strength, yield strength, elongation at break, and Young's modulus were determined from the tensile stress-strain curves, while flexural strength and flexural modulus were determined from the flexural stress-strain curves. The impact strength was determined from the notched Izod impact tests.

3.2. Water Absorption Test

Water immersion tests were performed according to ASTM D570. At least six specimens of each formulation were pre-conditioned in a 50 °C oven for 24 h to remove all moisture. All the samples were cooled at room temperature for 30 min and then weighed to the nearest 0.001 g. The conditioned samples were then immersed completely in a container of distilled water at 23 °C. The samples were weighed again to the nearest 0.001 g after 24 h, seven days, 21 days, and 35 days until the increase in weight for 3 consecutive weighing averages less than 1%.

3.3. Microscopy

The fractured surfaces of the tensile specimens were mounted vertically on sample holder to expose the composite's cross-section perpendicular to the injection molding flow direction. The cross-sections of samples were observed using a FEI Quanta FEG 250 field emission SEM (Thermo Fisher Scientific, (Waltham, MA, USA). A 10 keV beam was used and backscatter electron images were collected. The low vacuum mode was used with 60 Pa water vapor environment. The working distance was set at approximately 10 mm.

3.4. Statistical Analysis

Results were analyzed by ANOVA using SAS (version 9.2; SAS Institute Inc., Cary, SC, USA) and XLStat (version 2013.4; Addinsoft, New York, NY, USA). A significance level of $\alpha = 0.05$ with Tukey adjustment for multiple comparisons was used to determine significant differences.

4. Results and Discussion

4.1. Mechanical Properties

4.1.1. Strength and Elongation

Figure 1 shows the tensile strength and strain at maximum load for the composites. The 30% glass fiber control sample exhibited the highest tensile strength (64% more than unfilled control) while the unfilled control exhibited the highest elongation at the ultimate strength. Contrary to cellulose being a high elongation fiber [6], the 30% cellulose control exhibited the lowest elongation at the ultimate strength. The significant reduction in both tensile strength and elongation may have been caused by agglomeration of cellulose fibers at high loading levels as well as poor adhesion between fiber and matrix. This result is in agreement with Sukri et al., who reported that extension was reduced by 63% when kenaf fiber loading levels were increased from 0 to 30% in rPP/rPA6 blends [17]. Santos et al. also reported that the tensile strength and extension of PA6 and Curaua fiber composites (with fiber content from 20 to 40%) was lower compared to the PA6 control sample [18].

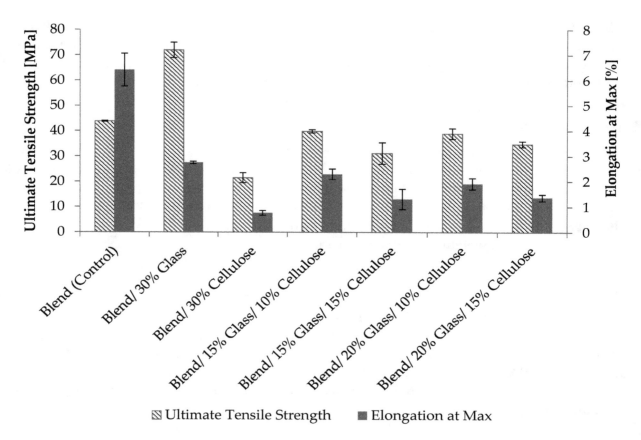

Figure 1. Tensile strength and elongation at maximum load of unfilled polymer blends and cellulose + glass fiber hybrid reinforced composites.

No significant differences were observed in tensile strength and elongation of composites containing 15 or 20% glass fibers. It is speculated that the effect of the glass fiber reinforcement in these composites is maximized near 15% loading level. In more detail, strength as a function of glass content asymptotically approaches a maximum level around 15% filler levels. The increment of filler ratio could also cause more fiber ends to act as stress generators during deformation of composites and this results in causing earlier fiber-matrix debonding and crack development during deformation [27,28]. This assumption is supported by a study by Mishra et al. in which the tensile strength of sisal/glass hybrid polyester composites did not increase with the addition of glass fibers beyond 5.7 wt%. Similarly, the author reported that the tensile strength of a pineapple leaf/glass hybrid polyester composite decreased by about 10% when loaded with more than 12.9% glass fiber [9]. This result is also in line with the studies by Franciszczak et al. in which, addition of 14 vol%. glass fiber with PP gave merely 27% further increase in composite's tensile strength [28]. In general, as the cellulose content in the composites increased, the tensile strength and elongation was reduced. While cellulose is a high elongation fiber [6], this reduction may be attributed to the decreased strength of cellulose compared to glass fiber. The glass fiber, relatively stiff, failed first which caused the transfer of the applied load to the cellulose fibers. The hybrid composite exhibited medium elongation (higher compared to the cellulose control but lower compared to the glass control) compared to the individual fiber reinforced composites [12].

The results of flexural strength and impact strength for PP, PA6, PPgMA, cellulose, and glass fiber composites are shown in Figure 2. The 30% glass fiber control and 30% cellulose control samples exhibited the largest and smallest flexural and impact strengths, respectively. Results from composites containing 15% glass fiber suggest that increasing cellulose loading-levels reduced the flexural and impact strength of the hybrid composites. The impact strength of fiber-reinforced composites depends on many factors, including the nature of the constituents, fiber/matrix adhesion, construction, and geometry of the composites, and test conditions. It is possible that cellulose fiber agglomeration resulted in the formation of stress concentrations, reducing the energy needed to initiate crack propagation.

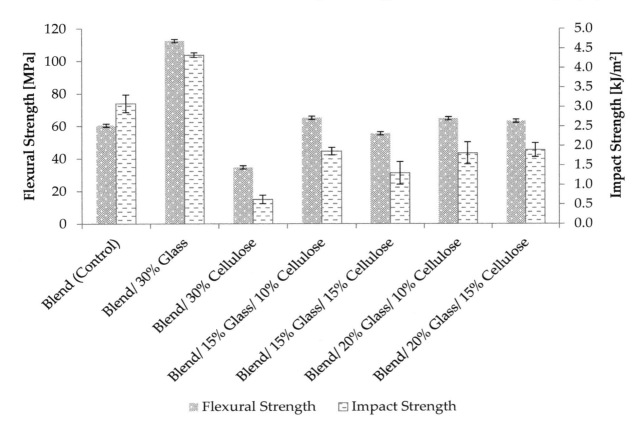

Figure 2. Flexural and impact strength of unfilled polymer blends and cellulose and glass fiber reinforced composites.

Additionally, replacement of cellulose fiber with glass fiber resulted in an increase in flexural and impact strength. For example, 15% glass + 15% cellulose composites (30% total fiber content) exhibited lower flexural and impact strength compared to 20% glass + 10% cellulose composites (30% total fiber content) (Figure 2). This is consistent with previous studies that reported an increase in impact strength with glass fiber content. For example, Misra et al. observed a 34% increase in impact strength by the addition of 8.5% glass fiber to sisal-fiber-reinforced polyester composites. In a three-point flexure test, failure occurred as a result of bending failure and shear failure [9]. It is possible that higher glass fiber content prevents shearing, resulting in an increase in flexural strength.

4.1.2. Stiffness

Figure 3 shows the Young's and flexural moduli for the composites. The 30% glass composites exhibited the highest moduli of all tested composites. It is also important to note that the hybrid composite of glass and cellulose fiber results in higher flexural strength than the tensile strengths implying the composite has better response to compression stress than tensile. In general, Young's and flexural moduli increased with increasing cellulose and glass fiber content and these results are in agreement with the Rule of Hybrid Mixture (ROHM). This result is in agreement with Kahl et al., who reported the increase in tensile modulus with increasing content of glass fibers in the PP/glass fiber and cellulose composite due to the higher modulus of glass fiber [13]. Several mechanisms, such as compression, shearing *etc.*, take place together during the stiffness test [29]. The significant increase in stiffness is due to the reinforcement effect of the cellulose and the increased resistance to shearing with the addition of glass fiber. This result is supported by a study of Sukri et al. who studied rPP, rPA6 and kenaf fiber by varying the fiber content from 10 to 30% [17]. Lei et al. also found a 50% increase in modulus by adding 30 wt% of bagasse to recycled high density polyethylene (rHDPE) [30].

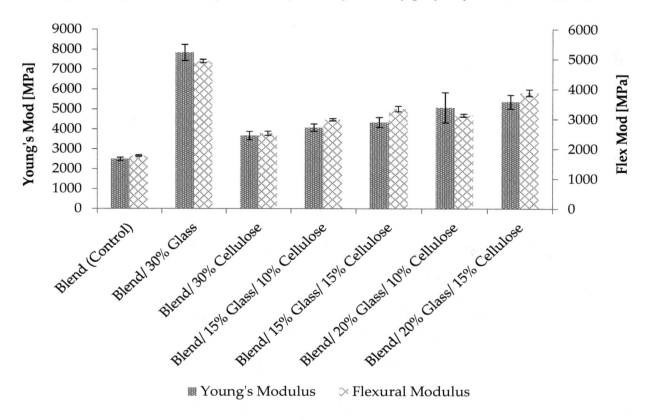

Figure 3. Young's modulus and flexural modulus of unfilled polymer blends and cellulose and glass fiber reinforced composites.

4.2. Water Absorption

The water absorption of PA6/ PP/ PPgMA composites is shown in Figure 4. The highest and lowest water absorptions during a 3-week period was exhibited by 30% cellulose control (5.5%) and 30% glass fiber (2.89%), respectively. The hydrophilic OH groups on the surface of the cellulose crystallites or in the amorphous region may be available for bonding with water if there is no crosslinking with other OH groups. The water bonding at the amorphous region and the free water in the cellulose cavities cause an increase in absorption [31]. Among the composites, there was no significant difference in water absorption with increasing cellulose content. However, increasing the composite's glass fiber content from 15 to 20% significantly reduced water absorption. The water enters through the interface and can diffuse through the porous structure of the fibers. Water penetration and diffusion mainly occur at the fiber–matrix interface and through the fibers via capillary mechanism [31]. When glass fiber was incorporated, the water uptake decreased, as the diffusion of water is not possible through glass fiber, as shown in Figure 4. In more detail, it is seen that in general, water absorption is inversely proportional to glass content. Similar observation has been seen by Joseph et al. who reported that the incorporation of a small amount of glass fiber (12%) increased the resistance of banana/PF composites to water sorption very effectively [12].

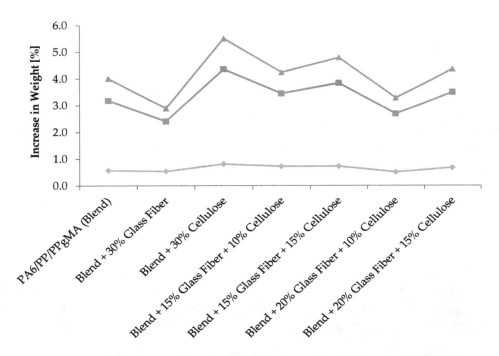

Figure 4. Water absorption of unfilled polymer blends and cellulose and glass fiber reinforced composites.

A summary of the mechanical results is shown in Table 2. Incorporation of glass fibers increased the strength and moduli of the PA6/PP/PPgMA blend composites. The cellulose control group reduced the strength, elongation, stiffness, and increased water absorption of the polymer composites. On average, the composites with glass/cellulose fiber mixtures showed only increased Young's modulus and flexural modulus of the composite and a reduction in tensile strength and elongation compared to the unfilled control. It has no effect on the flexural, impact strength and water absorption (for 10 and 15% only). The addition of glass and cellulose fibers improved the stiffness of the polymer blend, but all composites showed lower strength and elongation compared to the unfilled polymer blend. In this study, 15% glass + 10% cellulose composites exhibited the best balance in properties. Sreekala et al. analyzed the tensile, elongation, and stiffness properties of phenol formaldehyde reinforced with glass and oil palm fibers composites as a function of fiber composition. The author reported that 40 wt% fiber loading, composites with 0.74 volume fraction of oil palm fiber (29.6% oil palm fiber, 10.4% glass fiber) exhibited the highest tensile properties among the hybrid composites as excellent dispersion of the fibers and increased oil palm fiber/glass compatibility occurs at this composition [29].

Table 2. Details of mechanical properties for PP, PA6, and hybrid fibers (cellulose and glass fibers) composites. Samples containing the combination of one or both fibers were compared to their unfilled control, lacking the hybrid fibers. Boxes labelled (+) exhibited property improvement, white boxes with (o) experienced no significant property change, and boxes (-) experienced property degradation.

Properties		Single Filler Composites		Dual Filler Composites
		Glass Fiber	Cellulose	Glass Fiber + Cellulose
Strength	Tensile	+	-	-
	Flex	+	-	o
	Impact	+	-	o
Elongation	Tensile	-	-	-
Stiffness	Young's Modulus	+	-	+
	Flexural Modulus	+	-	+
Absorptivity	Water Absorption	o	+	+

4.3. Morphological Properties

Scanning Electron Microscopy

SEM images in Figure 5 show morphological differences between the control polymer blends and composite containing the blend and the glass and cellulose fibers. The micrograph of Figure 1a illustrates the well dispersion of glass fibers in the polymer blend matrix. Figure 1b shows cellulose fibers within a polymer blend matrix (cellulose fibers are circled). The matrix surrounding the cellulose fibers is cracked, suggesting that adhesion between the fiber and matrix is weak and this is in line with the low tensile strength achieved in the results. The micrographs of blend + 15% glass fiber + 10% cellulose fiber suggests that the area of the cellulose fibers is surrounded with glass fibers, which effectively reinforces the fiber and matrix together, leading to good interfacial adhesion between the two and this in result brings the tensile property of this formulation close to the one of neat blend formulation.

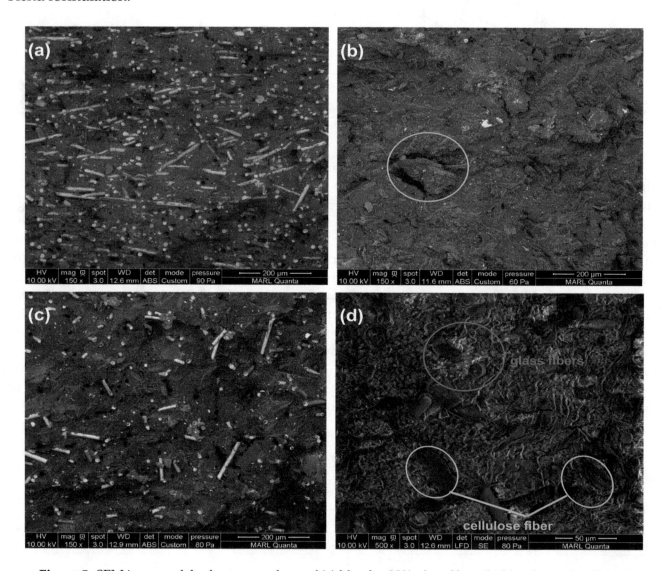

Figure 5. SEM images of the fracture surfaces of (**a**) blend + 30% glass fiber; (**b**) blend + 30% cellulose fiber (circled); and (**c,d**) blend + 15% glass fiber + 10% cellulose fiber at 150× and 500× respectively.

5. Conclusions

This study determined the effects of adding cellulose and glass fibers on the mechanical, morphological, and water absorption properties of PP/PA6/PPgMA composites. In general, the mechanical properties were enhanced by addition of glass fibers alone: tensile strength, Young's modulus, flexural modulus, flexural strength, and Izod impact strength were increased at 30% glass fiber load. Even though the composites with hybrid (cellulose and glass) fibers did not perform better than the control PP/ PA6 blend, increasing the cellulose content from 10 to 20% increased the Young's and flexural modulus and water absorption, while flexural strength and elongation was reduced. In addition, when the glass fiber content was increased from 15 to 20% in the composite, the flexural strength, Young's modulus and impact strength improved as well. There are no significant differences observed in the different loading levels from 10 to 30%. The 15% glass + 10% cellulose fiber composite showed the best properties among the composites with hybrid (cellulose and glass) fibers as the modulus and percentage elongation at break was the highest. Effect of fiber reinforcement on thermal properties and effect of compatibilizer should also be studied in the future to learn more about the thermal stability, crystallinity, compatibility, and to improve the properties of these composites.

Author Contributions: Conceptualization, A.K. and D.M.; methodology, C.A. and A.L.; software, C.A.; validation, C.A., A.L., A.K., D.G., and R.M.; formal analysis, C.A.; investigation, C.A. and A.L.; resources, D.M.; writing—original draft preparation, C.A.; writing—review and editing, C.A., A.L., A.K., D.G., and R.M.; visualization, A.L.; supervision, A.L., A.K., D.M., D.G., and R.M.; project administration, A.L., A.K.

References

1.   U.S. Environmental Protection Agency. Recycling and Reuse: End-of Life-Vehicles and Extended Producer Responsibility: European Union Directive. 2016. Available online: https://archive.epa.gov/oswer/international/web/html/200811_elv_directive.html. (accessed on 20 February 2018).

2.   Holbery, J.; Houston, D. Natural-fibre-reinforced polymer composites in automotive applications. *J. Miner. Met. Mater. Soc.* **2006**, *58*, 80–86. [CrossRef]

3.   Huiqin, W. 2011 engineering plastics sector to see increase of 10.93%. 2012. Available online: http://en.ce.cn/Insight/201202/01/t20120201_23034917.shtml. (accessed on 18 February 2018).

4.   Kiziltas, A.; Erbas Kiziltas, E.; Boran, S.; Gardner, D.J. Micro-and nanocellulose composites for automotive applications. In Proceedings of the Society of Plastics Engineers-13th Annual Automotive Composites Conference and Exhibition, ACCE, Novi, MI, USA, 11–13 September 2013; Volume 1, pp. 402–414.

5.   Njuguna, J.; Wambua, P.; Pielichowski, K.; Kayvantash, K. Natural Fibre-Reinforced Polymer Composites and Nanocomposites for Automotive Applications. In *Cellulose Fibers: Bio- and Nano-Polymer Composites*; Kalia, S., Kaith, B.S., Kaur, I., Eds.; Springer: Heidelberg, German, 2011; Volume 70, pp. 661–700.

6.   Saba, N.; Tahir, P.M.; Jawaid, M. A review on potentiality of nano filler/natural fiber filled polymer hybrid composites. *Polymers* **2014**, *6*, 2247–2273. [CrossRef]

7.   Ahmad, F.; Choi, H.S.; Park, M.K. A review: Natural fiber composites selection in view of mechanical, light weight, and economic properties. *Macromol. Mater. Eng.* **2015**, *300*, 10–24. [CrossRef]

8.   Yashas Gowda, T.G.; Sanjay, M.R.; Subrahmanya Bhat, K.; Madhu, P.; Senthamaraikannan, P.; Yogesha, B. Polymer matrix-natural fiber composites: An overview. *Cogent Eng.* **2018**, *5*, 1446667. [CrossRef]

9.   Mishra, S.; Mohanty, A.K.; Drzal, L.T.; Misra, M.; Parija, S.; Nayak, S.K.; Tripathy, S.S. Studies on mechanical performance of biofibre/glass reinforced polyester hybrid composites. *Compos. Sci. Technol.* **2003**, *63*, 1377–1385.

10. Kalaprasad, G.; Joseph, K.; Thomas, S. Influence of Short Glass Fiber Addition on the Mechanical Properties of Sisal Reinforced Low Density Polyethylene Composites. *J. Compos. Mater.* **1997**, *31*, 509–527. [CrossRef]

11. Langhorst, A.E.; Burkholder, J.; Long, J.; Thomas, R.; Kiziltas, A.; Mielewski, D. Blue-agave fiber-reinforced polypropylene composites for automotive applications. *BioResources* **2018**, *13*, 820–835. [CrossRef]

12. Joseph, S.; Sreekala, M.S.; Koshy, P.; Thomas, S. Mechanical properties and water sorption behavior of phenol–formaldehyde hybrid composites reinforced with banana fiber and glass fiber. *J. Appl. Polym. Sci.* **2008**, *109*, 1439–1446. [CrossRef]

13. Kahl, C.; Feldmann, M.; Sälzer, P.; Heim, H.P. Advanced short fiber composites with hybrid reinforcement and selective fiber-matrix-adhesion based on polypropylene – Characterization of mechanical properties and fiber orientation using high-resolution X-ray tomography. *Compos. Part A Appl. Sci. Manuf.* **2018**, *111*, 54–61. [CrossRef]

14. Karian, H.G. *Handbook of Polypropylene and Polypropylene Composites*, 2nd ed.; CRC press: Boca Raton, FL, USA, 2003.

15. Ha, C.S.; Park, H.D.; Cho, W.J. Compatibilizers for Recycling of the Plastic Mixture Wastes. II. the Effect of a Compatibilizer for Binary Blends on the Properties of Ternary Blends. *J. Appl. Polym. Sci.* **2000**, *76*, 1048–1053. [CrossRef]

16. Langhorst, A.; Kiziltas, A.; Mielewski, D.; Lee, E. Selective dispersion and compatibilizing effect of cellulose filler in recycled PA6/PP blends. Presented at SPE ACCE, Novi, MI, USA, September 2015.

17. Sukri, S.M.; Suradi, L.; Arsad, A.; Rahmat, A.R.; Hassan, A. Green composites based on recycled polyamide-6/recycled polypropylene kenaf composites: mechanical, thermal and morphological properties. *J. Polym. Eng.* **2012**, *32*, 291–299. [CrossRef]

18. Santos, P.A.; Spinacé, M.A.S.; Fermoselli, K.K.G.; De Paoli, M.A. Polyamide-6/vegetal fiber composite prepared by extrusion and injection molding. *Compos. Part A Appl. Sci. Manuf.* **2007**, *38*, 2404–2411. [CrossRef]

19. Jannerfeldt, G.; Booch, L.; Manson, J.A.E. Influence of hyperbranched polymers on the interfacial tension of polypropylene/polyamide-6 blends. *J. Polym. Sci. Part B Polym. Phys.* **1999**, *37*, 2069–2077. [CrossRef]

20. Pinheiro, L.A.; Hu, G.-H.; Pessan, L.A.; Canevarolo, S.V. In-line measurements of the morphological parameters of PP/PA6 blends during extrusion in the transient mode. *Polym. Eng. Sci.* **2008**, *48*, 806–814. [CrossRef]

21. Hietaoja, P.; Heino, M.; Vainio, T.; Seppala, J. Compatibilization of PP/PBT and PP/PA6 blends with a new oxazoline-functionalized polypropylene. *Polym. Bull.* **1996**, *37*, 353–359. [CrossRef]

22. Roeder, J.; Oliveira, R.V.B.; Gonçalves, M.C.; Soldi, V.; Pires, A.T.N. Polypropylene/polyamide-6 blends: influence of compatibilizing agent on interface domains. *Polym. Test.* **2002**, *21*, 815–821. [CrossRef]

23. Shi, D.; Ke, Z.; Yang, J.h.; Gao, Y.; Wu, J.; Yin, J. Rheology and Morphology of Reactively Compatibilized PP/PA6 Blends. *Macromolecules* **2002**, *35*, 8005–8012. [CrossRef]

24. Fu, S.-Y.; Lauke, B.; Li, R.K.Y.; Mai, Y.-W. Effects of PA6,6/PP ratio on the mechanical properties of short glass fiber reinforced and rubber-toughened polyamide 6,6/polypropylene blends. *Compos. Part B Eng.* **2005**, *37*, 182–190. [CrossRef]

25. Benderly, D.; Siegmann, A.; Narkis, M. Polymer encapsulation of glass filler in ternary PP/PA-6/glass blends. *Polym. Compos.* **1996**, *17*, 86–95. [CrossRef]

26. Abbacha, N.; Fellahi, S. Synthesis of PP-g-MAH and evaluation of its effect on the properties of glass fibre reinforced nylon 6/polypropylene blends. *Macromol. Symp.* **2002**, *178*, 131–138. [CrossRef]

27. Akil, H.M.; Omar, M.F.; Mazuki, A.A.M.; Safiee, S.Z.; Ishak, A.M.; Abu Bakar, A. Kenaf fiber reinforced composites: A review. *Mater. Des.* **2011**, *32*, 4107–4121. [CrossRef]

28. Franciszczak, P.; Kalniņš, K.; Błędzki, A.K. Hybridisation of man-made cellulose and glass reinforcement in short-fibre composites for injection moulding–Effects on mechanical performance. *Compos. Part B Eng.* **2018**, *145*, 14–27. [CrossRef]

29. Sreekala, M.S.; George, J.M.; Kumaran, G.; Thomas, S. The mechanical performance of hybrid phenol-formaldehyde-based composites reinforced with glass and oil palm fibres. *Compos. Sci. Technol.* **2002**, *62*, 339–353. [CrossRef]

30. Lei, Y.; Wu, Q.; Yao, F.; Xu, Y. Preparation and properties of recycled HDPE/natural fiber composites. *Compos. Part A Appl. Sci. Manuf.* **2007**, *38*, 1664–1674. [CrossRef]

31. Kabir, M.M.; Wang, H.; Lau, K.T.; Cardona, F. Chemical treatments on plant-based natural fibre reinforced polymer composites: An overview. *Compos. Part B Eng.* **2012**, *43*, 2883–2892. [CrossRef]

4

Toughness Enhancement of PHBV/TPU/Cellulose Compounds with Reactive Additives for Compostable Injected Parts in Industrial Applications

Estefanía Lidón Sánchez-Safont [1], Alex Arrillaga [2], Jon Anakabe [2], Luis Cabedo [1] and Jose Gamez-Perez [1,*]

[1] Polymers and Advanced Materials Group (PIMA), Universitat Jaume I, 12071 Castellón, Spain; esafont@uji.es (E.L.S.-S.); lcabedo@uji.es (L.C.)

[2] Leartiker S. Coop., Xemein Etorbidea 12A, 48270 Markina-Xemein, Spain; aarrillaga@leartiker.com (A.A.); janakabe@leartiker.com (J.A.)

* Correspondence: gamez@uji.es

Abstract: Poly(3-hydroxybutyrate-co-3-valerate), PHBV, is a bacterial thermoplastic biopolyester that possesses interesting thermal and mechanical properties. As it is fully biodegradable, it could be an alternative to the use of commodities in single-use applications or in those intended for composting at their end of life. Two big drawbacks of PHBV are its low impact toughness and its high cost, which limit its potential applications. In this work, we proposed the use of a PHBV-based compound with purified α-cellulose fibres and a thermoplastic polyurethane (TPU), with the purpose of improving the performance of PHBV in terms of balanced heat resistance, stiffness, and toughness. Three reactive agents with different functionalities have been tested in these compounds: hexametylene diisocianate (HMDI), a commercial multi-epoxy-functionalized styrene-co-glycidyl methacrylate oligomer (Joncryl® ADR-4368), and triglycidyl isocyanurate (TGIC). The results indicate that the reactive agents play a main role of compatibilizers among the phases of the PHBV/TPU/cellulose compounds. HMDI showed the highest ability to compatibilize the cellulose and the PHBV in the compounds, with the topmost values of deformation at break, static toughness, and impact strength. Joncryl® and TGIC, on the other hand, seemed to enhance the compatibility between the fibres and the polymer matrix as well as the TPU within the PHBV.

Keywords: biopolyester; compatibilizer; cellulose; elastomer; toughening; biodisintegration; heat deflection temperature

1. Introduction

Nowadays, the use of plastics is widely extended in almost all production fields, such as packaging, electronics, automotive, household, etc., and the market is dominated by the so-called commodities, traditional oil-based plastics. The growing concern over the environmental problems involved with petroleum-based polymers related to their non-renewable origin and poor biodegradability is leading the industry to replace current materials with biodegradable alternatives [1]. Therefore, researchers have been looking for alternatives that may be more environmentally sustainable, especially in short- and medium-term applications, such as packaging. Within this context, biopolyesters have received great attention, especially those that are bio-sourced and biodegradable, as a way to overcome some of the waste management issues [2].

Among the different commercially available biopolyesters, one of the most promising candidates to replace commodities is the poly(3-hydroxybutyrate-co-3-valerate) (PHBV) [3–6]. PHBV is a bacterial thermoplastic biopolyester from the polyhydroxyalcanoates family that possesses physical properties comparable to conventional polyolefins, high static mechanical performance [7], and relatively high thermal resistance [8], while being fully biodegradable. However, two important drawbacks of PHBV are its low impact toughness and its cost, which is still quite high [9,10]. These disadvantages are a serious handicap for its use in applications in rigid packaging parts, for instance, that could be obtained by injection moulding.

One of the most promising eco-friendly approaches to reduce the manufacturing costs of PHBV while maintaining its biodegradability and sustainability is the development of natural fibre-based polymer composites. Indeed, it also improves its mechanical performance in terms of stiffness as well as thermal resistance [11]. On the other hand, in order to enhance the toughness of PHBV, several attempts have been reported in the literature, some of them related to blending with other polymers such as poly(butylene adipate-co-terephthalate) (PBAT) [12,13], polybutylene succinate (PBS) [14,15], or polycaprolactone (PCL) [12] or by the addition of impact modifiers such as ethylene vinyl acetate [16], epoxidized natural rubber [13,17], or thermoplastic polyurethane (TPU) [12,18–20], showing in all cases great improvements in elongation at break.

In this work, we proposed the use of a purified α-cellulose fibres and a thermoplastic polyurethane (TPU) with the purpose of improving the performance of PHBV in terms of balanced heat resistance, stiffness, and toughness without compromising biodisintegrability in composting conditions.

However, previous works have shown that the interaction of these fillers with PHA matrices was not very strong, resulting in low toughness and tensile strength [21,22]. Nonetheless, some strategies to improve the chemical affinity between the cellulose and other polyesters have been used in order to increase the reinforcement effect of the cellulose, such as fibre treatments or use of compatibilizers (reviewed by [11,23–26]).

From an industrial point of view, reactive extrusion is a convenient, cost-effective approach to improve the interfacial adhesion of the different phases via an in situ reaction during melt processing [27]. Within this objective, three reactive agents have been tested: (a) hexametylene diisocianate (HMDI); (b) (Joncryl® ADR-4368), a commercial multi-epoxy-functionalized styrene-acrylic oligomer; and (c) triglycidyl isocyanurate (TGIC) (Figure 1). These reactive agents possess three different functional groups that could potentially react with the hydroxyl groups present at the cellulose surface and the ones from the alcohol and carboxylic acid groups at the polymer chain ends [28]. Some reports have been found in the literature about the use of diisocyanates as compatibilizers in biopolyester/fibre composites [29–31], PHBV/polylactic acid (PLA) blends [32], and PLA/TPU blends [33,34], showing good improvements in interfacial adhesion. Hao et al. showed improved interfacial adhesion in PLA/sisal fibre composites using Joncryl® [35] and Nanthananon et al. reported similar improvements in PLA/eucalyptus fibre systems [36]. Furthermore, the use of Joncryl® has also been proved efficient in the compatibilization of POM/TPU blends [37]. TGIC was successfully used to compatibilize polylactide/sisal fibre biocomposites [38].

In this work, the combined effect of TPU, cellulose fibres, and the use of three different reactive agents (HMDI, Joncryl®, and TGIC) is explored in order to improve the interfacial adhesion and compatibility of PHBV, TPU, and cellulose through reactive extrusion. This strategy is aimed at building a ternary system that will overcome the handicaps of PHBV that prevent its usage in injection-moulded applications in terms of cost, toughness, and thermal resistance.

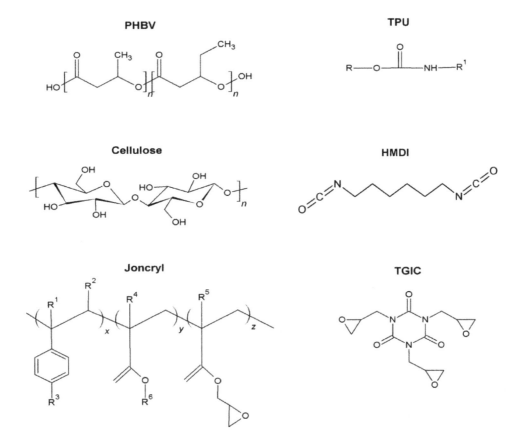

Figure 1. Chemical structures of materials used in this study.

2. Results and Discussion

2.1. Preparation of Compounds and Analysis of Their Processability

PHBV/TPU/Cellulose triple systems with different content of additives (TPU and cellulose) and reactive agents (HMDI, Joncryl®, and TGIC) were prepared by a co-rotating twin-screw extruder in the proportions described in Table 1.

The melt flow index is a useful tool to predict the processability of materials in industrial equipment such as injection moulding and gives an indirect measurement of melt viscosity, as it is indirectly proportional to viscosity. Figure 2 represents the melt flow index values of neat PHBV and the compounds (PHBV/30T/10C and PHBV/30T/30C) with 0, 0.3, 0.5, and 1 phr reactive agents content.

As seen in Figure 2, the addition of TPU and cellulose significantly decreases the melt fluidity of PHBV, especially for the highest cellulose content. This increment in melt viscosity is typical in fibre-based composites because of the increased shear produced by the restricted chain mobility induced by the fibres [39]. The addition of the different reactive agents in 0.3 phr leads to a further drastic reduction of the melt fluidity. As the reactive agent content increases, the MFI values decrease slightly, except in the case of TGIC, where the compounds have similar melt indexes. Among the three reactive agents, the highest reduction in MFI is found in the compositions with HMDI. This reduction in fluidity with the incorporation of the reactive agents is indicative of some reactivity with the components of the system and can be related to a compatibilization between the fibres and the polymers and/or between the PHBV and the TPU [27].

With respect to the processability, the reduced fluidity of the compositions with the reactive agents led to increased injection pressure values. However, despite the low MFI values of the compounds, the injected samples were successfully obtained without any change in the processing parameters with respect to neat PHBV.

Table 1. List of compounds and their composition.

Sample	TPU	Cellulose	HMDI	Joncryl®	TGIC
			(phr) *		
Neat PHBV	-	-	-	-	-
PHBV/30T/10C **	30	10	-	-	-
PHBV/30T/10C-0.3HMDI	30	10	0.3	-	-
PHBV/30T/10C-0.5HMDI	30	10	0.5	-	-
PHBV/30T/10C-1HMDI	30	10	1	-	-
PHBV/30T/10C-0.3Joncryl	30	10	-	0.3	-
PHBV/30T/10C-0.5Joncryl	30	10	-	0.5	-
PHBV/30T/10C-1Joncryl	30	10	-	1	-
PHBV/30T/10C-0.3TGIC	30	10	-	-	0.3
PHBV/30T/10C-0.5TGIC	30	10	-	-	0.5
PHBV/30T/10C-1TGIC	30	10	-	-	1
PHBV/30T/30C ***	30	30	-	-	-
PHBV/30T/30C-0.3HMDI	30	30	0.3	-	-
PHBV/30T/30C-0.5HMDI	30	30	0.5	-	-
PHBV/30T/30C-1HMDI	30	30	1	-	-
PHBV/30T/30C-0.3Joncryl	30	30	-	0.3	-
PHBV/30T/30C-0.5Joncryl	30	30	-	0.5	-
PHBV/30T/30C-1Joncryl	30	30	-	1	-
PHBV/30T/30C-0.3TGIC	30	30	-	-	0.3
PHBV/30T/30C-0.5TGIC	30	30	-	-	0.5
PHBV/30T/30C-1TGIC	30	30	-	-	1

* phr refers to the 100 unit weight PHBV matrix; ** PHBV/30T/10C corresponds to 71.4 wt % PHBV, 21.4 wt % TPU, and 7.2 wt % cellulose; *** PHBV/30T/30C corresponds to 62.5 wt % PHBV, 18.8 wt % TPU, and 18.8 wt % cellulose.

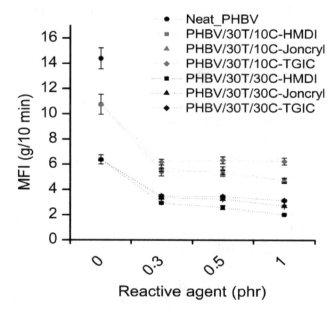

Figure 2. Melt flow index (MFI) of neat PHBV, PHBV/30T/10C, and PHBV/30T/30C with 0, 0.3, 0.5, and 1 phr reactive agent content.

2.2. Characterization

The morphology of the PHBV/TPU/cellulose triple systems was analysed by scanning electron microscopy (SEM). Micrographs of PHBV/30T/10C and PHBV/30T/30C without reactive agents and with 1 phr of HMDI, Joncryl®, and TGIC are depicted in Figures 3 and 4, respectively.

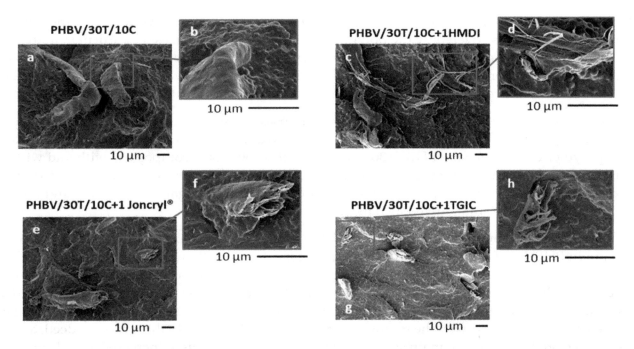

Figure 3. SEM micrographs of PHBV/30T/10C (**a,b**) and PHBV/30T/10C with 1 phr of HMDI (**c,d**), Joncryl® (**e,f**), and TGIC (**g,h**).

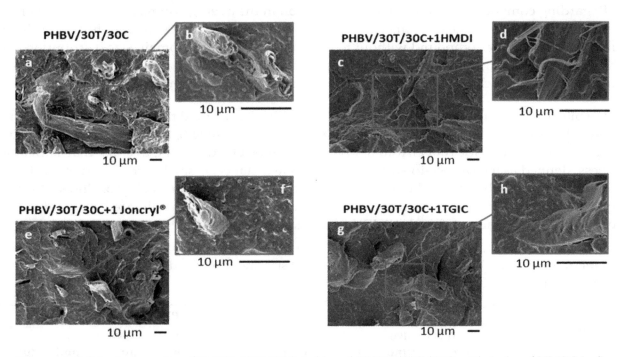

Figure 4. SEM micrographs of PHBV/30T/30C (**a,b**) and PHBV/30T/30C with 1 phr of HMDI (**c,d**), Joncryl® (**e,f**), and TGIC (**g,h**).

With respect to the fillers, in samples without reactive agent (Figures 3a and 4a), a good distribution of the fibres was observed, indicating good compounding and certain affinity between the fibres and the polymer matrix probably due to the formation of hydrogen bonds between the –OH groups of the fibre surface and PHBV [21,40,41]. However, some detachment of the fibres is also observed, indicating that the hydrogen bonding type is not enough to provide a strong adhesion between these phases. Regarding the fibre distribution, no remarkable differences were observed with the addition of the different reactive agents (Figures 3 and 4). Nevertheless, regarding the fibre–matrix interface, with the presence of the reactive agents the fibres seem to be well trapped by the polymer matrix, as broken fibres covered by the polymer were observed in all cases. In particular, in the case of HMDI (Figure 3c,d and Figure 4c,d), the fibres appear broken in the longitudinal direction, and defibrillation was observed, indicating a cohesive failure. These observations suggest the strongest adhesion between the cellulose fibres and the PHBV, indicating a compatibilization effect.

With respect to the polymeric matrix, all the compositions present a drop in matrix morphology, where the disperse phase is the TPU, as shown in Figures 5 and 6. These figures show the SEM images of the polymeric matrix for the PHBV/30T/10 and PHBV/30T/30C composites, with and without the reactive agents. The droplet size distributions of the dispersed phase are also included in the aforementioned figures. The average domain size (d), the estimated ligament distance, and the d10, d50, and d90 values are summarized in Table 2.

According to the measurements performed, the average domain size (d) of the TPU is 0.416 and 0.420 μm in PHBV/30T/10C and PHBV/30T/30C, respectively. Although some detachment of TPU is observed, the small size of the dispersed phase domains indicates a certain affinity between the phases. With the incorporation of the reactive agents the average TPU droplet size was reduced, as shown in Table 2. The highest droplet size reductions were obtained in the compounds containing the highest amount of cellulose with the reactive agents TGIC and Joncryl®. In these cases, indeed, a slight dependence on the average domain size (d) as the reactive agent content increases was observed in compounds with 10 phr of Cellullose, but not in those with 30 phr Cellulose.

Regarding compounds with HMDI, the reduction in the average domain size (d) is lower with respect to the other reactive agents. In fact, the compound PHBV/30T/10C with 1 phr HMDI shows a similar value of d as the compound without reactive agents, with the domain size distribution being slightly displaced to bigger sizes, presenting the highest d90 value among all compounds. Nevertheless, the matrix ligament thickness of this system is in the same range as in the rest of the composites.

Despite the differences in the size distributions, the droplet size and the estimated matrix ligament thickness (T) are quite small for all compositions. As reported by Wu [42], the matrix ligament thickness plays an important role for rubber toughening in polymer blends. If the average matrix ligament thickness, defined as the surface-to-surface interparticle distance, is smaller than a critical value, the blend will be tough, whereas on the contrary the blend will be brittle. It can be concluded, from this analysis, that the use of TGIC and Joncryl® reactive agents produced an enhanced compatibilization effect on the TPU domains within the PHBV/cellulose matrix.

Table 2. Estimated d10, d50 and d90, average droplet size and ligament distance values of the TPU dispersed phase.

(phr)	PHBV/30T/10C					PHBV/30T/30C				
	d10 (μm)	d50 (μm)	d90 (μm)	d (μm)	T (μm)	d10 (μm)	d50 (μm)	d90 (μm)	d (μm)	T (μm)
0	0.17	0.34	0.79	0.42	0.13	0.17	0.35	0.77	0.42	0.13
0.3 HMDI	0.14	0.31	0.79	0.40	0.12	0.09	0.20	0.57	0.27	0.08
0.5 HMDI	0.11	0.21	0.51	0.26	0.08	0.09	0.18	0.76	0.31	0.10
1 HMDI	0.11	0.26	1.01	0.43	0.13	0.09	0.21	0.77	0.33	0.10
0.3 Joncryl	0.09	0.20	0.52	0.26	0.08	0.07	0.14	0.38	0.19	0.06
0.5 Joncryl	0.09	0.17	0.43	0.22	0.07	0.08	0.14	0.48	0.21	0.07
1 Joncryl	0.09	0.16	0.39	0.20	0.06	0.07	0.14	0.36	0.18	0.06
0.3 TGIC	0.10	0.21	0.57	0.28	0.09	0.07	0.14	0.43	0.20	0.06
0.5 TGIC	0.09	0.21	0.52	0.26	0.08	0.08	0.14	0.39	0.20	0.06
1 TGIC	0.08	0.17	0.54	0.25	0.08	0.09	0.17	0.42	0.21	0.07

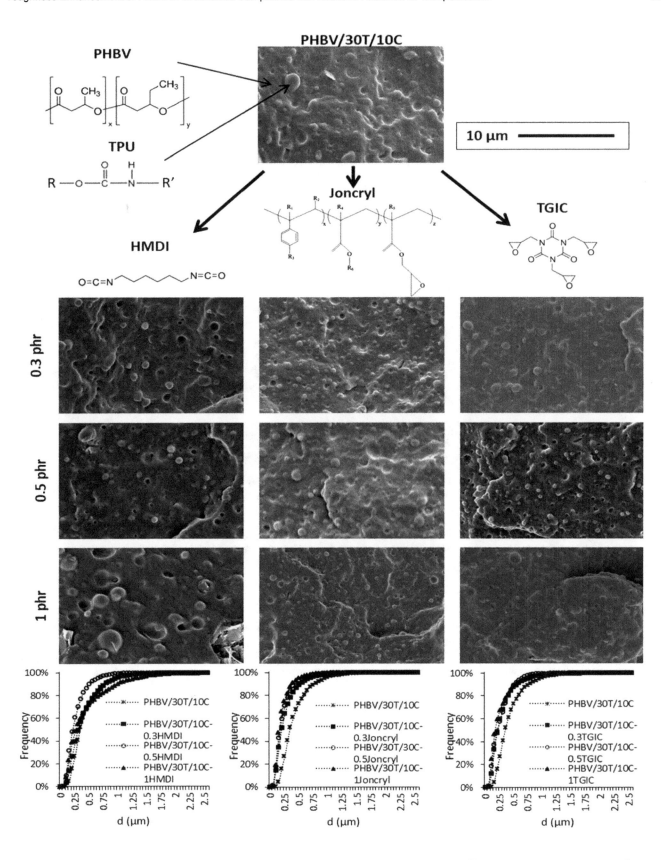

Figure 5. SEM images of the PHBV/30T/10C composites with the different reactive agents and cumulative frequency droplet size histograms of the dispersed phase.

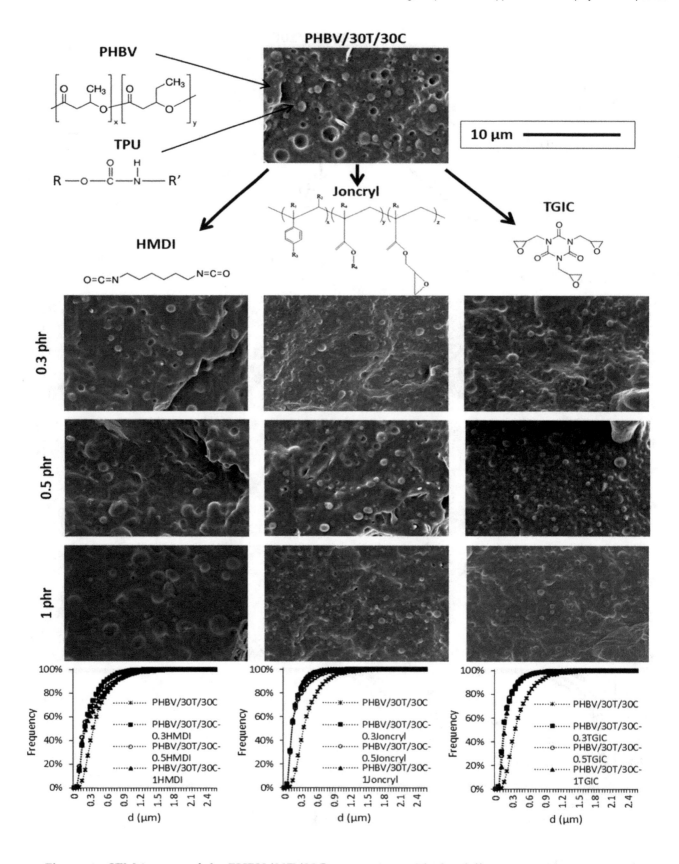

Figure 6. SEM images of the PHBV/30T/30C composites with the different reactive agents and cumulative frequency droplet size histograms of the dispersed phase.

2.2.1. Mechanical Properties

Tensile tests up to failure were conducted in order to study the mechanical properties of neat PHBV and the compounds with and without reactive agents (HMDI, Joncryl®, and TGIC). The Young's modulus, tensile yield strength, and elongation at break of the different compositions are shown in Figure 7. The representative strain–stress curves of neat PHBV and PHBV/TPU/cellulose composites with and without the highest level of reactive agents (1 phr) are also represented for the sake of clarity.

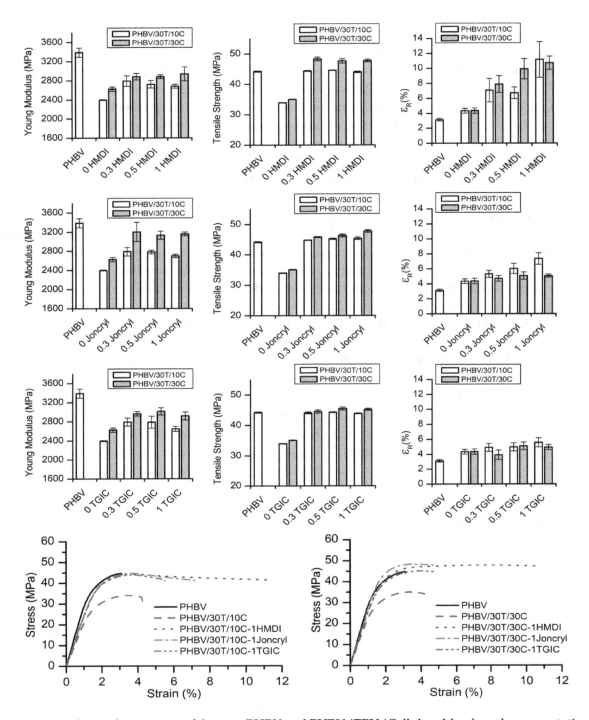

Figure 7. Mechanical properties of the neat PHBV and PHBV/TPU/Cellulose blends and representative stress–strain curves of neat PHBV, PHBV/30T/10C, and PHBV/30T/30C systems with and without 1 phr reactive agents.

PHBV presents a typical stiff and brittle mechanical performance, with high values of elastic modulus and tensile strength and low elongation at break (<5%). With respect to neat PHBV, the incorporation of TPU (30 phr) and cellulose (10 or 30 phr) leads to a reduction in the rigidity (about 25%) and the tensile strength (about 20%) and an enhancement in elongation at break (ca. 40%) and static toughness (25% and 32%, respectively) [43]. This increase in elongation at break is related with both the good distribution and small droplet size of the dispersed elastomeric phase (TPU) [19,20]. On the other hand, although there is a certain affinity among the phases, this limited interaction is not enough to ensure an efficient load transfer to the cellulose fibres. Without strong adhesion, the fibres detach at low deformation values, lowering the tensile strength of the compound and acting as stress concentrators for premature material failure [44]. With the addition of the reactive agents to the compounds, the elastic modulus and the tensile strength clearly increase with respect to the PHBV/TPU/cellulose without them. These parameters are strongly influenced by the matrix–fibre interaction and their improvement indicates a better load transfer due to an enhanced adhesion [30]. For the 10 phr cellulose compounds, the different reactive agents show a similar impact in these parameters, regardless of their content, supposing an improvement in the elastic modulus of about 15% and an increase in the tensile strength of around 30%. For the 30 phr cellulose compounds, the highest rise in the tensile modulus was obtained with the addition of Joncryl® (20% vs. ca. 10% for HMDI and TGIC). On the other hand, the tensile strength was improved by around 40%, 30%, and 35% with the HMDI, Joncryl®, and TGIC, respectively, reaching that of neat PHBV.

These results reveal that the three tested reactive agents are effective at improving the interfacial adhesion of the cellulose with the polymeric matrix and are in accordance with the SEM observations and the MFI values, pointing to an increased interaction among the phases [45]. This conclusion is in agreement with some other works that have been reported in the literature, on biopolyester–fibre composites compatibilized with diisocyanates [29–31], Joncryl® [35,36], or TGIC [38].

The biggest difference among the tested reactive agents with respect to their influence on the mechanical performance of the compounds is in the elongation at break. In all cases, this parameter was improved with respect to both neat PHBV and the compound without reactive agents. In Figure 6, looking at any PHBV/30T/10C compounds, as the reactive agent addition increases, the elongation at break rises too. On the other hand, in the case of the PHBV/30T/30C compounds, only the compounds with HMDI show an increase of elongation at break as the reactive agent content increases. This difference may point that the role of HMDI may not be the same as TGIC or Joncryl®.

In fact, the compounds with the highest TGIC level (1 phr) show an increase in elongation at break of ca. 28% with respect to the uncompatibilized PHBV/30T/10C system and 13% for PHBV/30T/30C. Similarly, the addition of 1 phr Joncryl® improved elongation at break by 70% in PHBV/30T/10C and 16% in PHBV/30T/30C. However, the compounds with 1 phr of HMDI showed an extraordinary enhancement of the elongation at break; the elongation at break was improved by 160% and 150% for compounds with 10 and 30 phr cellulose, respectively. Moreover, the static toughness (calculated from the area below the stress–strain curve) was enhanced by 320% and 340% with respect to the compound without reactive agents, and 420% and 450% with respect to neat PHBV.

2.2.2. Impact Resistance

Figure 8 summarizes the values obtained from unnotched and notched Charpy's impact tests, along with the static toughness from tensile tests of neat PHBV and the compounds with and without the reactive agents.

It is known that PHBV is very brittle and therefore it presents very low values of resilience in both unnotched and notched Charpy's impact tests and low static toughness, as shown in Figure 7. In this figure it can be observed that the compounds with TPU and cellulose clearly show an improvement in toughness resistance in the case of unnotched impact tests (Figure 8a,d,g) [43]. This improvement could be attributed to the positive role that the elastomeric TPU phase plays in absorbing impact energy [46]. However, concerning the notched tests (Figure 8b,e,h), there is no such increase in the impact energy absorbed, probably because there is a preferred crack propagation pathway through the matrix/fibre interfaces, where the adhesion is not very strong, as previously pointed out when discussing the variations in the elastic modulus of the compounds.

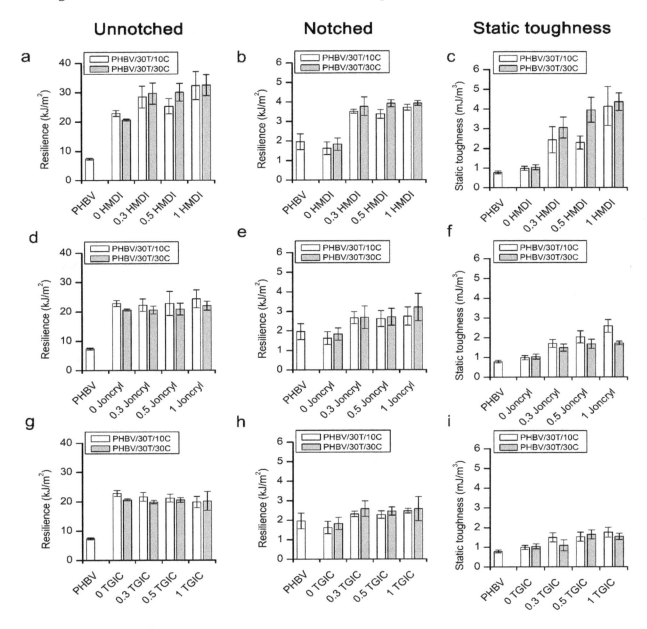

Figure 8. Charpy's impact results for unnotched specimens (**a,d,g**), notched specimens (**b,e,h**) and static toughness from the area below the strain–stress curve (**c,f,i**).

Nevertheless, the addition of reactive agents significantly improves the notched impact properties of the composites, which is in agreement with the static toughness determined from the area below the stress–strain curves (Figure 8c,f,i). This enhancement of the matrix–fibre interface adds to the effect of the small droplet size of the elastomeric phase that implies a low ligament thickness [42,47] and to by the enhanced interfacial interactions between TPU and PHBV [37], resulting in higher impact resistance in the presence of a notch.

When analysing the influence of the different reactive agents, the greatest increase in impact strength is obtained for composites with HMDI. For these composites, the absorbed impact energy was highly improved in both unnotched and notched impact tests, as well as in the static toughness (Figure 8a–c). Attending to the SEM micrographs of the impact fractured surfaces, with the HMDI addition (Figures 3 and 4) most of the fibres appear broken at their longitudinal direction, thus indicating a cohesive failure that confirms the presence of a very strong interface. This was not the case in compounds with the addition of TGIC or Joncryl®.

In polymer matrix composites, when there is a weak interface between the second phase and the polymeric matrix, the detachment of the particles during tensile deformations leads to the formation of flaws and voids at the interface of the fibre and the matrix. Those voids can coalesce and act as either crack initiators or provide a fast propagation crack pathway, which eventually leads to the premature failure of the material [13,48]. With a stronger particle–matrix adhesion, the possibility of growth and merge of those internal flaws is reduced, so there is an effective load transfer between the two phases, improving the fracture toughness [30]. It can be said that when the shear strength at the particle–matrix interface is higher than the shear yielding of any of the phases, plastic deformation of any of them can occur, thus increasing the energy absorbed. Thus, reactive agents can play different roles, increasing the adhesion between the PHBV/TPU, PHBV/cellulose and TPU/cellulose interfaces.

The impact performance of the compounds with HMDI stands out over the other ones. In this case, it seems that HMDI strongly increases the adhesion between PHBV and cellulose, which results in a synergetic effect with the addition of the TPU. The well-dispersed elastomeric phase decreases the yield strength of the polymer matrix and the strong interaction between the polymer and the fibres allows effective load transfer without producing flaws at the interfaces. Moreover, the exceptional mechanical performance of these compositions in terms of elongation at break also suggests that HMDI could play a positive role in enhancing the interfacial adhesion between PHBV and TPU. Indeed, diisocyanates have demonstrated effectiveness in improving the compatibility of biopolyester/TPU blends, as has been reported by Dogan et al. [33,34].

TGIC and Joncryl®, according to this reasoning, would not be so effective at enhancing the cellulose/PHBV interface, thus showing limited values of impact resistance, especially in the presence of a notch.

2.2.3. Heat Deflection Temperature HDT-A

The thermal resistance of neat PHBV and PHBV/TPU/cellulose composites was evaluated by means of heat deflection temperature (HDT-A) measurements. The results are grouped in Table 3.

Table 3. HDT-A values for neat PHBV and PHBV/TPU/cellulose systems with and without reactive agents.

Sample		HDT-A (°C)		
		0.3 phr	0.5phr	1 phr
PHBV	108 ± 1			
PHBV/30T/10C	94 ± 3			
PHBV/30T/10C + HMDI		93 ± 1	98 ± 3	95 ± 4
PHBV/30T/10C + Joncryl		95 ± 1	98 ± 3	97 ± 1
PHBV/30T/10C + TGIC		90 ± 3	90 ± 1	90 ± 3
PHBV/30T/30C	96 ± 1			
PHBV/30T/30C + HMDI		97 ± 2	99 ± 2	100 ± 1
PHBV/30T/30C + Joncryl		98 ± 4	97 ± 3	94 ± 1
PHBV/30T/30C + TGIC		104 ± 3	99 ± 3	99 ± 1

The PHBV presents a relatively high thermal resistance, showing an HDT-A value of 108 °C, in agreement with previously reported values [8,49,50]. The HDT values obtained for PHBV/30T/10C and PHBV/30T/30C are 94 and 96 °C, respectively. In spite of the relatively high content of the elastomeric additive (30 phr), the thermal resistance is not that much lower. This is due to the positive role played by the cellulose fibres in terms of reinforcement. As is widely reported in literature, in fibre-based polymer composites the restricted mobility of polymer chains in the presence of fibres leads to an increase in the dimensional stability and, thus, higher temperatures are required to deform them [50].

The use of reactive agents did have a significant influence on HDT values, but a trend was not seen with variation on their relative content. For the compounds with the lowest cellulose content (10 phr), HDT values ranged between 90 °C (TGIC) and 98 °C (Joncryl®), compared with a value of 94 °C for the compound without reactive agents. On the other hand, for the PHBV/30T/30C compounds, the thermal resistance was in almost all cases improved with HDT-A values around 100 °C (especially HMDI and TGIC), being the HDT value for the compound without reactive agents 96 °C.

These results are in agreement with the improved rigidity of the samples in the presence of reactive agents. Indeed, since there is no dependence of the HDT value on increasing the content of reactive agents, crosslinking reactions among the polymer chains can be discarded, thus indicating that the effect of the reactive agents is only a consequence of the compatibilization of the different phases. The increase in the HDT of fibre-based composites is therefore related to the reinforcement of the cellulose fibres [51] and, along with the use of the studied reactive agents, allows for enhancing the toughness and mechanical performance of PHBV without drastically decreasing its thermal resistance.

2.2.4. Biodisintegration in Composting Conditions

To explore the influence of cellulose content and the different reactive agents on the compostability of the PHBV/TPU/cellulose ternary systems, biodisintegration tests were conducted according to the ISO 20200 standard. The disintegration (weight loss) level over composting time is represented in Figure 9.

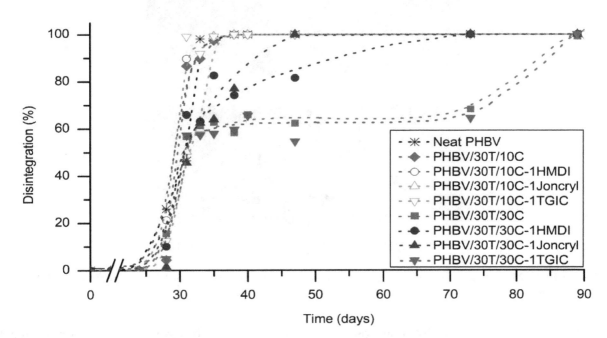

Figure 9. Disintegration in composting conditions of neat PHBV and PHB/TPU/cellulose systems with and without 1 phr reactive agents.

In general, all the compositions studied can be considered biodisintegrable in composting conditions according to ISO 20200. As shown in Figure 9, the PHBV disintegration process starts after an incubation period of 28 days. At this time the disintegration rate drastically increases to achieve total disintegration at 38 days of composting, in accordance with previous works [20,52,53]. No differences in the biodisintegration rate were detected for the PHBV/TPU/cellulose composites containing 10 phr cellulose, independent of the presence of reactive agents or the reactive agent type.

When the cellulose content was increased, the biodisintegration rate was, oddly, significantly reduced. To understand this occurrence, it must be taken into account how the samples were prepared and how biodisintegration takes place. For the composting tests, the specimens were obtained by hot pressing. Under the hot pressing conditions, the formation of a percolation mesh of interconnected cellulose fibres is favoured due to the high fibre–fibre affinity of the cellulose. This cellulose mesh is partially covered by TPU, which possess a low biodisintegration rate [20] with respect to PHBV and cellulose. Then, during the incubation time, a biofilm is formed at the surface of the testing specimen and the microbial advance occurs from the surface to the bulk, preferentially through the PHBV phase, as it is deduced by the stabilization of the disintegrated mass at around 60 wt % (approximately, the PHBV weight content) after 35 days of composting. We think that the TPU droplets, which take longer to biodisintegrate and are quite sticky at high temperature and moisture content, cover the fibres, limiting the access of the microbial advance to the cellulose.

This phenomenon causes a slowdown in the biodisintegration rate, but when the microorganisms have access to cellulose the weight loss rises rapidly and total disintegration is achieved within 90 days of composting. To validate this hypothesis, similar samples to those used for biodisintegration were placed for Soxhlet extraction of the PHBV phase with chloroform, and the resulting morphology analysed by SEM (Figure 10). TPU can be seen covering the fibres, partially confirming this reasoning.

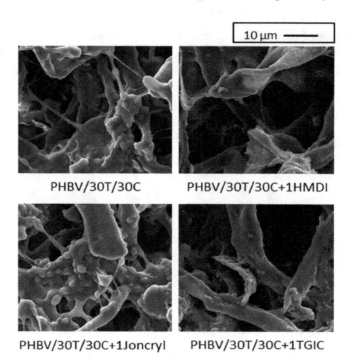

Figure 10. SEM micrographs of PHBV/30T/30C samples with and without 1 phr reactive agents after Soxhlet extraction.

Furthermore, PHBV/30T/30C composites showed different behaviour depending on the reactive agent added. The composition with TGIC presented a similar trend to the composition without reactive agent (that is, a slowdown at 60% weight loss); the ones with HMDI presented a fast biodisintegration of about 80% after 33 days of composting, reaching complete disintegration at day 73; and the composition with Joncryl® was totally degraded after 47 days. These differences in the biodisintegration rates among PHBV/30T/30C composites with different reactive agents could be influenced by the interactions of the reactive agents with the fibres, the PHBV, and the TPU. It is hypothesized that when there is a high interaction between the cellulose and the PHBV (promoted by the reactive agents), the microorganisms can access the cellulose more easily and the biodisintegration is completed earlier. On the contrary, when the PHBV–cellulose interaction is weak, the TPU can be easily located at the fibres surface, hindering the microbial advance. When looking at the SEM pictures of the compounds after Soxhlet extraction (Figure 10), there is more polymer covering the fibres in the case of no reactive agent or TGIC addition than in the case of the compounds with HMDI and Joncryl®, supporting the aforementioned reasoning.

3. Experimental

3.1. Materials

Poly(3-hydroxybutyrate-co-3-hydroxyvalerate) (PHBV) commercial grade with 3 wt % valerate content was purchased from Tianan Biologic Material Co. (Ningbo, China) in pellet form (ENMAT Y1000P). Thermoplastic polyurethane (TPU) Elastollan® 890 A 10FC was supplied by BASF (Ludwigshafen, Germany). Purified alpha-cellulose fibre grade (TC90) (alpha-cellulose content >99.5%) from CreaFill Fibers Corp. (Chestertown, MD, USA) was used. The reactive agents hexamethylene diisocyanate (HMDI) and triglycidyl isocyanurate (TGIC) were supplied by Sigma-Aldrich (Spain) and the Joncryl® 4368 was purchased from BASF (Ludwigshafen, Germany).

3.2. Sample Preparation

The PHBV and TPU used in this study were dried at 80 °C for at least 6 h in a DESTA DS06 HT dehumidifying dryer and the cellulose was dried in a lab oven (Memmert universal oven U, Schwabach, Germany) at 90 °C for a minimum of 16 h prior to the blending step, whilst the three reactive agents (HMDI, Joncryl®, and TGIC) were used as received.

PHBV/TPU/cellulose triple systems with different content of additives (TPU and cellulose) and reactive agents (HMDI, Joncryl®, and TGIC) (see Table 1) were prepared in a Labtech LTE (Samutprakarn, Thailand) (Ø = 26 mm, L/D ratio = 40) co-rotating twin-screw extruder. The temperature profile was set at 145/155/160/170 °C from hopper to nozzle, the rotation speed was 250 rpm, and the feeding speed was about 5 kg/h. All the components were manually dry-mixed before extrusion except HMDI, which was dispensed at the feeding zone by means of a peristaltic pump (Watson Marlow 120 S/R, Sondika, Spain). The extruded material was cooled in a water bath and pelletized (MAAG PRIMO S pelletizer, Stuttgart, Germany).

Material pellets were dried again at 80 °C for 8 h (DESTA DS06 HT) before the injection process. Standardized tensile specimens (ISO-527 Type 1A) were injection-moulded in a DEMAG IntElect 100 T injection moulding machine (Schwaig, Germany) with an injection temperature of 185 °C at the nozzle. A holding pressure of 600 bars was applied for 12 s, followed by 40 s of cooling time. For the sake of comparison, neat PHBV was also processed under identical conditions.

Prior to any characterization, all the samples were annealed at 80 °C for 48 h in order to obtain equivalent crystallinity and mechanical performance to aged samples.

3.3. Characterization

The melt flow index (MFI) of the different compounds was measured in a Tinius Olsen MP600 (Surrey, England) melt flow indexer according to the ISO 1133 standard. The tests were performed at 185 °C and 2.16 kg load.

The morphology of the PHBV/30T/10C and PHBV/30T/30C triple systems with and without reactive agents (HMDI, Joncryl® and TGIC) was examined by scanning electron microscopy (SEM) using a high-resolution field-emission JEOL 7001F microscope (Japan). The fracture surfaces from impact-fractured specimens were previously coated by sputtering with a thin layer of Pt. From selected representative SEM images (at 2500× magnification), the diameters of the droplets corresponding to the dispersed phase were measured using Fiji® software (ImageJ 1.51j8). The number of droplets measured in all cases was higher than 600. From the individual measures, the following parameters were determined: the average droplet size (d) and the droplet size distribution parameters d10, d50, and d90 (corresponding to the size where 10%, 50%, and 90% of the droplets are included, respectively). The matrix ligament thickness (T) was also estimated, according to Wu's equation [42]:

$$T = d\left[\left(\frac{\pi}{6\varphi_r}\right)^{\frac{1}{3}} - 1\right], \tag{1}$$

where d is the average domain size of the dispersed phase and φ_r is the volume fraction of the dispersed phase, determined as follows:

$$\varphi_r = \frac{\rho_m w_r}{(\rho_r w_m + \rho_m w_r)}, \tag{2}$$

where ρ_m and ρ_r are the densities of the matrix and dispersed phases, respectively, and w_m and w_r are their weight fractions.

Tensile tests were conducted on ISO-527 type 1A injection-moulded specimens in a Hounsfield H25K universal testing machine (Surrey, England) equipped with a 25 kN load cell according to the ISO-527-1:2012 standard.

Notched and unnotched Charpy impact tests were carried out by means of an ATS faar IMPats-15 (Segrate, Italy) impact pendulum with a 4 J hammer according to the ISO 179 standard. Samples were cut from injection-moulded bars.

Heat deflection temperature (HDT) analyses were performed using a Deflex 687-2 (Barcelona, Spain). A heating rate of 120 °C/h was used with an applied load of 1.8 MPa in accordance with Method A of ISO 75 standard. The temperature was recorded until the sample deflects 0.35 mm.

Biodisintegration tests were carried out with samples ($15 \times 15 \times 0.2$ mm^3) obtained from hot-pressed plates (180 °C, 5 min, and ca. 40 bar). Tests were performed according to the ISO 20200 standard [54]. Solid synthetic waste was prepared by mixing 10% of activated mature compost (VIGORHUMUS H-00, purchased from Buras Profesional, S.A., Girona, Spain), 40% sawdust, 30% rabbit feed, 10% corn starch, 5% sugar, 4% corn seed oil, and 1% urea. The water content of the mixture was adjusted to 55%. The samples were placed inside mesh bags to simplify their extraction and allow the contact of the compost with the specimens, and then were buried in compost bioreactors at 4–6 cm depth. Bioreactors were incubated at 58 °C. The aerobic conditions were guaranteed by periodically mixing the synthetic waste and adding water according to the standard requirements. Three replicates of each sample were removed from the boxes at different composting times for analysis. Samples were washed with water and dried under a vacuum at 40 °C until a constant mass. The disintegration degree was calculated by normalizing the sample weight to the initial weight with Equation (3):

$$D = \frac{m_i - m_f}{m_i} \times 100 \tag{3}$$

where m_i is the initial dry mass of the test material and m_f is the dry mass of the test material recovered at different incubation stages. Moreover, the morphology of the films prepared for the composting tests was analysed by SEM after the Soxhlet extraction with chloroform of the PHBV phase.

4. Conclusions

In this study three reactive agents used in reactive extrusion (HMDI, Joncryl®, and TGIC) were tested in PHBV/TPU/cellulose for injection moulding applications that require biodisintegration in composting conditions. The influence of the cellulose content, the reactive agent type and the reactive agent content, were analysed. It was observed that the incorporation of TPU and cellulose in PHBV led to a reduction in the tensile elastic modulus and tensile strength, but an enhancement in elongation at break, with an overall increase in static toughness attributed to the toughening effect of the TPU. However, the addition of the reactive agents to the compounds resulted in a rise in the tensile strength and elastic modulus up to values close to or higher than neat PHBV and an increase in the value of strain at break with respect to the compounds without reactive agents.

In terms of impact resistance, the addition of the reactive agents improved the toughness of the compounds in notched and unnotched configurations. Furthermore, even though the TPU in the compounds causes a decrease in the thermal strength with respect to neat PHBV, the addition of cellulose up to 30 phr with the reactive agents was able to moderate this drop.

Those results indicate that the reactive agents play a main role as compatibilizers among the phases of the PHBV/TPU/cellulose compounds. HMDI showed the highest ability to compatibilize the cellulose and the PHBV in the compounds, with the topmost values of deformation at break and static toughness. Joncryl® and TGIC, on the other hand, seemed to enhance the compatibility between the fibres and the polymer matrix as well as the TPU within the PHBV.

The findings of this work point to a route to modify the properties of PHBV (and PHAs in general) through blending with reactive agents, which can help to overcome some of the difficulties that these materials encounter in standard applications.

Author Contributions: Formal analysis, E.L.S.-S.; Funding acquisition, L.C. and J.G.-P.; Investigation, E.L.S.-S.; Methodology, E.L.S.-S., A.A., J.A., L.C. and J.G.-P.; Resources, A.A. and J.A.; Supervision, J.G.-P.; Writing—original draft, E.L.S.-S.; Writing—review & editing, A.A., J.A., L.C. and J.G.-P.

Acknowledgments: The authors acknowledge financial support for this research from the Ministerio de Economia y Competitividad (AGL2015-63855-C2-2-R) and the Pla de Promoció de la Investigació de la Universitat Jaume I (UJI-B2016-35). The authors thank Servicios Centrales de Instrumentación (SCIC) of Universitat Jaume I for the use of SEM. We are also grateful to Raquel Oliver and Jose Ortega for experimental support.

References

1. Plastics Europe-Association of Plastics Manufacturers. Plastics-the Facts 2017: Analysis of European Plastics Production, Demand and Waste Data. Available online: http://www.plasticseurope.org/en/resources/publications/plastics-facts-2017 (accessed on 2 February 2018).

2. Rujnić-Sokele, M.; Pilipović, A. Challenges and opportunities of biodegradable plastics: A mini review. *Waste Manag. Res.* **2017**, *35*, 132–140. [CrossRef] [PubMed]

3. Cunha, M.; Fernandes, B.; Covas, J.A.; Vicente, A.A.; Hilliou, L. Film blowing of PHBV blends and PHBV-based multilayers for the production of biodegradable packages. *J. Appl. Polym. Sci.* **2016**, *133*. [CrossRef]

4. Laycock, B.; Halley, P.; Pratt, S.; Werker, A.; Lant, P. The chemomechanical properties of microbial polyhydroxyalkanoates. *Prog. Polym. Sci.* **2013**, *38*, 536–583. [CrossRef]

5. Albuquerque, P.B.S.; Malafaia, C.B. Perspectives on the production, structural characteristics and potential applications of bioplastics derived from polyhydroxyalkanoates. *Int. J. Biol. Macromol.* **2018**, *107*, 615–625. [CrossRef] [PubMed]

6. Wang, Y.; Yin, J.; Chen, G.Q. Polyhydroxyalkanoates, challenges and opportunities. *Curr. Opin. Biotechnol.* **2014**, *30*, 59–65. [CrossRef] [PubMed]

7. Pilla, S. *Handbook of Bioplastics and Biocomposites Engineering Applications*; John Wiley & Sons: Hoboken, NJ, USA, 2011; ISBN 9780470626078.

8. Peelman, N.; Ragaert, P.; Ragaert, K.; De Meulenaer, B.; Devlieghere, F.; Cardon, L. Heat resistance of new biobased polymeric materials, focusing on starch, cellulose, PLA, and PHA. *J. Appl. Polym. Sci.* **2015**, *132*. [CrossRef]

9. Keskin, G.; Kızıl, G.; Bechelany, M.; Pochat-Bohatier, C.; Öner, M. Potential of polyhydroxyalkanoate (PHA) polymers family as substitutes of petroleum based polymers for packaging applications and solutions brought by their composites to form barrier materials. *Pure Appl. Chem.* **2017**, *89*, 1841–1848. [CrossRef]

10. Bugnicourt, E.; Cinelli, P.; Lazzeri, A.; Alvarez, V. Polyhydroxyalkanoate (PHA): Review of synthesis, characteristics, processing and potential applications in packaging. *Express Polym. Lett.* **2014**, *8*, 791–808. [CrossRef]

11. Väisänen, T.; Haapala, A.; Lappalainen, R.; Tomppo, L. Utilization of agricultural and forest industry waste and residues in natural fiber-polymer composites: A review. *Waste Manag.* **2016**, *54*, 62–73. [CrossRef] [PubMed]

12. Jost, V.; Miesbauer, O. Effect of different biopolymers and polymers on the mechanical and permeation properties of extruded PHBV cast films. *J. Appl. Polym. Sci.* **2018**, *135*, 46153. [CrossRef]

13. Zhang, K.; Misra, M.; Mohanty, A.K. Toughened sustainable green composites from poly(3-hydroxybutyrate-co-3-hydroxyvalerate) based ternary blends and miscanthus biofiber. *ACS Sustain. Chem. Eng.* **2014**, *2*, 2345–2354. [CrossRef]

14. Chikh, A.; Benhamida, A.; Kaci, M.; Pillin, I.; Bruzaud, S. Synergistic effect of compatibilizer and sepiolite on the morphology of poly(3-hydroxybutyrate-co-3-hydroxyvalerate)/poly(butylene succinate) blends. *Polym. Test.* **2016**, *53*, 19–28. [CrossRef]

15. Ma, P.; Hristova-Bogaerds, D.G.; Lemstra, P.J.; Zhang, Y.; Wang, S. Toughening of PHBV/PBS and PHB/PBS Blends via In situ Compatibilization Using Dicumyl Peroxide as a Free-Radical Grafting Initiator. *Macromol. Mater. Eng.* **2012**, *297*, 402–410. [CrossRef]

16. El-Taweel, S.H.; Khater, M. Mechanical and Thermal Behavior of Blends of Poly(hydroxybutyrate-co-hydroxyvalerate) with Ethylene Vinyl Acetate Copolymer. *J. Macromol. Sci. Part. B* **2015**, *54*, 1225–1232. [CrossRef]

17. Adams, B.; Abdelwahab, M.; Misra, M.; Mohanty, A.K. Injection-Molded Bioblends from Lignin and Biodegradable Polymers: Processing and Performance Evaluation. *J. Polym. Environ.* **2017**, 1–14. [CrossRef]

18. Wang, S.; Chen, W.; Xiang, H.; Yang, J.; Zhou, Z.; Zhu, M. Modification and Potential Application of Short-Chain-Length Polyhydroxyalkanoate (SCL-PHA). *Polymers* **2016**, *8*, 273. [CrossRef]

19. González-Ausejo, J.; Sánchez-Safont, E.; Cabedo, L.; Gamez-Perez, J. Toughness Enhancement of Commercial Poly (Hydroxybutyrate-co-Valerate) (PHBV) by Blending with a Thermoplastic Polyurethane (TPU). *J. Multiscale Model.* **2016**, *7*, 1640008. [CrossRef]

20. Martínez-Abad, A.; González-Ausejo, J.; Lagarón, J.M.; Cabedo, L. Biodegradable poly(3-hydroxybutyrate-co-3-hydroxyvalerate)/thermoplastic polyurethane blends with improved mechanical and barrier performance. *Polym. Degrad. Stab.* **2016**, *132*, 52–61. [CrossRef]

21. Bhardwaj, R.; Mohanty, A.K.; Drzal, L.T.; Pourboghrat, F.; Misra, M. Renewable resource-based green composites from recycled cellulose fiber and poly(3-hydroxybutyrate-co-3-hydroxyvalerate) bioplastic. *Biomacromolecules* **2006**, *7*, 2044–2051. [CrossRef] [PubMed]

22. Sánchez-Safont, E.L.; Aldureid, A.; Lagarón, J.M.; Gámez-Pérez, J.; Cabedo, L. Biocomposites of different lignocellulosic wastes for sustainable food packaging applications. *Compos. Part. B Eng.* **2018**, *145*, 215–225. [CrossRef]

23. Satyanarayana, K.G.; Arizaga, G.G.C.; Wypych, F. Biodegradable composites based on lignocellulosic fibers—An overview. *Prog. Polym. Sci.* **2009**, *34*, 982–1021. [CrossRef]

24. Pereira, P.H.F.; Rosa, M.D.F.; Cioffi, M.O.H.; Benini, K.C.C.D.C.; Milanese, A.C.; Voorwald, H.J.C.; Mulinari, D.R. Vegetal fibers in polymeric composites: A review. *Polímeros* **2015**, *25*, 9–22. [CrossRef]

25. Wei, L.; McDonald, A. A Review on Grafting of Biofibers for Biocomposites. *Materials* **2016**, *9*, 303. [CrossRef] [PubMed]

26. Misra, M.; Pandey, J.K.; Mohanty, A.K. *Biocomposites: Design and Mechanical Performance*; Elsevier Inc.: Amsterdam, The Netherlands, 2015; ISBN 9781782423942.

27. Muthuraj, R.; Misra, M.; Mohanty, A.K. Biodegradable compatibilized polymer blends for packaging applications: A literature review. *J. Appl. Polym. Sci.* **2018**, *135*, 45726. [CrossRef]

28. Stenstad, P.; Andresen, M.; Tanem, B.S.; Stenius, P. Chemical surface modifications of microfibrillated cellulose. *Cellulose* **2008**, *15*, 35–45. [CrossRef]

29. Jiang, L.; Chen, F.; Qian, J.; Huang, J.; Wolcott, M.; Liu, L.; Zhang, J. Reinforcing and Toughening Effects of Bamboo Pulp Fiber on Poly(3-hydroxybutyrate-co-3-hydroxyvalerate) Fiber Composites. *Ind. Eng. Chem. Res.* **2010**, *49*, 572–577. [CrossRef]

30. Zarrinbakhsh, N.; Mohanty, A.K.; Misra, M. Improving the interfacial adhesion in a new renewable resource-based biocomposites from biofuel coproduct and biodegradable plastic. *J. Mater. Sci.* **2013**, *48*, 6025–6038. [CrossRef]

31. Anderson, S.; Zhang, J.; Wolcott, M.P. Effect of Interfacial Modifiers on Mechanical and Physical Properties of the PHB Composite with High Wood Flour Content. *J. Polym. Environ.* **2013**, *21*, 631–639. [CrossRef]

32. González-Ausejo, J.; Sánchez-Safont, E.; Lagarón, J.M.; Balart, R.; Cabedo, L.; Gámez-Pérez, J. Compatibilization of poly(3-hydroxybutyrate-co-3-hydroxyvalerate)–poly(lactic acid) blends with diisocyanates. *J. Appl. Polym. Sci.* **2017**, *134*, 1–11. [CrossRef]

33. Dogan, S.K.; Reyes, E.A.; Rastogi, S.; Ozkoc, G. Reactive compatibilization of PLA/TPU blends with a diisocyanate. *J. Appl. Polym. Sci.* **2014**, *131*. [CrossRef]

34. Dogan, S.K.; Boyacioglu, S.; Kodal, M.; Gokce, O.; Ozkoc, G. Thermally induced shape memory behavior, enzymatic degradation and biocompatibility of PLA/TPU blends: "Effects of compatibilization". *J. Mech. Behav. Biomed. Mater.* **2017**, *71*, 349–361. [CrossRef] [PubMed]

35. Hao, M.; Wu, H.; Qiu, F.; Wang, X. Interface Bond Improvement of Sisal Fibre Reinforced Polylactide Composites with Added Epoxy Oligomer. *Materials* **2018**, *11*, 398. [CrossRef] [PubMed]

36. Nanthananon, P.; Seadan, M.; Pivsa-Art, S.; Hiroyuki, H.; Suttiruengwong, S. Biodegradable polyesters reinforced with eucalyptus fiber: Effect of reactive agents. In *AIP Conference Proceedings*; AIP Publishing: Melville, NY, USA, 2017; p. 70012.

37. Tang, W.; Wang, H.; Tang, J.; Yuan, H. Polyoxymethylene/thermoplastic polyurethane blends compatibilized with multifunctional chain extender. *J. Appl. Polym. Sci.* **2013**, *127*, 3033–3039. [CrossRef]

38. Hao, M.; Wu, H. Effect of in situ reactive interfacial compatibilization on structure and properties of polylactide/sisal fiber biocomposites. *Polym. Compos.* **2018**, *39*, E174–E187. [CrossRef]

39. Ferrero, B.; Fombuena, V.; Fenollar, O.; Boronat, T.; Balart, R. Development of natural fiber-reinforced plastics (NFRP) based on biobased polyethylene and waste fibers from Posidonia oceanica seaweed. *Polym. Compos.* **2015**, *36*, 1378–1385. [CrossRef]

40. Hameed, N.; Guo, Q.; Tay, F.H.; Kazarian, S.G. Blends of cellulose and poly(3-hydroxybutyrate-co-3-hydroxyvalerate) prepared from the ionic liquid 1-butyl-3-methylimidazolium chloride. *Carbohydr. Polym.* **2011**, *86*, 94–104. [CrossRef]

41. Mechanical and biodegradation performance of short natural fibre polyhydroxybutyrate composites. *Polym. Test.* **2013**, *32*, 1603–1611.

42. Wu, S. Phase structure and adhesion in polymer blends: A criterion for rubber toughening. *Polymer* **1985**, *26*, 1855–1863. [CrossRef]

43. Sánchez-Safont, E.L.; Arrillaga, A.; Anakabe, J.; Gamez-Perez, J.; Cabedo, L. PHBV/TPU/Cellulose compounds for compostable injection molded parts with improved thermal and mechanical performance. *J. Appl. Polym. Sci.*. submitted.

44. Seggiani, M.; Cinelli, P.; Mallegni, N.; Balestri, E.; Puccini, M.; Vitolo, S.; Lardicci, C.; Lazzeri, A. New Bio-Composites Based on Polyhydroxyalkanoates and Posidonia oceanica Fibres for Applications in a Marine Environment. *Materials* **2017**, *10*, 326. [CrossRef] [PubMed]

45. Yatigala, N.S.; Bajwa, D.S.; Bajwa, S.G. Compatibilization improves physico-mechanical properties of biodegradable biobased polymer composites. *Compos. Part. A Appl. Sci. Manuf.* **2018**, *107*, 315–325. [CrossRef]

46. Wang, S.; Xiang, H.; Wang, R.; Peng, C.; Zhou, Z.; Zhu, M. Morphology and properties of renewable poly(3-hydroxybutyrate-co-3-hydroxyvalerate) blends with thermoplastic polyurethane. *Polym. Eng. Sci.* **2014**, *54*, 1113–1119. [CrossRef]

47. Margolina, A.; Wu, S. Percolation model for brittle-tough transition in nylon/rubber blends. *Polymer* **1988**, *29*, 2170–2173. [CrossRef]

48. Nagarajan, V.; Misra, M.; Mohanty, A.K. New engineered biocomposites from poly(3-hydroxybutyrate-co-3-hydroxyvalerate) (PHBV)/poly(butylene adipate-co-terephthalate) (PBAT) blends and switchgrass: Fabrication and performance evaluation. *Ind. Crops Prod.* **2013**, *42*, 461–468. [CrossRef]

49. Muthuraj, R.; Misra, M.; Mohanty, A.K. Reactive compatibilization and performance evaluation of miscanthus biofiber reinforced poly(hydroxybutyrate-co-hydroxyvalerate) biocomposites. *J. Appl. Polym. Sci.* **2017**, *134*. [CrossRef]

50. Rossa, L.V.; Scienza, L.C.; Zattera, A.J. Effect of curauá fiber content on the properties of poly(hydroxybutyrate-co-valerate) composites. *Polym. Compos.* **2013**, *34*, 450–456. [CrossRef]

51. Buchdahl, R. Mechanical properties of polymers and composites—Vols. I and II, Lawrence, E. Nielsen, Marcel Dekker, Inc.; New York, 1974, Vol. I 255 pp. Vol. II 301 pp. Vol. I $24.50, Vol. II $28.75. *J. Polym. Sci. Polym. Lett. Ed.* **1975**, *13*, 120–121. [CrossRef]

52. González-Ausejo, J.; Sanchez-Safont, E.; Lagaron, J.M.; Olsson, R.T.; Gamez-Perez, J.; Cabedo, L. Assessing the thermoformability of poly(3-hydroxybutyrate-co-3-hydroxyvalerate)/poly(acid lactic) blends compatibilized with diisocyanates. *Polym. Test.* **2017**, *62*, 235–245. [CrossRef]

53. Sánchez-Safont, E.L.; González-Ausejo, J.; Gámez-Pérez, J.; Lagarón, J.M.; Cabedo, L. Poly(3-Hydroxybutyrate-co-3-Hydroxyvalerate)/Purified Cellulose Fiber Composites by Melt Blending: Characterization and Degradation in Composting Conditions. *J. Renew. Mater.* **2016**, *4*, 123–132. [CrossRef]

54. ISO. *Determinación del Grado de Desintegración de Materiales Plásticos Bajo Condiciones de Compostaje Simuladas en un Laboratorio*; UNE-EN ISO UNE-EN ISO 20200; ISO: Geneva, Switzerland, 2006.

5

Exploring the Structural Transformation Mechanism of Chinese and Thailand Silk Fibroin Fibers and Formic-Acid Fabricated Silk Films

Qichun Liu [1,2], Fang Wang [1,*], Zhenggui Gu [2], Qingyu Ma [3] and Xiao Hu [4,5,6,*]

1   Center of Analysis and Testing, Nanjing Normal University, Nanjing 210023, China; 40021@njnu.edu.cn
2   School of Chemistry and Materials Science, Nanjing Normal University Jiangsu, Nanjing 210023, China; guzhenggui@njnu.edu.cn
3   School of Physics and Technology, Nanjing Normal University, Nanjing 210023, China; maqingyu@njnu.edu.cn
4   Department of Physics and Astronomy, Rowan University, Glassboro, NJ 08028, USA
5   Department of Biomedical Engineering, Rowan University, Glassboro, NJ 08028, USA
6   Department of Molecular and Cellular Biosciences, Rowan University, Glassboro, NJ 08028, USA
*   Correspondence: wangfang@njnu.edu.cn (F.W.); hu@rowan.edu (X.H.)

Abstract: Silk fibroin (SF) is a protein polymer derived from insects, which has unique mechanical properties and tunable biodegradation rate due to its variable structures. Here, the variability of structural, thermal, and mechanical properties of two domesticated silk films (*Chinese and Thailand B. Mori*) regenerated from formic acid solution, as well as their original fibers, were compared and investigated using dynamic mechanical analysis (DMA) and Fourier transform infrared spectrometry (FTIR). Four relaxation events appeared clearly during the temperature region of 25 °C to 280 °C in DMA curves, and their disorder degree (f_{dis}) and glass transition temperature (T_g) were predicted using Group Interaction Modeling (GIM). Compared with *Thai* (Thailand) regenerated silks, *Chin* (Chinese) silks possess a lower T_g, higher f_{dis}, and better elasticity and mechanical strength. As the calcium chloride content in the initial processing solvent increases (1%–6%), the T_g of the final SF samples gradually decrease, while their f_{dis} increase. Besides, SF with more non-crystalline structures shows high plasticity. Two α- relaxations in the glass transition region of tan δ curve were identified due to the structural transition of silk protein. These findings provide a new perspective for the design of advanced protein biomaterials with different secondary structures, and facilitate a comprehensive understanding of the structure-property relationship of various biopolymers in the future.

Keywords: silk fibroin; glass transition; DMA; FTIR; stress-strain

---

1. Introduction

Silk is a biopolymer with perfect biocompatibility and tunable biodegradability due to its unique protein compositions and structures [1–5]. In the past few decades, silk has been developed into variable biomaterials including tubes, sponges, hydrogels, fibers and thin films, and combined with various functional nanomaterials to provide unique properties that can be applied to biomedical, electrical, or material engineering [6–11].

Generally, different material fabrication methods can affect the multi-step structural transitions and physical properties of silk fibroin materials. For example, Philips et al. [12] compared the dissolution of silk fibroin using different ionic liquids, and demonstrated that 1-butyl-3-methylimidazolium chloride, 1-butyl-2,3-dimethylimidazolium, and 1-ethyl-3-

methylimidazolium were able to disrupt the hydrogen bonding in silk fibroin fibers. By controlling drying rate, Lu et al. [13] were able to prepare water-insoluble silk films from 9.3 mol/L LiBr aqueous solution. Tian et al. [14] added poly epoxy materials, such as formaldehyde, glutaraldehyde, and epoxy compounds into silk fibroin. They suggested that the flexibility of silk materials can be improved through the epoxy compounds, which also acted as crosslinking agents for silk fibroin proteins.

Dynamic Mechanical Analysis (DMA) is one of thermal analysis techniques, which is an advanced technique for measuring the viscoelastic change of polymeric materials during their structural relaxation. [15–20] Juan et al. [21] investigated the effect of temperature and thermal history on the mechanical properties of native silkworm and spider dragline silks by dynamic mechanical thermal analysis (DMTA). Their results showed that the DMA storage modulus and loss tangent of silk materials depend on their different chemical and physical processing methods. Wang et al. [22] also explored the variability of individual as-reeled *A. pernyi* silk fibers using DMTA. They suggested that different polar solvents could affect the tensile properties and structure of silk fibers during the quasi-static tensile tests in ethanol, air, methanol, or water. Porter et al. [23,24] assumed that spider silk's stiffness and strength attributed to the high cohesive energy density of hydrogen bonding, and the toughness attributed to the high energy absorption during post-yield deformation. Furthermore, they found that silk strength was associated with the peculiar molecular and nanoscale structure of its morphology. Kawano et al. [25] measured *Nafion* silk films with different types of solvent and cations using DMA in the controlled force mode. Their results demonstrated that silk elasticity decreased with the increase of water, methanol, ethanol, or ethanol/water mixture content in the *Nafion* film, and also decreased with increasing temperature and cation substitutions (Li^+, Na^+, K^+, Cs^+ and Rb^+).

Besides, Step-scan Differential Scanning Calorimetry (SSDSC) is a relatively new technique which is another thermal analysis technique under temperature modulation, where the temperature program comprises a periodic succession of short, heating rates, and isothermal steps; thus, the measured heat flow contains contributions which arise from the heat capacity and those due to physical transformations or chemical reactions. The total heat flow can be separated into the reversing and non-reversing components, because the reversing component is only observed on the heating part of the cycle and the non-reversing only on the isothermal. Since both the heat capacity equilibration and DSC equilibration are rapid, the C_p calculation is said to be independent of kinetic processes [26]. Therefore, through SSDSC, the "reversing heat capacity", which represents the reversible heat effect of samples within the temperature range of the modulation, such as the specific heat of samples during the glass transition region, can be measured and calculated. Hu et al. [3] used temperature modulated DSC (TMDSC) to eliminate the non-reversing thermal phenomena of the sample and measure the reversing thermal properties of the silk-tropoelastin samples. Sheng et al. [27] characterized the heat capacity, phase contents and transitions of PLA scaffold using SSDSC approach.

Bombyx Mori silkworms are domestically raised silkworms that can produce white silk fibers (e.g., from China (*Chin* silk)), or yellow silk fibers (e.g., from Thailand (*Thai* silk)), due to their different geographical and growing environments [28,29]. Derived from *Bombyx mori* cocoons, silk fibroin (SF) is a fibrous protein consisting of repeating glycine-alanine or glycine-serine peptides responsible for beta-sheet crystal structures mixed with amorphous regions [3–11]. Different silkworm species have different amino acid compositions and therefore have different crystallinity [5,23,24]. The environmental climate can also affect the mulberry leaves. Various silkworm leaves or foods may lead to differences in their cocoons, such as the color and strength. In our previous work, Indian *Antheraea mylitta*, *Philosamia ricini*, *Antheraea assamensis*, Thailand and Chinese *Bombyx mori* mulberry (*Thai, Chin*) silk films have been successfully regenerated from the aqueous solution [30]. Moreover, it was found that Chin and Thai silk fibroin films can be regenerated through a calcium chloride-formic acid ($CaCl_2$/FA) solution system [31]. It was demonstrated that Ca^{2+} ions could interact with the silk structure, and change their glass transition temperature, specific heat, and thermal stability [32]. In this work, the DMA technique was, for the first time, used to explore and compare the structure and mechanical property of these two kinds of silk fibers (Thai, Chin), and also combined with SSDSC and FTIR technologies to investigate these properties and transformation mechanism of their protein films regenerated from the FA solution with a changing $CaCl_2$ content (1%~6%). In addition, a theory developed by Porter et al. [24] Group Interaction Modeling (GIM), was used to investigate and verified the relationship between the stability and structure of regenerated silk materials during the glass transition temperature region (T_g). This work also explained the impact of $CaCl_2$ content to the dynamic mechanical properties of two domesticated silks comprehensively. These comparative studies are important for the design of advanced silk-based materials with tunable structures and properties.

2. Results and Discussion

2.1. Dynamic Mechanical Analysis of the Degumming Silk Fiber

Figure 1 shows the storage modulus E' and loss factor tan δ curves of CRS and TRS natural fibers with the change of temperature at five frequencies (1 Hz, 2 Hz, 5 Hz, 10 Hz, and 20 Hz), respectively. Four peaks were observed in E' curves of Chin silk fibroin fiber sample under all frequencies (Figure 1a), which were assigned to the protein relaxation of γ, β, α_c and α at 25 °C to 280 °C, respectively. As the frequency increases from 1 Hz to 20 Hz, the transition peak moves slightly to a higher temperature, since the time of the molecular segment relaxation and movement is inversely proportional to the frequency intensity. According to the equivalent principle of time and temperature, the increased frequency is equivalent to the shortened relaxation time of the material. Therefore, when the material is tested at a higher frequency, the transition peak of the segment movement could move to a higher temperature [33,34]. Different frequencies of 1 Hz to 20 Hz have the same effect of dynamic thermomechanical property on the silk fibroin. Therefore, in the remaining studies, we will only discuss experimental phenomena with a frequency of 1 Hz. In the 1 Hz force-controlled E' curve (solid line), the γ-relaxation endear at about 41.64 °C due to the molecular motion of the silk protein side chain and the initial evaporation of free water molecules from silk. The β-relaxation at 96.12 °C could not be precisely assigned, but may be related to the molecular motion of silk fibroin after complete evaporation of water, or to the pendant group of the silk polymers (e.g., Ardhyananta et al. [35] has pointed out that the pendant group of the polysiloxanes could affect the thermal and mechanical properties). This can be confirmed by previous findings [36,37] that regenerative silk usually contains 5–10% (w/w) bound water, which significantly affect the thermal properties of silk. The water content of our samples in this work was around 6 wt.% (Table 1) measured by thermogravimetric analysis (TG), which has been discussed previously [29]. The α-relaxation at about 235.34 °C is associated with the glass transition of the silk protein noncrystalline structure due to the segmental motion of the silk protein backbone. Notably, Um et al. [38] pointed out that an α_c-relaxation above 260 °C might occur after the α-relaxation in silk proteins fabricated from the aqueous solution. In our present work, this relaxation appeared around 269.92 °C for CRS and 272.39 °C for TRS. Besides, during the heating scan, the tan δ curves also show clearly three peaks (39.91 °C, 92.72 °C, and 214.25 °C), corresponding to the γ, β, and α-relaxation (Figure 1b). However, the peak of α_c-relaxation did not appear obviously in the tan δ curve. The same phenomena can be also found from the TRS fiber sample. For the TRS fiber, three transition events can be observed at 43.82 °C, 109.07 °C, and 238.48 °C in the E' curve at 1 Hz (Figure 1c, solid line), which correspond to γ, β, α relaxations, respectively. The final transition peak at 272.39 °C (Figure 1c) is belonged to α_c-relaxation. Furthermore, in the tan δ curve (Figure 1d, solid line), γ, β and α-relaxation peaks were observed at 42.11 °C, 100.55 °C, and 218.92 °C in the range of 25~280 °C, respectively.

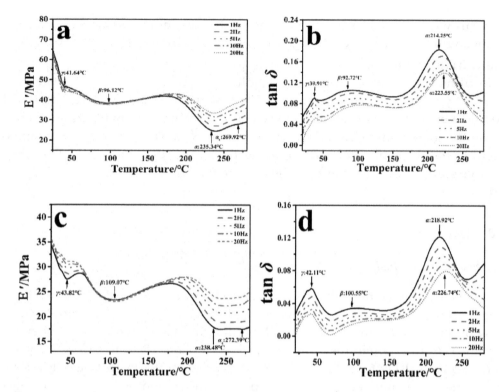

Figure 1. Dynamic mechanical analysis of the storage modulus (E') curves of Chin silk fibroin fibers (**a**) and Thai silk fibroin fibers (**c**); and tan δ curves of Chin silk fibroin fibers (**b**) and Thai silk fibroin fibers (**d**) at five different frequencies: 1 Hz (solid line), 2 Hz (dash line), 5 Hz (dot line), 10 Hz (dash dot line), and 20 Hz (short dot line).

Table 1. The water content of different SF film samples by Thermogravimetric Analysis (TG) *.

Sample	Water Content/%	Sample	Water Content/%
CRS	9.70	TRS	7.26
CSF-1.0	3.23	TSF-1.0	1.26
CSF-1.5	2.97	TSF-1.5	1.04
CSF-2.0	2.05	TSF-2.0	1.10
CSF-3.0	1.79	TSF-3.0	1.20
CSF-4.0	1.35	TSF-4.0	0.52
CSF-6.0	1.16	TSF-6.0	0.51

* All of the numbers have an error bar of ± 3%.

Table 2 compared E' and tan δ of these two kinds of regenerated silk fibers (CRS and TRS) in the relationship of γ, β, α, and α_c under the condition of 1Hz frequency. During the γ-relaxation, CRS sample has a storage modulus of 18.33 MPa and tan δ of 10.45 at 41.64 °C, while TRS sample has a lower storage modulus of 7.48 MPa and a loss factor of 7.21 at 43.82 °C. In the storage modulus E' curve, the peak temperatures of the CRS sample (41.64 °C, 96.12 °C, 235.34 °C, and 269.92 °C) were 2.18~12.95 °C lower than the TRS sample (43.82 °C, 109.07 °C, 238.48 °C, and 272.39 °C) (Table 2). In tan δ curve, the peak temperatures of the CRS sample (39.91 °C, 92.72 °C, and 214.25 °C) were also 2.20~4.67 °C lower than those of TRS sample (42.11 °C, 100.55 °C, and 218.2 °C) (Table 2). Moreover, the storage modulus E' of TRS sample under the 1 Hz (7.48 MPa, 11.71 MPa, 23.54 MPa, and 25.73 MPa) were also lower than those of the CRS sample (18.33 MPa, 26.28 MPa, 40.20 MPa and 38.34 MPa) during the γ, β, α and α_c transitions, respectively (Table 2). These results indicated that *Chin* white silk fiber (CRS) can dehydrate more easily and have more disorder in its structure than the yellow *Thai* TRS fiber, and *Chin* silk molecular chains can be moved more easily when heated. Besides, it will possess a higher degree of viscous deformation, stronger damping, and faster energy dissipation than the TRS sample. Meanwhile, this might also imply that the CRS fiber has more elasticity and stiffness.

Table 2. The experimental parameters of degummed Chinese (CRS) and Thailand (TRS) *B. Mori* silk fibers obtained from DMA analysis *

Sample	CRS/TRS			
Attribution	γ	β	α	α_c
E'/MPa	18.33/7.48	26.28/11.71	40.20/23.54	38.34/25.73
Tan δ	10.45/7.21	6.24/4.72	23.49/19.36	N/A
$T_{E'}$/°C	41.64/43.28	96.12/109.07	235.34/238.48	269.92/272.39
$T_{tan\delta}$/°C	39.91/42.11	92.72/100.55	214.25/218.92	N/A
$\Delta T_{E'}$/°C	2.18	12.95	3.14	2.47
$\Delta T_{tan\delta}$/°C	2.20	7.83	4.67	N/A

* E' and tan δ represent the storage modulus and integral loss factor of CRS and TRS samples at 1 Hz frequency respectively in DMA tensile mode. $T_{E'}$ and $T_{tan\delta}$ represent the peak temperatures for the storage modulus and the loss factor curves, which are corresponding to the protein relaxation of γ, β, α and α_c. $\Delta T_{E'}$ and $\Delta T_{tan\delta}$ represent the peak temperature differences between TRS and CRS samples at γ, β, or α-relaxation regions. The E' and Tan δ values have an error bar of ±0.3, the $T_{E'}$ and $T_{tan\delta}$ values have an error bar of ±0.5 °C.

Born et al. [39] pointed out that the thermally induced vitreous transition of silk fibroin was proposed to the derive non-cooperative or cooperative movements of the skeleton segments in the non-crystalline or disordered regions of silk structure, when the intermolecular forces pass through a maximum or the intermolecular rigidity tends to zero. The transition condition is known as Born's elastic instability criterion, which focuses on the stiffness or mobility of the bonds perpendicular to the axis of interaction instead of the bonds along the axis interaction. Quantitatively, Porter's Group Interaction Modeling (GIM) theory provided a relationship between the properties and structure of polymeric materials, and the expression between structural parameters and T_g can be presented in Equation (1):

$$T_g{}^c = 0.224y + 0.0513\frac{E_{coh}}{N} \tag{1}$$

where the $T_g{}^c$ is the theoretical glass transition temperature at a reference rate of 1 Hz, which can be written in terms of several parameters: (1) the temperature of skeletal mode vibrations, y; (2) the cohesive energy, E_{coh}; and (3) the skeletal degrees of freedom, N [24].

The degrees of freedom N, in the GIM frame is defined as the number of normal vibration skeletons at the axis of the polymer backbone. For detailed calculation of E_{coh} for each peptide base, reference can be made to Porter's work [24]. Wang et al. [22] and Guan et al. [40] studied the cohesive energy E_{coh} and degree of freedom N of each group of China silk from *Bomby Mori cocoon* according to the data in Table 3. *B. mori* silk' E_{coh} was calculated as: E_{coh} = 24.3 (contribution of the peptide base) + 0 × 47.5% (contribution of glycine -H) + 4.5 × 31.7% (Alanine contribution -CH3) + 10.8 × 15.8% (Serine -CH2-OH contribution) + 35.8 × 5% (tyrosine contribution -CH2-Ph-OH) = 29.2 (kJ·mol^{-1}).

Table 3. Calculated GIM parameters based on peptide group contributions and amino acid (AA) sequences for *B Mori* silks [24].

Peptide	Structure	E_{coh}/kJ·mol^{-1} (without H-Bond)	Degrees of Freedom N	Molar Fraction as Counted in AA Sequence *B Mori*
Peptidebase(-R group)	-C-CO-NH-	24.3	6	0
Glycine(G)	-H	0	0	47.5%
Alanine(A)	-CH3	4.5	2	31.7%
Serine(S)	-CH2-OH	10.8	3	15.8%
Glutamine(Q)	-CH2-CH2-CO-NH2	28.8	5	0
Tyrosine(Y)	-CH2-Ph-OH	35.8	4	5%
Leucine(L)	-CH2-C(CH3)2	18	4	0
Arginine(R)	-CH2-CH2-CH2-NH-C(NH2)2	45	7	0
Average	*B. Mori*	29.2	6	/

While the CRS and TRS samples in our study were grown in different regions (China and Thailand) and have different thermal properties [28–30], where they all came from *Bombyx Mori* silkworm species. Therefore, we considered that these two kinds of silk have the same cohesive energy E_{coh} and degree of freedom N.

Meanwhile, in our research, the experimental value of T_g was defined as the temperature from fitted gaussian peak position on the tan δ curve during the glass transition, as shown in Figure 1b,d and Table 4. Using the GIM method and the structural parameters in Table 3, we have E_{coh} = 29.2 kJ·mol^{-1}, N = 6, and y = 241 °C. The value of y was common to all structures due to the same average group molecular weight [23]. Therefore, the theoretical value of $T_g{}^c$ of CRS or TRS fiber sample was calculated to 82.4 °C, without considering the contribution from hydrogen-bonds. This theoretical value $T_g{}^c$ was much lower than that of DMA observation (about 218 °C). If one hydrogen bond per peptide group was taken, the calculated result T_g from Equation (1) would become 157.6 °C, which is close to the lower limit of the silk's experimental T_g. Guan et al. [40] argued that the molecular structures responsible for the glass transition of silk do not have a singular form, but a probability spectrum with several favored forms, e.g., one or two hydrogen bonds per peptide. Thus, the experimental T_g temperatures of 214.25 °C for Chin silk fibroin fibers in Figure 1b and of 218.92 °C for Thai silk fibroin fibers in Figure 1d are the results of the averaged hydrogen-bonding density contributed by hydrogen-bonding forms with different probabilities, respectively. Vollrath et al. [23] believed that if one or two hydrogen bonds were adopted, the molecular structure in the silk would have a 70% chance of 2 hydrogen bonds (H-bonds). Hydrogen bond energy took 10 kJ·mol^{-1} as an average of N-H···O and N-H···N forms, respectively. The higher T_g implied more hydrogen bonds existed among amide groups of silks, through which highly oriented molecular structure and the number of hydrogen bonds have impact on the cohesive energy directly. In *B. Mori* silk sample, if two hydrogen bonds per peptide were counted, an additional energy of 20 kJ·mol^{-1} would be added to the peptide base value of 29.2 kJ·mol^{-1}, which gives the final average E_{coh} of 49.2 kJ·mol^{-1} for each characteristic segment. As a result, the theoretical value $T_g{}^c$ of silk is 243.1 °C through Equation (1) calculation, which is close to the upper limit of experimental temperature of T_g at 214.25 °C in tan δ curve and at 235.34 °C in E' curve for CRS fiber sample, while at 218.92 °C in tan δ curve, and at 238.48 °C in E' curve for TRS fiber sample.

Table 4. GIM parameters used for T_g and the degree of structural disorder f_{dis} calculations b*

Sample	Group	H-Bonds	E_{coh}/kJ·mol^{-1}	N	$T_g{}^c$/°C	T_g/°C	Tan δ^c	Tan δ	f_{dis}
Chin	$G_{0.475}A_{0.317}S_{0.158}Y_{0.005}$	1	39.2	6	157.6	214.25	56	23.49	0.63
		2	49.2	6	243.1		70		0.51
Thai	$G_{0.475}A_{0.317}S_{0.158}Y_{0.005}$	1	39.2	6	157.6	218.92	56	19.36	0.52
		2	49.2	6	243.1		70		0.42

* The number of H-bonds per peptide group is 1 or 2. Cohesive energy E_{coh} is the sum of energy from hydrogen bonds and the peptide base. N is the degrees of freedom. $T_g{}^c$ is the theoretical glass transition temperature calculated from Equation (1), in which y is set as 241 °C for all cases. T_g represents the experimental glass transition temperature from tan δ curve at α relaxation. The theoretical Tan δ^c was calculated from Equation (2), which represents the energy dissipation for 100% degrees of structural disorder. Tan δ is the integral loss factor at α relaxation in DMA curve, and f_{dis} is the predicted degree of structural disorder by using Equation (3).

Tan δ curves from DMA can be further used to determine the structural change of regenerated silk. As previously introduced, the order-disorder distribution can avoid the complicated assignments of secondary structures, which could be used to effectively predict the macroscopic properties of silk materials. First, a structural parameter, f_{dis}, is defined as the degree of structural disorder, which is the molar fraction of the non-crystalline structures that are responsible for glass transition. The degree of structural disorder approximates an averaged structural parameter of heterogeneous nano-structures in a mean-field homogeneous micro and macroscopic morphology. For these two kinds of silk, their f_{dis} values obtained from amino acid sequence analysis are listed in Table 4.

Equation (2) from GIM model described the quantitative relationship of cumulative tan δ (over the transition temperature range) with the structural parameters of the interactive group, E_{coh} and N. A quick calculation of tan δ^c using Equation (2) for *B. Mori* silk is shown in Table 4, which appeared much greater than the experimental values of tan δ. Therefore, for semi-crystalline silks, the degree of structural disorder f_{dis} was introduced into the equation, and adapted as a new form, as presented in Equation (3). The function of factor f_{dis} is easy to understand as only the motions of the disordered structure could be activated during the glass transition and contribute to the experimentally measured tan δ.

The coefficient (2/3) in Equation (3) was used to correct the experimental effect of the uniaxial tensile mode in DMA measurement, because the molecular structures subjected to the static stress of the tensile direction could not be relaxed as the motions along this direction are restrained. As a result, the probability of molecular motions of the overall disordered structure through glass transition was reduced by one dimension, or a factor of 2/3.

$$\tan \delta^c = 0.0085 \times \frac{E_{coh}}{N} \tag{2}$$

$$\tan \delta = \frac{2}{3} \times f_{dis} \times 0.0085 \frac{E_{coh}}{N} \tag{3}$$

Equation (3) opened two avenues: First, it allowed the prediction of the cumulative loss tangent with a known degree of structural disorder. Second, it allowed the calculation of the degree of structural disorder from the theoretical tan δ^c. A quick calculation using GIM framework (Equation (2)) for *B. Mori* silk showed that tan δ^c was in the numerical range of 56–70, which represented the energy dissipation of 100% degrees of structural disorder. The apparent discrepancy between the experimental cumulative tan δ and the theoretical tan δ^c during the glass transition for native *B. Mori* silks clearly suggested that crystalline or ordered structure existed in our silk samples. This phenomenon was also mentioned in the work of Porter et al. [41–43] The degree of disorder f_{dis} for two silk fibers were also calculated individually by using both the number of hydrogen bonds and the cohesive energy from the Equation (3) (Table 4). Simultaneously, apparent discrepancy between the experimental tan δ and the theoretical tan δ^c in glass transition region of silks appeared by using the GIM framework. Guan et al. and Porter et al. also reported the same phenomenon [21,24]. By comparing the tan δ of these two kinds of regenerated *B. Mori* silk, the degree of disorder f_{dis} was deduced: 0.63 for one H-bond and 0.51 for two H-bonds in the CRS sample, and 0.52 for one H-bond and 0.42 for two H-bonds in the TRS sample, as listed in Table 4. The results showed that the disordered structure of silk fibroin had a significant effect on its glass transition. Additionally, the glass transition temperature (T_g) decreased with the degree of disorder (f_{dis}) increasing, since tan δ is directly associated with the f_{dis} as well as the ordered molecular structures in silk.

2.2. Structural Transformation of Chin and Thai Silk Protein Films

Glass transition temperature is a characterization temperature at which the chain segment of polymer molecules starts to move, which is related to the flexibility of polymer chains. In the glass transition region, when the semi-crystalline polymer material changes from the solid state to the flowing liquid state, the specific heat of the semi-crystalline polymer material undergoes a discontinuous mutation during the heating process [28,29]. Our previous studies on silk fibroin films by scanning electron microscopy (SEM) [30] and X-ray diffraction (XRD) [33] showed that a high $CaCl_2$ concentration can significantly reduce the silk fibril structure and micro-/nanoscale morphology in the silk film. Further, two small diffraction peaks appeared at 20.7° and 24.0° in the XRD curves of low $CaCl_2$ concentration sample (e.g., TSF-1.5), which is recognized as the silk I structure. This phenomenon implied that the β-sheet crystal content decreases with the increase of calcium chloride concentration, and higher calcium chloride concentrations may disrupt the hydrogen bonds between the silk fibroin molecular chains, which reduces the silk II content and increases the silk I content. Here, we will explore the structure transformation of silk proteins fabricated from CaCl$_2$–formic acid solution by using DMA, SSDSC, and FTIR results.

In general, the loss factors (tan δ) of silk fibroin membranes CSF-1.0, CSF-1.5, CSF-2.0, CSF-3.0, CSF-4.0, and CSF-6.0 all have three discontinuous events that corresponded to the protein relaxation of γ, β, and α, which have been discussed in the previous section. The γ relaxation peak around 50~60 °C with little shoulders in two side of curve implied the co-events of γ-relaxation (protein-water T_g) and the evaporation of mobile H$_2$O in this region [21]. The water evaporation completed around 80~140 °C, and the major molecular motion of pure silk fibroin (the α relaxation peak of tan δ curve) appeared in the range of 150–230 °C (Figure 2). The water content of various silk protein films samples contained around 0.5–3.23% bound water molecules measured by TG in our previous work [30]. Hu et al. [36]. focused on the interaction of the solid silk film with the intermolecular bound water molecules. The results showed that the silk film start to release the water molecules into air at 35 °C. As the temperature increased, more and more water evaporated and the weight of silk film decreased until about 160 °C. Above 160 °C, there is no more intermolecular water in the silk film. Based on these results, the water should have no contribution to α-relaxation. In addition, this peak also decreased gradually with the increasing of calcium chloride. To better understand the change during the T_g region, StepScan differential scanning calorimetry (SSDSC) measurement were also used to determine the heat capacity increment (ΔC_p^s) of silk samples during the T_g. It demonstrated that the glass transition temperature (T_g^s) of SF increased with the calcium chloride concentration decreasing, e.g., from 157.30 °C to 176.88 °C with the change of calcium chloride concentration from 6.0% to 1.0%, respectively. The heat capacity increment (ΔC_p^s) is directly proportional to the average chain mobility of proteins, which reflects the number of freely rotating bonds capable of changing the chain conformation [30]. The ΔC_p^s results summarized in Table 5 indicated that a non-crystalline structure exists in all samples and the SF-6.0 protein chains have the highest fraction of the non-crystalline structure, while the average chain mobility and non-crystalline fraction in the SF-1.0 sample are the lowest. Therefore, with the CaCl$_2$ concentration increasing, the T_g^s decreased while ΔC_p^s increased, which indicated that content of non-crystalline structures in SF is increasing with the increase of CaCl$_2$ content.

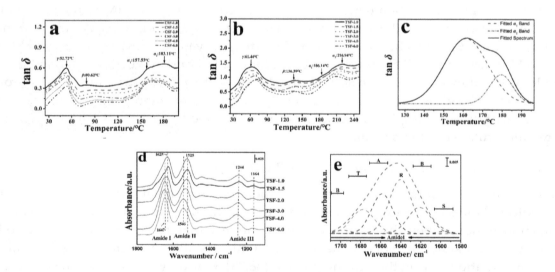

Figure 2. Tan δ curves of CSF (**a**) and TSF (**b**) from 1.0% (Solid); 1.5 (Dash); 2.0 (Dot); 3.0 (Dash Dot); 4.0 (Short Dot); 6.0 (Dash Dot Dot) CaCl$_2$/FA; (**c**) Curve fitting example of the T_g region (sample TSF-3.0). The fitted peaks are shown as Dash (α$_1$ Peak) and Dash Dot Dot (α$_2$ Peak); (**d**) FTIR absorbance spectra of TSF samples from different CaCl$_2$ conctration solutions in the range of 1100–1800 cm^{-1}; (**e**) Curve fitting example of the amide I region (sample TSF-3.0) in FTIR spectra. The fitted peaks were shown as short dash-dotted lines, and assigned as side chains (S), β-sheets (B), random coils (R), α-helixes (A), and turns (T).

Table 5. SSDSC and DMA analysis of CSF and TSF samples in their glass transition region [30] *.

Sample	$[CaCl_2]$/ wt %	$T_g{}^s$/°C	ΔC_p/ $J \cdot g^{-1} \cdot °C^{-1}$	$T_{g\text{-}\alpha 1}$/°C	Content in α_1 Region/%	Tan $\delta_{\text{-}\alpha 1}$	$f_{dis\text{-}\alpha 1}$	$T_{g\text{-}\alpha 2}$/°C	Content in α_2 Region/%	Tan $\delta_{\text{-}\alpha 2}$	$f_{dis\text{-}\alpha 2}$
CSF	1.0	176.88	0.1667	157.53	32.87	6.65	0.18	177.11	67.13	7.52	0.20
	1.5	172.65	0.1721	155.49	52.56	7.15	0.19	171.47	47.44	7.92	0.21
	2.0	169.04	0.1890	155.16	53.95	7.89	0.21	168.32	46.05	8.91	0.24
	3.0	167.42	0.1998	154.87	71.53	9.89	0.27	165.51	28.47	9.25	0.25
	4.0	163.21	0.2157	154.39	73.54	11.57	0.3	163.93	26.46	10.37	0.27
	6.0	157.30	0.2204	153.92	79.99	12.44	0.34	162.36	20.01	11.03	0.30
TSF	1.0	181.54	0.1641	186.14	54.70	4.34	0.12	216.94	45.30	4.02	0.10
	1.5	177.08	0.1643	184.54	65.51	5.73	0.15	212.01	34.49	4.97	0.13
	2.0	175.94	0.1644	184.31	71.14	6.52	0.18	209.27	28.86	5.56	0.14
	3.0	173.42	0.1972	183.69	75.24	7.22	0.19	206.45	24.76	6.25	0.16
	4.0	164.01	0.2136	183.08	75.72	8.93	0.23	204.85	24.28	7.67	0.20
	6.0	159.20	0.2137	182.47	76.65	10.17	0.26	200.82	23.35	8.93	0.23

* $[CaCl_2]$ is the concentration of calcium chloride in the solution. $T_g{}^s$ is the glass transition temperature of SF measured by SSDSC. $T_{g\text{-}\alpha 1}$ and $T_{g\text{-}\alpha 2}$ represent the peak temperatures of α_1-relaxtion and α_2-relaxtion from the tan δ curves, respectively. Their content values were obtained by fitting the tan δ curve in α_1 and α_2-relaxtion regions using Gaussian peaks. Tan $\delta_{\text{-}\alpha 1}$ and Tan $\delta_{\text{-}\alpha 2}$ are the integral loss factor at α relaxations. $f_{dis\text{-}\alpha 1}$ and $f_{dis\text{-}\alpha 2}$ are the predicted degree of structural disorder at α_1-relaxtion and α_2-relaxtion by the GIM model, respectively. The $T_g{}^s$, $T_{g\text{-}\alpha 1}$, and $T_{g\text{-}\alpha 2}$ have an error bar of ± 0.5 °C. The Content in α_1 region and Content in α_2 region have an error bar of $\pm 3\%$. The Tan $\delta_{\text{-}\alpha 1}$, Tan $\delta_{\text{-}\alpha 2}$, $f_{dis\text{-}\alpha 1}$ and $f_{dis\text{-}\alpha 2}$ have an error bar of ± 0.05.

In our previous study [30], we found that the concentration of $CaCl_2$/FA could significantly affect the secondary structures of silk. Since DMA technique is more sensitive than the SSDSC technique for the glass transition measurement, two relaxation peaks (α_1 and α_2) can be found in the glass transition region of tan δ curve from 125 °C to 195 °C (Figure 2). For the CSF-1.0 sample, the two transition peaks appeared at 157.53 °C ($T_{g\text{-}\alpha 1}$) and 177.11 °C ($T_{g\text{-}\alpha 2}$), which are associated with α_1 and α_2- relaxation (Figure 2a, solid line), respectively. For the TSF-1.0 sample, the α_1 and α_2-relaxation transitions appeared at 186.14 °C ($T_{g\text{-}\alpha 1}$) and 216.94 °C ($T_{g\text{-}\alpha 2}$), respectively. To quantify the percentage of these two peaks, the tan δ curves were curve fitted using Gaussian peaks in the glass transition region of 125–195 °C.

Figure 2c showed an example of curve fitted tan δ curves from the TSF-3.0 sample (dashed lines). Table 5 summarized the fitted percentage of each peak in T_g region for all silk samples. It shows that the TSF-1.0 film has 54.70% α_1 relaxion at 186.14 °C ($T_{g\text{-}\alpha 1}$), and 45.30% α_2-relaxation at 216.94 °C ($T_{g\text{-}\alpha 2}$). While the TSF-6.0 sample has 76.65% α_1 relaxion at 182.47 °C and 23.35% α_2 relaxion at 200.82 °C. This suggests that as the concentration of calcium chloride increased, both peaks shifted to lower temperatures (Table 5), but the content of α_1 relaxion increased, while the content of α_2 relaxion decreased. To better understand the secondary structures of our materials, a FTIR deconvolution method was performed to quantify the percentage of the secondary structures in all silk samples [28]. Figure 2d showed the main characteristics of protein structures of all six silk films in the FTIR spectra. With the concentration of $CaCl_2$ increasing, the center of the absorption band in Amide I region shifted gradually from 1625 cm^{-1} (beta-sheet structure, TSF-1.0) to 1647 cm^{-1} (random coil structure, TSF-6.0), while the peak in Amide II region also shifted gradually from 1525 cm^{-1} (TSF-1.0) to 1546 cm^{-1} (TSF-6.0). Moreover, for the Amide-III and FTIR fingerprinting regions, Thai silk sample showed same characteristic peaks and two obvious peaks at 1244 cm^{-1} and 1164 cm^{-1} [30]. Figure 2e shows an example of the Amide I region of TSF-3.0 sample with the fitted FTIR peaks shown as dashed lines. The peak positions and their related secondary structures were assigned as side chains (S), β-sheets (B), random coil (R), α-helix (A), and turns (T) [30]. We found that the content of helix structures in TSF samples could increase significantly by increasing the concentration of $CaCl_2$. For example, with the increase of $CaCl_2$ concentration, the percentage of β-sheet structures gradually decreased from 26.60% in TSF-1.0 film to 7.07% in the TSF-6.0 sample, while the percentage of random coils decreased from 56.07% to 46.43%, while α-helixes content increased from 8.59% to 28.39%, and the percentage of turns also increased from 7.11% to 17.74%. The similar trend of α-relaxation temperatures and the content of secondary structures in the silk samples were also observed in CSF samples. Table 6 summarized the percentage of secondary structures for each CSF and TSF sample. These results implied that α_1 and α_2- relaxation events in tan δ curve are associated with the change of protein secondary structures. Um et al. [38] found that there is α_c-relaxation after the major α-relaxation in water-regenerated silk materials, which is related to the cooling crystallization of silk proteins. In our silk samples regenerated from $CaCl_2$/FA system, with the decrease of β-sheets in the silk films, the α_1 content increased, while α_2 percentage decreased, and both of the α_1 and α_2 peaks shifted to a lower temperature. Therefore, we can claim that random coils or other non-crystalline structures may be contributed to the transition of α_1-relaxation which was equaled to the α-relaxation, while the change of α-helix structure may be related to α_2-relaxation which was the α_c-relaxation found previously [38].

Table 6. Percentage of secondary structures in CSF and TSF samples obtained by FTIR structural analysis [28].

[CaCl₂] (wt %)	CSF					TSF				
	β-Sheet/%	Turns/%	Side Chains%	α-Helix/%	Random Coils/%	β-Sheet/%	Turns/%	Side Chains/%	α-Helix/%	Random Coils/%
1.0	23.32	6.03	1.33	10.07	59.25	26.60	7.11	1.63	8.59	56.07
1.5	20.25	7.59	1.07	11.39	59.70	24.76	8.02	1.15	9.96	56.11
2.0	13.56	12.05	0.91	13.78	59.70	14.70	14.31	1.19	12.54	57.26
3.0	8.09	14.13	0.85	18.79	58.14	10.92	16.61	0.94	18.22	53.31
4.0	8.17	15.41	0.63	25.66	50.13	8.25	17.05	0.79	21.83	52.08
6.0	6.83	15.89	0.57	30.26	46.45	7.07	17.74	0.37	28.39	46.43

Similarly, DMA experiment also showed that $T_{g\text{-}\alpha1}$ and $T_{g\text{-}\alpha2}$ of the CSF samples were lower than those of the TSF samples (Figure 2 and Table 5) at the same calcium chloride content. Both FTIR and DMA results indicate that lower β-sheet structures and higher non-crystalline structures can be obtained in silk films regenerated from the solution with higher calcium chloride concentration.

Base on the above work, the relationship between degree of disorder f_{dis} and the T_g were discussed and the influence of calcium chloride on structure of these silk samples was further explored in this section. It was supposed that the amount of hydrogen bonding effects of calcium chloride on different silk fibroin films were identical (H-bonds = 1; $N = 6$; $E_{\text{coh}} = 39.2$ kJ·mol^{-1}, $y = 241$ °C), the degree of disorder $f_{\text{dis-}\alpha1}$ and $f_{\text{dis-}\alpha2}$ of two kinds of regenerated silk fibroin were calculated from GIM Equation (3) and is shown in Table 5, as well as the tan $\delta_{\text{-}\alpha1}$, tan $\delta_{\text{-}\alpha2}$, $T_{g\text{-}\alpha1}$, and $T_{g\text{-}\alpha2}$ from DMA test. The increasing trend of $f_{\text{dis-}\alpha1}$ and $f_{\text{dis-}\alpha2}$ indicated that more concentration of calcium chloride could induce the β-sheet transform into the random coil in the secondary structure of silk fibroin and more non-crystalline structure form. Simultaneously, in the same concentration of calcium chloride, CSF samples showed more disorder degree than that of TSF samples, e.g., 0.20 for CSF-1.0 sample, 0.10 for TSF-1.0 sample at α_1-relaxation. The results suggested that more β-sheet and crystalline structures in TSF sample than those in CSF sample.

2.3. Stress-Strain Study of CSF and TSF

The stress-strain curve gives information about the Young modulus (slope at the origin), yield point, break point, and recovery behavior of polymeric films. Meanwhile, the stress-strain curve analysis can also provide information on polymer structure, degree of crosslinking, degree of crystallization, processing conditions, and viscoelastic properties of polymer [44–46]. The stress-strain curve of the CSF sample (Figure 3a) shows that as the calcium chloride concentration increases from 1.0 to 6.0 mass%, the initial slope decreases from 7.86 to 1.83 and its stress decreases from 14.91 to 11.93 MPa at the yield point, respectively. In addition, under the stress of 10 MPa, their strains were 1.21%, 2.15%, 2.47%, 3.39%, 4.17%, and 5.88%, respectively (Table 7). Compared with the trend of change in CSF samples under the initial slope and yield stress, the trend of change in TSF samples is the same (Figure 3b). These results reveal that increasing the content of calcium chloride can increase the elongation of the sample, while its elasticity and stiffness decrease due to the formation of dominated random coils and other non-crystalline structures in silk samples [33]. Besides, with the same concentration of calcium chloride and the same stress loaded on the sample, the initial slope of CSF sample was higher than that of the TSF sample. For example, the initial slope of the

CSF-6.0 sample is 1.83 under the 10 MPa, which was higher than the initial slope of 1.52 from the TSF-6.0 sample.

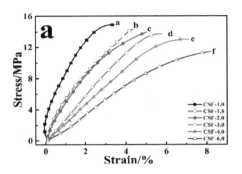 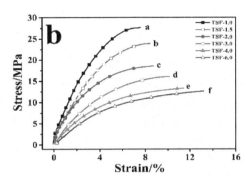

Figure 3. The stress-strain curves of CSF (**a**) and TSF (**b**) samples regenerated from 1.0% (solid square), 1.5% (hollow square), 2.0% (solid circle), 3.0% (hollow circle), 4.0% (solid triangle), and 6.0% (hollow triangle) $CaCl_2$/FA solutions.

Table 7. Mechanical properties of CSF and TSF samples measured by DMA *.

[CaCl$_2$]/wt %	CSF			TSF		
	Initial Slope	Yield Stress/MPa	Strain/%	Initial Slope	Yield Stress/MPa	Strain/%
1.0	7.86 ± 0.10	14.91 ± 1.57	1.21 ± 0.08	7.37 ± 0.24	14.28 ± 1.29	1.86 ± 0.07
1.5	7.11 ± 0.32	14.30 ± 1.38	2.15 ± 0.15	6.94 ± 0.19	13.33 ± 1.37	2.27 ± 0.10
2.0	5.48 ± 0.55	13.98 ± 1.55	2.47 ± 0.13	3.57 ± 0.36	12.45 ± 1.56	2.50 ± 0.05
3.0	2.96 ± 0.37	13.66 ± 1.29	3.39 ± 0.16	2.70 ± 0.27	11.20 ± 1.28	3.93 ± 0.12
4.0	2.16 ± 0.86	12.97 ± 1.37	4.17 ± 0.09	1.80 ± 0.58	10.17 ± 1.35	5.45 ± 0.27
6.0	1.83 ± 0.27	11.43 ± 0.98	5.88 ± 0.17	1.52 ± 0.49	8.93 ± 1.88	6.96 ± 0.16

* [CaCl$_2$] is the concentration of calcium chloride in the formic acid solution system. The initial slope is the ratio of stress to strain, which represents the elasticity of SF sample. The yield stress is the stress at the yield point in stress-strain curve. The strain value is the material elongation ratio under the stress force of 10 MPa. Every sample have to do five experiment times. Their errors or deviations were shown in each column after symbol '±'.

Similarly, the strain of the TSF-6.0 sample was 6.96%, which was greater than 5.88% of the CSF-6.0 sample. And the yield stress of CSF-6.0 sample was 11.43 MPa, which was higher than 8.93 MPa of the TSF-6.0 sample. These results indicate that the CSF sample has better elasticity than the TSF sample, while the TSF sample has better plasticity than the CSF sample. These results also imply that the average chain mobility of silk films in TSF samples is higher than that in CSF samples. Therefore, at the same $CaCl_2$ concentration, the TSF sample would contain more non-crystalline structures than the CSF sample, which has been proved by FTIR structural analysis in Table 6.

3. Materials and Methods

3.1. Materials and Preparation

The Chinese (Dandong Qiyue Trade co., LTD.) and Thailand *Bombyx mori* silk cocoons (Queen Sericulture Center, Nakornratchasima, Thailand) were first degummed into Chin regenerated silk fibroin fibers (CRS) and Thai regenerated silk fibroin fibers (TRS) according to a previously reported procedure [47]. Different amounts of calcium chloride (AR, purity 96%, West Gansu Chemical Plant, Shantou, Guangdong) were mixed with formic acid (AR, purity 88%, West Gansu Chemical Plant, Shantou, Guangdong) to form 1.0, 1.5, 2.0, 3.0, 4.0, and 6.0 mass% $CaCl_2$/FA solutions. Then, the degummed silk fibroin (SF) fibers were quickly dissolved into these $CaCl_2$/FA solutions at room temperature. The final solutions were immediately cast onto a Teflon mold (50 mm × 20 mm × 10 mm) and dried at room temperature to form SF films. After being washed (30 mins in running water) and vacuum dried to fully remove $CaCl_2$/FA solvents, the final Chin SF (CSF) and Thai SF (TSF) films were obtained. CSF and TSF samples were named with the numbers (−1.0,

−1.5, −2.0, −3.0, −4.0, and −6.0) after the sample names to indicate the initial concentration of $CaCl_2$ (1.0, 1.5, 2.0, 3.0, 4.0, and 6.0 mass%) in the solution.

3.2. Thermal and Mechanical Analyses

Sample with a dimension of $5.0 \times 2.0 \times 1.0 mm^3$ was subjected to the mechanical analysis by using a Dynamic Mechanical Analyzer (Perkin-Elmer Diamond DMA, Waltham, MA, USA). Experiment was proceeded under a temperature range from 25 °C to 280 °C with a heating rate of 2 °C min^{-1} and at 1, 2, 5, 10, and 20 Hz frequencies simultaneously. The glass transition temperature (T_g) of sample was taken from the temperature at which the maximum peak of the tangent δ was exhibited. Furthermore, at the atmosphere temperature (25 °C), the stress-strain property of silk fibroin film was tested by DMA in a tensile controlled model. The tensile force increased from 5 mN to 2000 mN at a lifting speed of 50 mN min^{-1} until the sample was broken. For all experiments, at least three samples were measured under the same test conditions to check the consistency of the experimental results. Dynamic mechanical analysis (DMA) refers to the technique of measuring the dynamic modulus and mechanical loss of a specimen and its relation to temperature or frequency under programmed temperature and alternating stress. In general, a periodically varying (usually refers to the sine) force was applied to the sample to produce periodic changes in the stress. The sample will have a corresponding deformation behavior of the stress; thus, the mechanical modulus of the sample can be determined by the stress and deformation. For example, if a shear stress is applied, a shear G can be obtained, and if a type of stress is applied in tension or bending, Young's Modulus E can be obtained. DMA measurement modes include tension, compression, bending, single-cantilever, shear, and reverse. In most cases, the specimen undergoes a periodical variation of the mechanical vibration stress, which causes the corresponding vibrational strain. However, the specimen does not always respond instantaneously to the changing stress and the lags behind for a certain period of time. This mainly depends on the viscoelasticity of the specimen and the phase shift between applied stress and deformation. Therefore, sample modulus consists of both real and imaginary parts. The real part describes the sample response as the periodic stress, which is a measurement of the elasticity of the sample, called the storage modulus. The imaginary part describes the response phase shifted by 90°, which is a measurement of the mechanical energy converted to heat, called the loss modulus. The ratio of loss modulus (E'') and storage modulus (E') is called the loss factor, the phase shift tangent δ (tan δ), which is the express of sample damping performance. The storage modulus E' is proportional to the mechanical energy stored in the specimen during stress, while the loss modulus E'' represents the energy dissipated in the specimen during stress. The larger the loss modulus is, the more viscous the specimen is, and the stronger the damping of specimen is. The tan δ is independent of the geometry, therefore it can be accurately measured even if the geometry of the sample is not regular. In dynamic mechanical analysis, the force amplitude F_A and the displacement amplitude L_A are used to calculate the modulus of samples. For example, the tensile modulus or elastic modulus of the experiment E' can be obtained through the formula:

$$E' = \frac{\sigma}{\varepsilon} = \frac{F_A}{L_A} \frac{L_0}{A} = \frac{F_A}{L_A} g \tag{4}$$

$$\sigma = F_A / A \tag{5}$$

$$\varepsilon = L_0 / L_A \tag{6}$$

$$g = L_0 / A \tag{7}$$

where σ and ε are the stress and strain of the sample, respectively; F_A and L_A are the force amplitude and the displacement amplitude, respectively; L_0 is the original length of the sample; A is the unit area of the sample; g is the geometry efficiency; and F_A / L_A is the rigidity of the material [48].

Generally, materials with low storage modulus E' implies it will deform easily when applying load on it; and the tan δ represents the viscoelasticity of materials during the loading cycle. A high value of tan δ suggests a high degree of energy dissipation and a high degree of viscous deformation for this material.

Besides, the dried CRS and TRS samples were encapsulated in Al pans and heated in a differential scanning calorimeter under Step-Scan modulated (SSDSC, Diamond DSC, Perkin-Elmer, USA) at a heating rate of 3 °C/min^{-1} with a 3 °C step and isothermal time of 2 min with 25 mL min^{-1} purged dry nitrogen gas, and equipped with a refrigerated cooling system. Each sample was about 5 mg.

3.3. Fourier Transform Infrared Spectrometry

Fourier transform infrared spectra (FTIR) of silk protein film sample was obtained using a FTIR spectrometer (Nicolet-NEXUS 670, Nicolet, Madison, WI, USA), equipped with a deuterated triglycine sulfate detector and a multiple-reflection, horizontal MIRacle ATR attachment (OMNIT, using a Ge crystal, Madison, WI, USA). Spectra were recorded in the wavenumber range of 1800 to 1100 cm^{-1} with a resolution of 4 cm^{-1}, and 64 scans were applied for each measurement. Fourier self-deconvolution (FSD) of the IR spectra covering the Amide I region (1595–1705 cm^{-1}) was performed by the Nicolet Omnic software. Deconvolution was performed using Gauss line shape with a half-bandwidth of 25 cm^{-1} and a noise reduction factor of 0.3. FSD spectra were then curve-fitted by Gaussian peaks to measure the relative areas in the Amide I region.

3.4. Thermogravimetric Analysis

Thermogravimetric (TG) analysis (PerkinElmer Pyris 1, Waltham, MA, USA) was used to measure the change in the mass of the silk samples during temperature increase. The TG curves were obtained under a nitrogen atmosphere with a gas flow of 50 mL·min^{-1}. Samples of about 2–3 mg were heated from 30 to 600 °C with a heating rate of 10 °C·min^{-1}. The mass change percentages during heating were recorded.

4. Conclusions

Dynamic Mechanical Analysis (DMA) analysis is more sensitive than Differential Scanning Calorimetry (DSC), as short-range chain motion changes are easier to detect than heat capacity changes during the phase transitions of biopolymer materials. The various structures and mechanical properties of two kinds of domesticated silk fibroin (SF) films (Chinese and Thailand B. Mori) regenerated from formic acid-CaCl$_2$ solutions were investigated using DMA. Our study showed that by using the GIM model, the disorder degree (f_{dis}) of silk samples can be inferred from the cohesive energy (E_{coh}), the skeletal degree of freedom (N), and the loss factor (tan δ) at the glass transition region. Four disordered phase relaxation events were explored by DMA technique, with a focus on the glass transition and the degree of structural disorder. Our results illustrated that there are nearly two hydrogen bonds formed in each peptide group in the Bombyx Mori silk fibers. With the increase of calcium chloride concentration in the SF sample, the T_g of silk material decreases, which implied that the sample could contain more non-crystalline structures, such as random coils and helix. α_1 and α_2-relaxation events in DMA curves are both associated with the silk secondary structures. The random coils as well as other non-crystalline structures may be attributed to the change of α_1-relaxation, while α_2-relaxation could be directly associated to the α-helix to β-sheet transition. Moreover, at the same calcium chloride concentration, the CSF sample is more disordered than the TSF sample, which suggests that there are more β-sheet in the TSF sample than in the CSF sample. It also showed that the elasticity of the TSF sample is lower than that of the CSF sample, while their ductility is the opposite. Besides, SF samples prepared with lower concentrations of calcium chloride have higher elasticity, while SF samples prepared with high concentrations have better ductility. The effects of calcium chloride concentrations on the structure and mechanical properties of regenerated silk fibroin films were further investigated by comparing GIM theoretical model calculations with experimental results. These results provide us a new way to understand the structural changes and mechanical properties of different domesticated silk regenerated from acid-based solution system, which would be critical for engineering applications of silk materials. These results provide us with a new way to understand the structural changes and mechanical properties of different domesticated silk regenerated from acid-based solution system, which would be critical for engineering applications of silk materials in the future.

Author Contributions: Q.L., F.W. and X.H. conceived and designed the experiments; Q.L., F.W., Z.G., and Q.M. performed the experiments, Q.L., F.W. and X.H. analyzed the data and wrote the paper.

References

1. Bhardwaj, N.; Singh, Y.P.; Devi, D.; Kandimalla, R.; Kotoky, J.; Manda, B.B. Potential of silk fibroin/chondrocyte constructs of muga silkworm Antheraea assamensis for cartilage tissue engineering. *J. Mater. Chem. B* **2016**, *4*, 3670–3684. [CrossRef]

2. Yu, S.; Yang, W.; Chen, S.; Chen, M.; Liu, Y.; Shao, Z.; Chen, X. Floxuridine-loaded silk fibroin nanospheres. *RSC Adv.* **2014**, *4*, 18171–18177. [CrossRef]

3. Hu, X.; Wang, X.; Rnjak, J.; Weiss, A.S.; Kaplan, D.L. Biomaterials derived from silk-tropoelastin protein systems. *Biomaterials* **2010**, *31*, 8121–8131. [CrossRef] [PubMed]

4. Hu, X.; Shmelev, K.; Sun, L. Regulation of Silk Material Structure by Temperature-Controlled Water Vapor Annealing. *Biomacromolecules* **2011**, *12*, 1686–1696. [CrossRef] [PubMed]

5. Jao, D.; Xue, Y.; Medina, J.; Hu, X. Protein-Based Drug-Delivery Materials. *Materials* **2017**, *10*, 517. [CrossRef] [PubMed]

6. Kambe, Y.; Murakoshi, A.; Urakawa, H.; Kimura, Y.; Yamaoka, T. Vascular induction and cell infiltration into peptide-modified bioactive silk fibroin hydrogels. *J. Mater. Chem. B* **2017**, *5*, 7557–7571. [CrossRef]

7. Aytemiz, D.; Sakiyama, W.; Suzuki, Y.; Nakaizumi, N.; Tanaka, R.; Ogawa, Y.; Takagi, Y.; Nakazawa, Y.; Asakura, T. Small-diameter silk vascular grafts (3 mm diameter) with a double-raschel knitted silk tube coated with silk fibroin sponge. *Adv. Healthc. Mater.* **2013**, *2*, 361–368. [CrossRef] [PubMed]

8. Shahbazi, B.; Taghipour, M.; Rahmani, H.; Sadrjavadi, K.; Fattahi, A. Preparation and characterization of silk fibroin/oligochitosan nanoparticles for siRNA delivery. *Colloids Surf. B* **2015**, *136*, 867–877. [CrossRef] [PubMed]

9. Tesfaye, M.; Patwa, R.; Kommadath, R.; Kotecha, P.; Katiyar, V. Silk nanocrystals stabilized melt extruded poly (lactic acid) nanocomposite films: Effect of recycling on thermal degradation kinetics and optimization studies. *Thermochim. Acta* **2016**, *643*, 41–52. [CrossRef]

10. Yin, Z.; Wu, F.; Xing, T.; Yadavalli, V.K.; Kundu, S.C.; Lu, S. A silk fibroin hydrogel with reversible sol-gel transition. *RSC Adv.* **2017**, *7*, 24085–24096. [CrossRef]

11. Hu, X.; Park, S.H.; Gil, E.S.; Xia, X.X.; Weiss, A.S.; Kaplan, D.L. The Influence of Elasticity and Surface Roughness on Myogenic and Osteogenic-Differentiation of Cells on Silk-Elastin Biomaterials. *Biomaterials* **2011**, *32*, 8979–8989. [CrossRef] [PubMed]

12. Phillips, D.M.; Drummy, L.F.; Conrady, D.G.; Fox, D.M.; Naik, R.R.; Stone, M.O. Dissolution and Regeneration of Bombyx mori Silk Fibroin Using Ionic Liquids. *J. Am. Chem. Soc.* **2004**, *126*, 14350–14361. [CrossRef] [PubMed]

13. Lu, Q.; Hu, X.; Wang, X.; Kluge, J.A.; Lu, S. Water-Insoluble Silk Films with Silk I Structure. *Acta Biomater.* **2010**, *6*, 1380–1387. [CrossRef] [PubMed]

14. Tian, L.; Chen, Y.; Min, S. Research on Cytotoxicity of Silk Fibroin Gel Materials Prepared with Polyepoxy Compound. *J. Biomed. Eng.* **2007**, *24*, 1309–1313.

15. Crawford, D.M.; Escarsega, J.A. Dynamic mechanical analysis of novel polyurethane coating for military applications. *Thermochim. Acta* **2000**, *357*, 161–168. [CrossRef]

16. Yin, B.; Hakkarainen, M. Core–shell nanoparticle–plasticizers for design of high-performance polymeric materials with improved stiffness and toughness. *J. Mater. Chem.* **2011**, *21*, 8670–8677. [CrossRef]

17. Khandaker, M.S.K.; Dudek, D.M.; Beers, E.P.; Dillard, D.A.; Bevan, D.R. Molecular modeling of the elastomeric properties of repeating units and building blocks of resilin, a disordered elastic protein. *J. Mech. Behav. Biomed. Mater.* **2016**, *61*, 110–121. [CrossRef] [PubMed]

18. Mahdi, E.M.; Tan, J.C. Dynamic molecular interactions between polyurethane and ZIF-8 in a polymer-MOF nanocomposite: Microstructural, thermo-mechanical and viscoelastic effects. *Polymer* **2016**, *97*, 31–43. [CrossRef]

19. Saba, N.; Jawaid, M.; Alothman, O.Y.; Paridah, M.T. A review on dynamic mechanical properties of natural fibre reinforced polymer composites. *Constr. Build. Mater.* **2016**, *106*, 149–159. [CrossRef]

20. Nalyanya, K.M.; Migunde, O.P.; Ngumbu, R.G.; Onyuka, A.; Rop, R.K. Thermal and dynamic mechanical analysis of bovine hide. *J. Therm. Anal. Calorim.* **2016**, *121*, 1–8. [CrossRef]

21. Guan, J.; Porter, D.; Vollrath, F. Thermally induced changes in dynamic mechanical properties of native silks. *Biomacromolecules* **2013**, *14*, 930–937. [CrossRef] [PubMed]

22. Wang, Y.; Guan, J.; Hawkins, N.; Porter, D.; Shao, Z. Understanding the variability of properties in Antheraea pernyi silk fibres. *Soft Matter* **2014**, *10*, 6321–6331. [CrossRef] [PubMed]

23. Vollrath, F.; Porter, D. Spider silk as a model biomaterial. *Appl. Phys. A* **2006**, *82*, 205–212. [CrossRef]

24. Porter, D.; Vollrath, F.; Shao, Z. Predicting the mechanical properties of spider silk as a nanostructured polymer. *Eur. Phys. J. E* **2005**, *16*, 199–206. [CrossRef] [PubMed]

25. Kawano, Y.; Wang, Y.; Palmer, R.; Aubuchon, A.; Steve, R. Stress-Strain Curves of Nafion Membranes in Acid and Salt Forms. *Polímeros* **2002**, *12*, 96–101. [CrossRef]

26. Pyda, M.; Wunderlich, B. Reversing and nonreversing heat capacity of poly (lactic acid) in the glass transition region by TMDSC. *Macromolecules* **2005**, *38*, 10472–10479. [CrossRef]

27. Sheng, S.J.; Hu, X.; Wang, F.; Ma, Q.Y.; Gu, M.F. Mechanical and thermal property characterization of poly-L-lactide (PLLA) scaffold developed using pressure-controllable green foaming technology. *Mat. Sci. Eng. C.-Mater.* **2015**, *49*, 612–622. [CrossRef] [PubMed]

28. Wang, F.; Wolf, N.; Rocks, E.M.; Vuong, T.; Hu, X. Comparative studies of regenerated water-based Mori, Thai, Eri, Muga and Tussah silk fibroin films. *J. Therm. Anal. Calorim.* **2015**, *122*, 1069–1076. [CrossRef]

29. Mazzi, S.; Zulker, E.; Buchicchio, J.; Anderson, B.; Hu, X. Comparative thermal analysis of Eri, Mori, Muga, and Tussar silk cocoons and fibroin fibers. *J. Therm. Anal. Calorim.* **2014**, *116*, 1337–1343. [CrossRef]

30. Wang, F.; Yu, H.Y.; Gu, Z.G.; Si, L.; Liu, Q.C.; Hu, X. Impact of calcium chloride concentration on structure and thermal property of Thai silk fibroin films. *J. Therm. Anal. Calorim.* **2017**, *130*, 1–9. [CrossRef]

31. Wang, F.; Chandler, P.; Oszust, R.; Sowell, E.; Graham, Z.; Ardito, W.; Hu, X. Thermal and structural analysis of silk–polyvinyl acetate blends. *J. Therm. Anal. Calorim.* **2017**, *127*, 923–929. [CrossRef]

32. Hu, X.; Lu, Q.; Sun, L.; Cebe, P.; Wang, X.Q.; Zhang, X.H.; Kaplan, D.L. Biomaterials from Ultrasonication-Induced Silk Fibroin—Hyaluronic Acid Hydrogels. *Biomacromolecules* **2010**, *11*, 3178–3188. [CrossRef] [PubMed]

33. Zhu, Z.; Jiang, C.; Cheng, Q.; Zhang, J.; Guo, S. Accelerated aging test of hydrogenated nitrile butadiene rubber using the time-temperature-strain superposition principle. *RSC Adv.* **2015**, *5*, 90178–90183. [CrossRef]

34. Butaud, P.; Placet, V.; Klesa, J.; Ouisse, M.; Foltête, E.; Gabrion, X. Investigations on the frequency and temperature effects on mechanical properties of a shape memory polymer (Veriflex). *Mech. Mater.* **2015**, *8*, 50–60. [CrossRef]

35. Ardhyananta, H.; Kawauchi, T.; Ismail, H.; Takeichi, T. Effect of pendant group of polysiloxanes on the thermal and mechanical properties of polybenzoxazine hybrids. *Polymer* **2009**, *25*, 5959–5969. [CrossRef]

36. Hu, X.; Kaplan, D.; Cebe, P. Effect of Water on Thermal Properties of Silk Fibroin. *Thermochim. Acta* **2007**, *461*, 137–144. [CrossRef]

37. Motta, A.; Fambri, L.; Migliaresi, C. Regenerated silk fibroin films: Thermal and dynamic mechanical analysis. *Chem. Phys.* **2002**, *203*, 1658–1665. [CrossRef]

38. Um, I.C.; Kim, T.H.; Kweon, H.Y.; Chang, S.K.; Park, Y.H. A comparative study on the dielectric and dynamic mechanical relaxation behavior of the regenerated silk fibroin films. *Macromol. Res.* **2009**, *17*, 785–790. [CrossRef]

39. Born, M. Thermodynamics of crystals and melting. *J. Chem. Phys.* **1939**, *7*, 591–603. [CrossRef]

40. Guan, J.; Wang, Y.; Mortimer, B.; Holland, C.; Shao, Z.; Porter, D.; Vollrath, F. Glass transitions in native silk fibres studied by dynamic mechanical thermal analysis. *Soft Matter* **2016**, *12*, 5926–5936. [CrossRef] [PubMed]

41. Porter, D.; Vollrath, F. Water mobility denaturation and the glass transition in proteins. *Biochim. Biophys. Acta* **2012** *1824*, 785–791. [CrossRef] [PubMed]

42.   Chen, F.; Porter, D.; Vollrath, F. Silk cocoon (Bombyx mori): Multi-layer structure and mechanical properties. *Acta Biomater.* **2012**, *8*, 2620–2627. [CrossRef] [PubMed]

43.   Fu, C.; Porter, D.; Chen, X.; Vollrath, F.; Shao, Z. Understanding the Mechanical Properties of Antheraea Pernyi Silk-From Primary Structure to Condensed Structure of the Protein. *Adv. Funct. Mater.* **2015**, *21*, 729–737. [CrossRef]

44.   Qin, G.; Hu, X.; Cebe, P.; Kaplan, D.L. Mechanism of resilin elasticity. *Nat. Commun.* **2012**, *3*, 1003–1013. [CrossRef] [PubMed]

45.   Roberts, D.R.T.; Holder, S.J. Mechanochromic systems for the detection of stress, strain and deformation in polymeric materials. *J. Mater. Chem.* **2011**, *21*, 8256–8268. [CrossRef]

46.   Hu, X.; Kaplan, D.; Cebe, P. Determining beta-sheet crystallinity in fibrous proteins by thermal analysis and infrared spectroscopy. *Macromolecules* **2006**, *39*, 6161–6170. [CrossRef]

47.   Yu, H.Y.; Wang, F.; Liu, Q.C.; Ma, Q.Y.; Gu, Z.G. Structure and Kinetics of Thermal Decomposition Mechanism of Novel Silk Fibroin Films. *Acta Phys.-Chim. Sin.* **2017**, *33*, 344–355.

48.   Sgreccia, E.; Chailan, G.F.; Khadhraoui, M.; Vona, M.L.D.; Knauth, P. Mechanical properties of proton-conducting sulfonated aromatic polymer membranes: Stress–strain tests and dynamical analysis. *J. Power Sources* **2010**, *195*, 7770–7775. [CrossRef]

6

Study on Thermal Decomposition Behaviors of Terpolymers of Carbon Dioxide, Propylene Oxide and Cyclohexene Oxide

Shaoyun Chen [1], Min Xiao [2], Luyi Sun [3] and Yuezhong Meng [2,*]

[1] College of Chemical Engineering and Materials Science, Quanzhou Normal University, Quanzhou 362000, China; chshaoy@qztc.edu.cn

[2] The Key Laboratory of Low-carbon Chemistry & Energy Conservation of Guangdong Province/State Key Laboratory of Optoelectronic Materials and Technologies, Sun Yat-Sen University, Guangzhou 510275, China; stsxm@mail.sysu.edu.cn

[3] Department of Chemical & Biomolecular Engineering and Polymer Program, Institute of Materials Science, University of Connecticut, Storrs, CT 06269, USA; luyi.sun@uconn.edu

* Correspondence: mengyzh@mail.sysu.edu.cn

Abstract: The terpolymerization of carbon dioxide (CO_2), propylene oxide (PO), and cyclohexene oxide (CHO) were performed by both random polymerization and block polymerization to synthesize the random poly (propylene cyclohexene carbonate) (PPCHC), di-block polymers of poly (propylene carbonate–cyclohexyl carbonate) (PPC-PCHC), and tri-block polymers of poly (cyclohexyl carbonate–propylene carbonate–cyclohexyl carbonate) (PCHC-PPC-PCHC). The kinetics of the thermal degradation of the terpolymers was investigated by the multiple heating rate method (Kissinger-Akahira-Sunose (KAS) method), the single heating rate method (Coats-Redfern method), and the Isoconversional kinetic analysis method proposed by Vyazovkin with the data from thermogravimetric analysis under dynamic conditions. The values of ln k vs. T^{-1} for the thermal decomposition of four polymers demonstrate the thermal stability of PPC and PPC-PCHC are poorer than PPCHC and PCHC-PPC-PCHC. In addition, for PPCHC and PCHC-PPC-PCHC, there is an intersection between the two rate constant lines, which means that, for thermal stability of PPCHC, it is more stable than PCHC-PPC-PCHC at the temperature less than 309 °C and less stable when the decomposed temperature is more than 309 °C. Pyrolysis-gas chromatography/mass spectrometry (Py-GC/MS) and thermogravimetric analysis/infrared spectrometry (TG/FTIR) techniques were applied to investigate the thermal degradation behavior of the polymers. The results showed that unzipping was the main degradation mechanism of all polymers so the final pyrolysates were cyclic propylene carbonate and cyclic cyclohexene carbonate. For the block copolymers, the main chain scission reaction first occurs at PC-PC linkages initiating an unzipping reaction of PPC chain and then, at CHC–CHC linkages, initiating an unzipping reaction of the PCHC chain. That is why the $T_{-5\%}$ of di-block and tri-block polymers were not much higher than that of PPC while two maximum decomposition temperatures were observed for both the block copolymer and the second one were much higher than that of PPC. For PPCHC, the random arranged bulky cyclohexane groups in the polymer chain can effectively suppress the backbiting process and retard the unzipping reaction. Thus, it exhibited much higher $T_{-5\%}$ than that of PPC and block copolymers.

Keywords: polycarbonate; thermal decomposition kinetics; TG/FTIR; Py-GC/MS

1. Introduction

Carbon dioxide (CO_2) is a nontoxic, nonflammable material that exists naturally in abundance. The use of CO_2 has attracted increasing interest in recent years and has been considered as an alternative approach to reduce the release of this greenhouse gas [1–5]. One good approach is using CO_2 to produce biodegradable polymeric materials. In 1969, Inoue et al. first observed that the copolymerization of carbon dioxide with epoxides could form aliphatic polycarbonates [1]. Since then, much work has been done to make CO_2 copolymerize with other monomers [2–5]. Poly (propylene carbonate) (PPC) made from carbon dioxide and propylene oxide is the main kind of CO_2-based copolymer that has been widely investigated [6–10]. In previous work, the high molecular weight alternating PPC was synthesized in very high yield. The PPCs exhibit good biodegradability [11] but show inferior thermal stability due to the flexible carbonate linkage in the backbone. It has also been found that PPC is easily decomposed to cyclic carbonate by the unzipping reaction, which is initiated by the free hydroxyl terminal groups [12,13]. In order to enhance the thermal properties of PPC, a third monomer cyclohexene oxide (CHO) copolymerizing with CO_2 and PO is considered as a profitable mean [14,15]. However, the thermal degradation kinetics of the terpolymer has not been investigated until now.

Generally, TG analysis is an effective method for studying thermal decomposition kinetics and provides information on a frequency factor, activation energy, and overall reaction order. Due to the insufficient analytical capability of the evolved gas mixture analysis, it is still not possible to provide sufficient information on the mechanism of thermal degradation. Therefore, the direct analysis of gas composition by continuous monitoring with thermogravimetric analysis/Fourier transform infrared spectrometry (TG/FTIR) has attracted more attention in the identification of gaseous products to study the pyrolysis mechanism [16–18]. Pyrolysis gas chromatography/mass spectrometry (Py-GC/MS) is widely used to evaluate the thermal decomposition behavior of polymers because of its high sensitivity, rapidity, and effective separation ability of complex compounds containing a pyrolysis product of similar compositions [19–21]. Therefore, the combination of TG/FTIR and Py-GC/MS has been widely used to study the thermal decomposition mechanism [22–25].

In this paper, CHO was introduced to copolymerize with CO_2 and PO to get the random copolymer, the di-block copolymer, and the tri-block copolymer. The thermal decomposition behaviors of the resultant polymers with different sequence structures are studied by the combination of Py-GC/MS and TG/FTIR techniques. In addition, the thermal degradation kinetic parameters are obtained by using the multiple heating rate method (Kissinger-Akahira-Sunose (KAS) [26–28]), the single heating rate method (Coats-Redfern method) [29,30], and the Isoconversional kinetic analysis method proposed by Vyazovkin [31–33]. This used the data from thermogravimetric analysis under dynamic conditions such as activation energy E, the pre-exponential factor A, and the rate constant k. They will provide theoretical basis of thermal stability for further application.

2. Results and Discussion

2.1. Thermal Decomposition Behavior

The synthesized copolymers containing polycarbonates and polyesters are terpolymers as Scheme 1. These have proven by our previous paper [15]. ^{1}H NMR spectrum of the purified terpolymer shown in Figures S1–S4 demonstrates the formation of ester and carbonate linkages. The proton resonances 1.3 [3H, CH_3], 4.2 [2H, CH_2CH], and 5.0 [1H, CH_2CH] correspond to CH_3, CH_2, and CH groups in the polycarbonate sequence. The peaks of 1.7 ppm, 2.1 ppm, and 4.6 ppm representing CH_2 of M-, O-cycloalkane ring, and CH link to carbonate indicated the cyclohexene carbonate unit in the block copolymer. The basic properties of the resultant polymers are shown in Table 1.

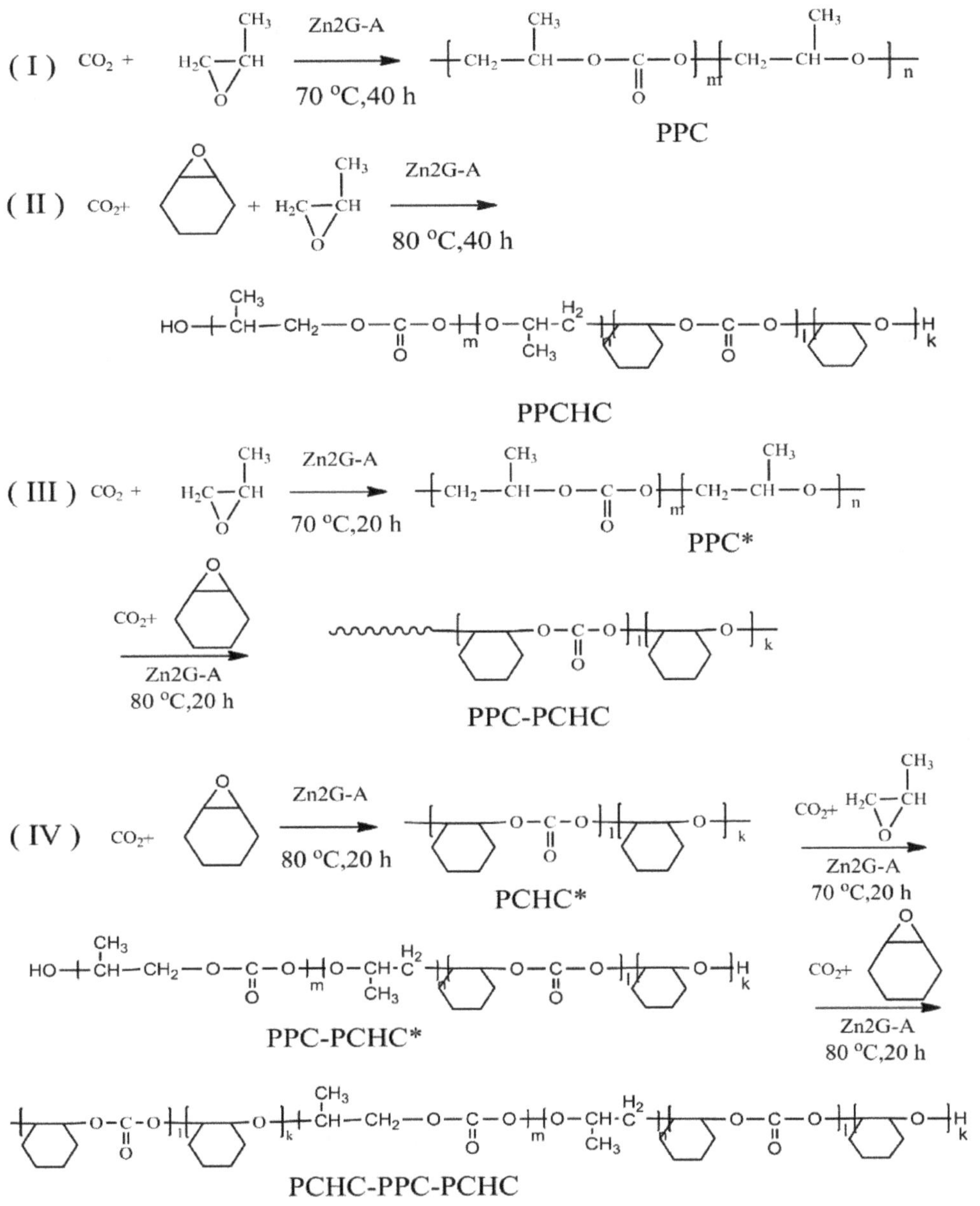

Scheme 1. The synthesized procedure of (I) PPC; (II) PPCHC; (III) PPC-PCHC; (IV) PCHC-PPC-PCHC copolymers.

Table 1. The properties of the resultant polymers.

Copolymer	Mn/Mw/PI [a]	$T_{-5\%}/T_{max}$ (°C) [c]	Composition (Molar Fraction %) [b]			
			f_{CO2}	f_{PC}	f_{CHC}	f_{PE}
PPC	$2.17 \times 10^5/3.78 \times 10^5/1.74$	255.7/278.2	48.9	48.9	-	2.2
PPCHC	$2.02 \times 10^5/7.13 \times 10^5/3.50$	281.0/313.4	47.9	36.2	11.7	4.2
PPC-PCHC	$2.97 \times 10^5/7.35 \times 10^5/2.47$	261.2/304.2, 342.7	48.0	35.1	12.9	4.0
PCHC-PPC-PCHC	$2.74 \times 10^5/7.88 \times 10^5/2.87$	275.2/305.8, 345.3	47.7	34.1	13.6	4.6

[a] Molecular weight was determined by GPC. [b] Determined by ^{1}H NMR spectroscopy. (see ESI Figure A1–A4) $f_{PC} = A_{4.2}/[(A_{4.2} + A_{4.6}) \times 2 + 0.8 \times A_{3.5}]$, $f_{CHC} = A_{4.6}/[(A_{4.2} + A_{4.6}) \times 2 + 0.8 \times A_{3.5}]$, $f_{CO2} = A_{4.2} + A_{4.6}/[(A_{4.2} + A_{4.6}) \times 2 + 0.8 \times A_{3.5}]$, $f_{PE} = 0.8 \times A_{3.5}/[(A_{4.2} + A_{4.6}) \times 2 + 0.8 \times A_{3.5}]$. [c] The heating rate is 10 °C/min.

In order to study the thermal degradation of the resultant polymers and to conclude the best way to improve the thermal stability of PPC, the non-isothermal kinetics of the thermal degradation of the resultant polymers were investigated with the Kissinger-Akahira-Sunose (KAS) method, the Coats-Redfern method, and the Isoconversional kinetic analysis method proposed by Vyazovkin.

TG-DTG curves of the polymers at different heating rates are presented in Figure 1. As the heating rate increased, thermal hysteresis became more and more evident. The thermal decomposition beginning temperature of the polymers also improved and the peak temperature moved to a higher temperature zone. Thermal decomposition of PPC and PPCHC were finished by one step and PPC-PCHC and PCHC-PPC-PCHC were mainly finished by two steps. Decomposition of four polymers at the heating rate 10 °C min^{-1} begin at 245.5 °C, 269.8 °C, 239.8 °C, and 257.2 °C, respectively. The maximum weight loss temperature from DTG curves for four polymers are 278.2 °C, 313.2 °C, 304.2 °C, and 305.8 °C, respectively.

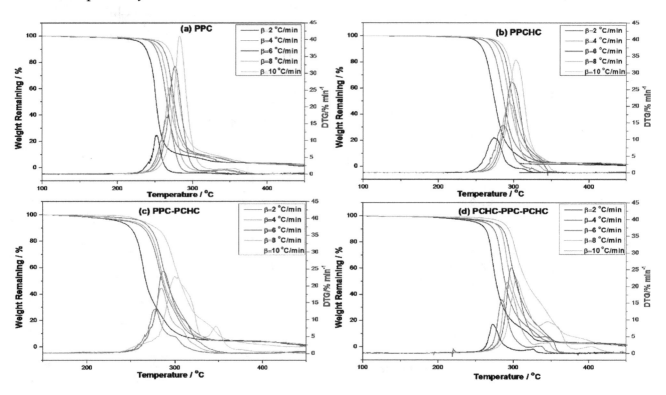

Figure 1. TG-DTG curves of the resultant polymers at different heating rates. (**a**) PPC; (**b**) PPCHC; (**c**) PPC-PCHC; (**d**) PCHC-PPC-PCHC.

In addition, the decomposition of four polymers are completed at ~350 °C. The KAS method has been employed to evaluate the activation energies of different polymers during thermal decomposition because of its good adaptability and validity for model-free approaches. At a constant value of conversion rate α, the plots of ln (β/T_{α}^2) versus $1/T_{\alpha}$. As shown in Figure S5, the plots of ln (β/T_{α}^2) versus $1/T$ at several heating rates obtain well fitted straight lines whose slopes allow the evaluation of the apparent activation energy. The distribution of the activation energy (Eα) for different polymers is presented in Figure 2. It can be seen that the effective activation energy varies with conversion, which is an indication of a complex mechanism of the decomposition reaction. The process of PPC, PPC-PCHC, and PCHC-PPC-PCHC begins with Eα~110 kJ mol^{-1} and then slightly increases to 130 kJ mol^{-1}. In addition, Eα for block polymers PPC-PCHC and PCHC-PPC-PCHC decrease sharply to ~90 kJ mol^{-1} and 75 kJ mol^{-1}. For random polymer PPCHC, the Eα remains stable ~140–150 kJ mol^{-1}. The average activation energies for PPC, PPCHC, PPC-PCHC, and PCHC-PPC-PCHC are 124.6 ± 6.8 kJ mol^{-1}, 145.5 ± 3.0 kJ mol^{-1}, 109.9 ± 7.6 kJ mol^{-1}, and 104.9 ± 16.2 kJ mol^{-1}, respectively. The variation tendency tells us that the Eα of the block copolymer PPC-PCHC and PCHC-PPC-PCHC have a similar trend during the process of thermal decomposition while the variation tendency of PPC and PPCHC were relatively stable.

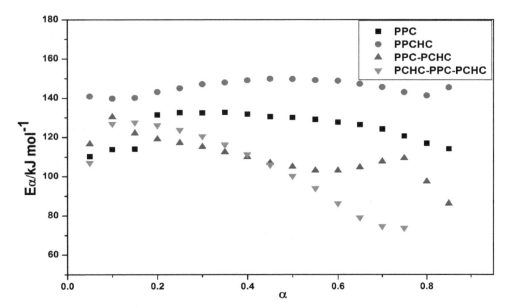

Figure 2. Values of the activation energy estimated by using the KAS method for the thermal decomposition of different polymers.

The model-fitting method (Coats-Redfern method) is suitable for the thermal decomposition of polymer during a certain pyrolysis stage in an entire thermal process. The Coats-Redfern method and 15 kinds of reaction models in solid-state reactions are adopted to focus on analyzing the pyrolysis kinetics of the polymers. The obtained kinetic parameters at different heating rates of polymers are presented in Table S1–S4. Then the iso-conversional pre-exponential factor values were evaluated by substituting the values of $E\alpha$ from the KAS method into the equation of the compensation effect in Equation (3) [31]. The compensation effect parameters a and b were determined by fitting the pairs of $\ln A_i$ and E_i from the Coats-Redfern method using 15 reaction models. The compensation effect parameters a and b were obtained by the average of five heating rates, which is shown in Figure S6. Therefore, the equations of compensation effect are $\ln A_\alpha = -3.23816 + 0.2244E_\alpha$ for PPC, $\ln A_\alpha = -3.5574 + 0.21475E_\alpha$ for PPCHC, $\ln A_\alpha = -2.9405 + 0.21602E_\alpha$ for PPC-PCHC, and $\ln A_\alpha = -3.0112 + 0.211782E_\alpha$ for PCHC-PPC-PCHC. The values of the pre-exponential factor as a function conversion for the thermal decomposition of different polymers are presented in Figure 3. The $\ln A_\alpha$ vs. α dependence exhibits similar facture as the one determined for the activation energy. The average preexponential factor for PPC, PPCHC, PPC-PCHC, and PCHC-PPC-PCHC are $24.7 \pm 1.5\,s^{-1}$, $27.7 \pm 0.6\,s^{-1}$, $20.8 \pm 1.6\,s^{-1}$, and $19.2 \pm 3.4\,s^{-1}$, respectively. Overall, the trend of activation energy values and pre-exponential factor values for decomposition is PCHC-PPC-PCHC < PPC-PCHC < PPC < PPCHC. However, the thermal stability of polymers determined by an increase activation energy values and a decrease pre-exponential factor values. In order to evaluate the overall effect of the activation energy and the pre-exponential factor on the kinetics of the process, the rate constant was calculated by using Equation (4) [32,33]. The values of $\ln k$ vs. T^{-1} for the thermal decomposition of four polymers are presented in Figure 4. From the results, decomposition of PPC, PPC-PCHC show higher values of the rate constant, so the thermal stability of PPC and PPC-PCHC are poorer than PPCHC and PCHC-PPC-PCHC. In addition, for PPCHC and PCHC-PPC-PCHC, there is an intersection between the two rate constant lines, which means that, for thermal stability, PPCHC is more stable than PCHC-PPC-PCHC at temperatures less than 309 °C and less stable when the temperature is more than 309 °C. The enhanced thermal stability of PPCHC is associated with an increase in the activation energy and the block polymer PPC-PCHC and PCHC-PPC-PCHC is mostly associated with a decrease in the pre-exponential factor. Above all, the random copolymerization of CHO, PO, and CO_2 is a better way to improve the thermal stability of PPC. We began with the idea that the tri-block polymer PCHC-PPC-PCHC had the same structure like SBS, which means its thermal stability was better than others. However, the results were not the same as predicted. In order to explain this, we studied the thermal degradation mechanisms of the polymers, which is discussed in the following part.

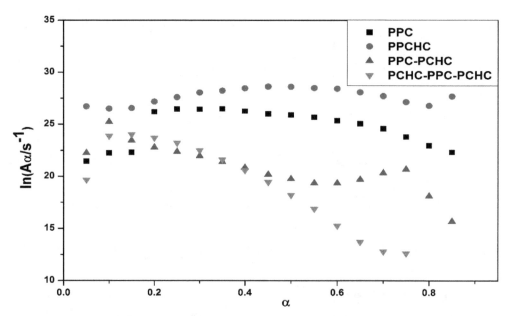

Figure 3. Pre-exponential factor as a function of conversion for the thermal decomposition of different polymers.

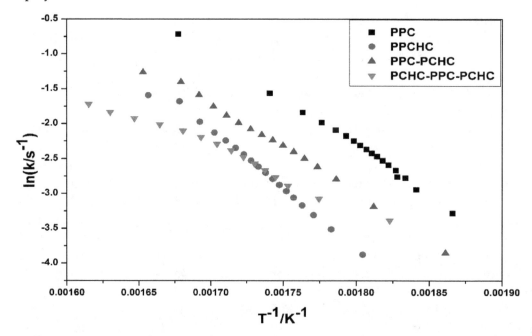

Figure 4. Rate constant as a function of reciprocal temperature for the thermal decomposition of different polymers.

2.2. Study the Thermal Decomposition Behavior Using Py-GC/MS and TG/FTIR

Thermal degradation kinetic parameters of the polymers were calculated by using thermogravimetric analysis. However, it could not provide enough information to analyze the thermal degradation mechanism. With the help of TG/IR technique, we can follow the dynamic process of the polymers decomposition. FTIR spectra (Figure S7) in three-dimensions for the decomposing dynamic process of polymers were obtained via this technique. The three-dimensional FTIR spectra of the pyrolysates derived from PPCHC, PPC-PCHC, and PCHC-PPC-PCHC are similar and the main IR absorbing peaks appear at 1109 cm^{-1}, 1863 cm^{-1}, and 2985 cm^{-1}, which indicates that the main pyrolysates of the block polymers are all cyclic carbonates.

According to previous research [34–37], the thermal decomposition behavior of PPC obeys two kinds of mechanism including the main chain scission reaction and the unzipping reaction. The unzipping reaction involves the backbiting of the terminal hydroxyl groups at the carbon of carbonate linkage leading to the formation of cyclic carbonate. From the FTIR spectra of pyrolysates at different temperatures shown in Figure 5, it can be seen that the peak stands for carbon dioxide is observed at 230 °C while the peaks stand for cyclic carbonates are observed at 250 °C for PPC, PPC-PCHC, and PCHC-PPC-PCHC and higher than 270 °C for PPCHC, which indicates the main chain scission reaction and the unzipping reaction during the decomposition process first occuring in PC-PC segments. In addition, the onset unzipping reaction temperature of the random polymer PPCHC is higher than that of PPC and block polymers. Comparing the intensity of the IR absorption peaks of the pyrolysates at a different temperature, we can see that large scale evolution of cyclic carbonates happens at a relatively lower temperature (about 300 °C) for PPC and higher temperature (about 350–400 °C) for PPCHC. While for the block copolymers, large scale evolution of cyclic carbonates happens at a wider temperature range (300–400 °C). It is supposed that the random polymer PPCHC probably has a large content of PC-CHC or CHC-PC linkages in the terpolymer, which means the steric hindrance of bulky cyclohexane groups can suppress the backbiting process and retard the unzipping reaction. Thus, the thermal stability of PPCHC is much better than PPC. While, for the block copolymers, the PC-PC block and CHC-CHC block exhibit similar thermal stability as PPC and PCHC, respectively, so the temperature range of large scale evolution of cyclic carbonate was wide and two maximum decomposition temperatures were observed, as shown in Table 1. The above supposition was further confirmed by Py-GC/MS, which provide a way to measure the thermal decomposition reactions in each stage by analyzing mixture-evolved gases. Based on the results of TG analysis, the pyrolysis temperatures were set at 200 °C, 250 °C, 300 °C, 350 °C, and 400 °C to obtain the formation curves of pyrolysates to analyze the degradation mechanisms. The chromatograms of pyrolysis products at different temperatures and their identifications are shown in Figure 6 and Table 2.

The pyrolysates of random copolymer PPCHC was not detected out at 300 °C and, at 350 °C, the pyrolysates are cyclic propylene carbonate (6.0–6.75 min) and cyclic cyclohexene carbonate (12.25–12.5 min). These phenomena confirms the large content of PC-CHC or CHC-PC linkages in the polymer chain. Due to the alternating PC and CHC main chain structure, the backbiting is suppressed by steric hindrance results in a retarded unzipping reaction and, once the unzipping reaction starts, the pyrolysates cyclic propylene carbonate and cyclic cyclohexene carbonate are produced at the same time.

For the di-block and tri-block copolymers, small peaks representing CO_2 (1.5–2.0 min) and cyclic propylene carbonate (6.0–6.75 min) appear when being pyrolyzed at 250 °C. The peak of cyclic cyclohexene carbonate appears at a pyrolysis temperature as high as 350 °C, which means that the main chain scission reaction first occurs at PC-PC linkages. This initiates the unzipping reaction of PC-PC linkages and then, at CHC-CHC linkages, initiates the unzipping reaction of the PCHC chain. In addition, this can explain why the degradation activation energies of the block polymers are lower at first and get higher with the decomposition conversion increasing. Based on above results and analysis, the possible decomposition pathways are given as Scheme 2. For the random copolymer PPCHC, the thermal decomposition behavior obeys the unzipping reaction mostly. In addition, for the block copolymers of PPC-PCHC and PCHC-PPC-PCHC, the thermal decomposition behavior obeys two kinds of mechanism including the main chain scission reaction and the unzipping reaction.

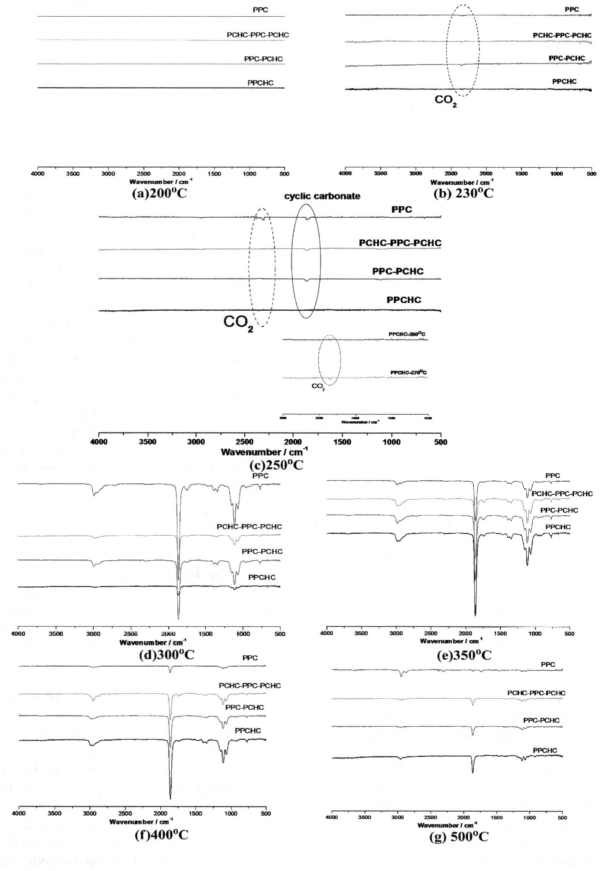

Figure 5. FTIR spectra of pyrolysates at (**a**) 200 °C; (**b**) 230 °C; (**c**)250 °C; (**d**) 300 °C; (**e**) 350 °C, (**f**) 400 °C, and (**g**) 500 °C.

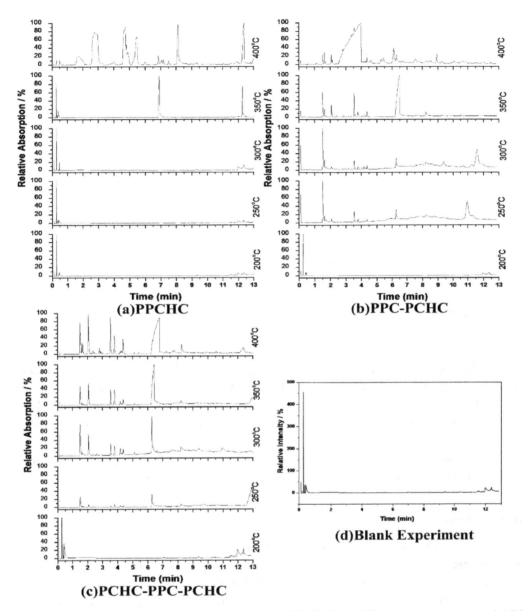

Figure 6. Chromatogram of pyrolysis products by Py-GC/MS at different temperatures :(**a**) PPCHC; (**b**)PPC-PCHC; (**c**)PCHC-PPC-PCHC; (**d**) Blank Experiment.

Table 2. Identification of pyrolysates in Py-GC/MS.

Compound	Retention Time (min)	Compound	Retention Time (min)
		CO_2	1.5–2 min
	4.0–4.5 min		4.5–5.0 min
	5.0–5.5 min		6.0–6.75 min
	7.5–8.0 min		12.25–12.5 min

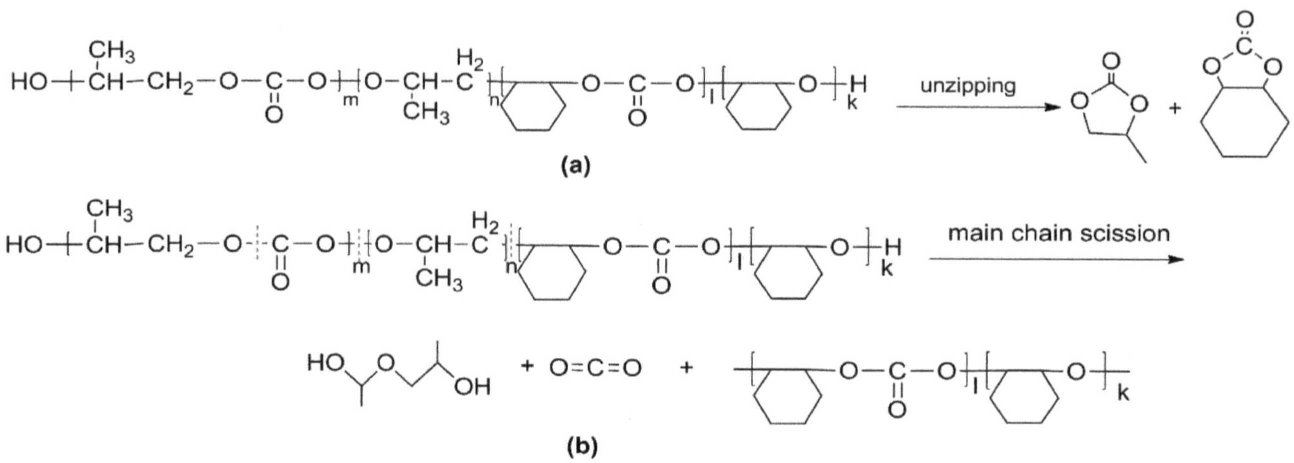

Scheme 2. Possible decomposition pathway of the polymers. (**a**) Unzipping of the polymer backbone and (**b**) main chain scission.

3. Materials and Methods

3.1. Materials

Carbon dioxide of a purity of 99.99% was used without further treatment. PO of a purity of 99.5% and CHO of a purity of 95.0% were refluxed over CaH_2 for 4 h and 24 h, respectively, and then distilled under dry nitrogen gas. Prior to use, they were stored over 4-Å molecular sieves. Other solvents and reagents such as ethanol and chloroform were of analytical grade and used without further purification.

Supported multi-component zinc dicarboxylate catalyst (Zn2G) was prepared according to previous work [38]. The catalyst was white powder with the Zn content of 11.6 wt %.

3.2. Preparation of Random Terpolymers and Block Copolymers

The di-block PPC-PCHC copolymerization was carried out using 'one pot, two steps' method in a 500 mL stainless steel autoclave equipped with a mechanical stirrer [15]. A multi-component catalyst (Zn2G) was introduced into the autoclave and the autoclave with the catalyst inside was dried for 24 h under vacuum at 80 °C and then cooled down to room temperature. Then the purified PO was immediately injected into the autoclave. The autoclave was pressurized to 5.2 MPa via a CO_2 cylinder and heated at 70 °C for 20 h. Following the evacuation of CO_2 and unreacted PO, CHO was introduced into the autoclave under an inert atmosphere. The autoclave was re-pressurized with 5.2 MPa of CO_2 and the reaction was performed at 80 °C for another 20 h. Then, the pressure in the autoclave was reduced to atmosphere to terminate the block copolymerization. Similarly, the tri-block polymer PCHC-PPC-PCHC was synthesized by using the 'one pot, three steps' method. The first step was the copolymerization of CO_2 with CHO, which is followed by the copolymerization of CO_2 with PO, and then the copolymerization of CO_2 with CHO again [15]. The random copolymer poly (propylene cyclohexene carbonate) (PPCHC) was prepared by terpolymerization of CO_2, PO, and CHO [14]. For the first 30 h, the temperature was set at 70 °C and, for the next 10 h, it was raised to 80 °C. PPC was made from PO and CO_2 while PCHC was a copolymer of CHO and CO_2. The resulting copolymers were separately dissolved in a proper volume of chloroform and 15 mL dilute HCl (5 wt %) was added to extract the catalyst residual from the product solution. The organic layer was then washed with distilled water three times. The viscous solution was concentrated to a proper concentration by using a rotary evaporator. Lastly, they were precipitated by being poured into vigorously stirred ethanol. The as-made copolymers were filtered and dried under vacuum at a temperature of 120 °C until a constant weight was obtained.

3.3. Determination of the Composition and the Molecular Weight of Polymers

^{1}H-NMR spectra of the block copolymer at room temperature using tetramethylsilane as an internal standard and D-chloroform (CDCl$_3$) as a solvent were recorded on a Bruker DRX-400 NMR spectrometer (Karlsruhe, Germany). Molecular weight (M$_w$ and M$_n$) of the resultant polymer product was measured by a gel permeation chromatography (GPC) system (Waters 515 HPLC Pump, Waters 2414 detector) (Milford, MA, USA) with a set of three columns (Waters Styragel 500, 10,000, and 100,000 Å) and chloroform (HPLC grade) as eluent. The GPC system was calibrated by a series of polystyrene standards with polydisperisties of 1.02.

3.4. Py-GC/MS Measurement

The Py-GC/MS measurements were carried out using a PYR-2A micro-tube furnace pyrolyser (Shimadzu-PYR2A) (Kyoto, Japan) coupled to an HP 6890 gas chromatograph linking to an HP 5973 quadruple mass spectrometer. The sample was heated in the PYR-2A compact in an oxygen-free furnace. Sample aliquots of about 1.00 mg were pyrolyzed by using a platinum coil attachment. This furnace's temperatures were set at 200 °C, 250 °C, 300 °C, 350 °C, and 400 °C, respectively. A quartz capillary column (SE-30; 15 m × 0.2 mm) (Kyoto, Japan) was used in the GC. The column temperature was initially held at 45 °C for 1 min and then programmed to 250 °C at a heating rate of 10 °C/min. The GC/MS interface was set at 280 °C. The pyrolysed products of polymers were directly injected and separated by gas chromatography using helium as an eluent gas and then characterized by the mass spectrometer. The mass spectra were recorded under electron impact ionization energy at 70 eV. The MS detector scanned from 29 to 350 m/z at a scan rate of 1.8 scan/s.

3.5. TG/FTIR Measurement

Thermogravimetric analysis (TGA) measurements were performed in a PerkinElmer Pyris Diamond TG/DTA analyzer (Waltham, MA, USA) under a protective nitrogen atmosphere. The temperature ranged from 50–500 °C with a heating rate of 2, 4, 6, 8, 10 °C/min, respectively. TG/IR analysis was performed with a TG/IR system, which combined with a PerkinElmer Pyris Diamond TG/DTA analyzer and a PerkinElmer Spectrum 100 FTIR spectrometer (Waltham, MA, USA). Samples of about 10 mg were pyrolysed in the TG analyzer and the evolved gases were led to the FTIR spectrometer directly through a connected heated gas line to obtain three dimensional FTIR spectra. The flow rate of N$_2$ is 10 mL/min. The aluminum pans are used for the samples. The temperature of the heated transfer line is 200 °C. The heating rate for taking TGA/FTIR spectra is 10 °C /min. The operating conditions of the FTIR had a frequency range of 4000–400 cm^{-1}, a resolution of 2.0 cm^{-1}, and a scan rate of 1.0 scan/s.

3.6. Thermal Decomposition Kinetics

The "International Confederation for Thermal Analysis and Calorimetry (ICTAC)" committee recommended that utilizing multiple heating rate programs leads to more reliable kinetic parameters with respect to the single heating rate program [31]. The Kissinger-Akahira-Sunose (KAS) method is an integral method by which the E can be obtained through the conversion values of reactant [26–28], which is represented by Equation (1) below.

$$ln\frac{\beta}{T^2} = ln\left(\frac{AE}{Rg(\alpha)}\right) - \frac{E}{RT} \tag{1}$$

Definitions of all the variables and parameters are the same with previous equations. Since the value of $ln\left(\frac{AE}{Rg(\alpha)}\right)$ is approximately constant when the values of α are the same at different β, the plot $ln\frac{\beta}{T^2}$ versus 1/T is approximately linear. Thus, by plotting $ln\frac{\beta}{T^2}$ against 1/T at certain conversion rates, the slope $-\frac{E}{RT}$ is calculated E.

The parameters of the compensation effect of the polymers' pyrolysis were obtained by the Coats-Redfern method. Fifteen kinds of frequently used reaction mathematical models are substituted into the Coats-Redfern equation. The Coats-Redfern equation is expressed by Equation (2) [29,30].

$$ln\frac{g(\alpha)}{T^2} = ln\frac{AR}{\beta E} - \frac{E}{RT} \qquad (2)$$

Substituting $g(\alpha)$ into Equation (2) and plotting $\ln\frac{g(\alpha)}{T^2}$ versus $1/T$, E and lnA of the different mathematical models can be calculated based on the slope (-E/R) and intercept ($ln\frac{AR}{\beta E}$).

The iso-conversional pre-exponential factor lnA and rate constant lnk are obtained by using the iso-conversional kinetic analysis method proposed by Vyazovkin [31–33]. The ln A is determined by the reaction feature of the reactant and it is independent of temperature. Therefore, the calculation is important for understanding the reaction feature. Since the compensation effect exists in E and A, Equation (3) is usually used to calculate ln A [31].

$$lnA = aE + b \qquad (3)$$

The compensation effect parameters a and b were determined by fitting the pairs of lnAi and Ei by 15 different models substituting into the Coats-Redfern method at each heating rate. Then, lnA can be determined for every conversion α by substituting the respective values of E from the KAS method into Equation (3) to obtain a dependence of lnA on α. LnA$_\alpha$ on α and Eα on α data were converted into the Arrhenius plot. Therefore, the rate constant for the thermal decomposition can be evaluated as Equation (4) [32,33].

$$lnk(T_\alpha) = lnA_\alpha - \frac{E_\alpha}{RT_\alpha} \qquad (4)$$

4. Conclusions

Thermal decomposition behaviors and degradation kinetic parameters of terpolymers with different sequence structures derived from CO_2, PO, and CHO were studied by the combination of Py-GC/MS and TG/IR techniques. In addition, the thermal degradation kinetic parameters were calculated by the Kissinger-Akahira-Sunose (KAS) method, the Coats-Redfern method, and the iso-conversional kinetic analysis method proposed by Vyazovkin with the data from thermogravimetric analysis under dynamic conditions. The average degradation activation energies of the polymers are PCHC-PPC-PCHC (104.9 ± 16.2 kJ mol^{-1}) < PPC-PCHC (109.9 ± 7.6 kJ mol^{-1} L) < PPC (124.6 ± 6.8 kJ/mol)) < PPCHC (145.5 ± 3.0 kJ mol^{-1}). The average pre-exponential factor ln A$_\alpha$ for PPC, PPCHC, PPC-PCHC, and PCHC-PPC-PCHC are 24.7 ± 1.5 s^{-1}, 27.7 ± 0.6 s^{-1}, 20.8 ± 1.6 s^{-1}, and 19.2 ± 3.4 s^{-1}, respectively. The rate constant values ln k vs. T^{-1} for the thermal decomposition of four polymers demonstrated that the thermal stability of PPC and PPC-PCHC are poorer than PPCHC and PCHC-PPC-PCHC and for PPCHC and PCHC-PPC-PCHC. There is an intersection between the two rate constant lines, which means that, for thermal stability, PPCHC is more stable than PCHC-PPC-PCHC at temperatures less than 309 °C and less stable when the temperature is more than 309 °C. The thermal degradation mechanism of the polymers was elucidated by IR-TG and Py-GC/MS techniques to be the main chain scissor reaction followed by the unzipping reaction. Due to large content of PC-CHC or CHC-PC linkages in the PPCHC, the steric hindrance of bulky cyclohexane groups restricted the unzipping reaction to some extent. The random copolymer showed one step decomposition with a T$_{-5\%}$ and T$_{max}$ as high as 281.0 °C and 313.4 °C, respectively. For the block polymers, the chain scission and unzipping reaction occurred first at the PPC block and then at the PCHC block. The CHC-CHC linkages could not restrict the PC-PCs unzipping reaction well, so the T$_{-5\%}$ of di-block and tri-block polymers are 261.2 °C and 275.2 °C, respectively, while two maximum decomposition temperatures were observed at 304.2–342.7 °C, and 305.8–345.3 °C, respectively. Lastly, we can conclude that random copolymerization of CHO, PO, and CO_2 is a better way to improve the thermal stability of PPC than block copolymerization.

Author Contributions: Y.M. conceive and designed the experiments. S.C., M.X., and L.S. conducted the experiments. Y.M., S.C., and M.X. analyzed the results and wrote the manuscript. All authors read and approved the final manuscript.

Abbreviations

TG/FTIR   Thermogravimetric analysis/ Fourier transform infrared spectrometry
Py-GC/MS   Pyrolysis-gas chromatography/mass spectrometry
PPC   poly (propylene carbonate)
PPCHC   Poly (propylene cyclohexene carbonate)
PPC-PCHC   Poly (propylene carbonate–cyclohexyl carbonate)
PCHC-PPC-PCHC Poly (cyclohexyl carbonate–propylene carbonate–cyclohexyl carbonate)

References

1. Inoue, S.; Koinuma, H.; Tsuruta, T. Copolymerization of carbon dioxide and epoxide. *J. Polym. Sci. Part B Polym. Lett.* **1969**, *7*, 287–292. [CrossRef]
2. Darensbourg, D.J.; Holtcamp, M.W. Catalysts for the reactions of epoxides and carbon dioxide. *Coord. Chem. Rev.* **1996**, *153*, 155–174. [CrossRef]
3. Meng, Y.Z.; Du, L.C.; Tiong, S.C. Effects of the structure and morphology of zinc glutarate on the fixation of carbon dioxide into polymer. *J. Polym. Sci. Part A Polym. Chem.* **2002**, *40*, 3579–3591. [CrossRef]
4. Coates, G.W.; Moore, D.R. Discrete metal-based catalysts for the copolymerization of CO_2 and epoxides: Discovery, reactivity, optimization, and mechanism. *Angew. Chem. Int. Ed.* **2004**, *43*, 6618–6639. [CrossRef] [PubMed]
5. Lu, X.B.; Ren, W.M.; Wu, G.P. CO_2 copolymers from epoxides: Catalyst activity, product selectivity, and stereochemistry control. *Acc. Chem. Res.* **2012**, *45*, 1721–1735. [CrossRef] [PubMed]
6. Czaplewski, D.A.; Kameoka, J.; Mathers, R.; Coats, G.W.; Craighead, H.G. Nanofluidic channels with elliptical cross sections formed using a nonlithographic process. *Appl. Phys. Lett.* **2003**, *83*, 4836–4838. [CrossRef]
7. Cao, M.; Xiao, M.; Lu, Y.; Meng, Y. Novel in situ preparation of crosslinked ethylene-vinyl alcohol copolymer foams with propylene carbonate. *Mater. Lett.* **2006**, *60*, 3286–3291.
8. Zeng, S.; Wang, S.; Xiao, M.; Meng, Y. Preparation and properties of biodegradable blend containing poly (propylene carbonate) and starch acetate with different degrees of substitution. *Carbohydr. Polym.* **2011**, *86*, 1260–1265. [CrossRef]
9. Chen, W.; Pang, M.; Xiao, M.; Wen, L.; Meng, Y. Mechanical, thermal, and morphological properties of glass fiber-reinforced biodegradable poly (propylene carbonate) composites. *J. Reinf. Plastics Compos.* **2010**, *29*, 1545–1550. [CrossRef]
10. Thorat, S.D.; Phillips, P.J.; Semenov, V.; Gakh, A. Physical properties of aliphatic polycarbonates made from CO_2 and epoxides. *J. Appl. Polym. Sci.* **2003**, *89*, 1163–1176. [CrossRef]
11. Kember, M.R.; Buchard, A.; Williams, C.K. Catalysts for CO_2/epoxide copolymerization. *Chem. Commun.* **2011**, *47*, 141–163. [CrossRef] [PubMed]
12. Li, X.H.; Meng, Y.Z.; Zhu, Q.; Tjong, S.C. Thermal decomposition characteristics of poly (propylene carbonate) using TG/IR and Py-GC/MS techniques. *Polym. Degrad. STable* **2003**, *81*, 157–165. [CrossRef]
13. Lu, X.L.; Zhu, Q.; Meng, Y.Z. Kinetic analysis of thermal decomposition of poly (propylene carbonate). *Polym. Degrad. STable* **2005**, *89*, 282–288. [CrossRef]
14. Wu, J.S.; Xiao, M.; He, H.; Wang, S.; Han, D.M.; Meng, Y.Z. Synthesis and characterization of high molecular weight poly (1, 2-propylene carbonate-co-1, 2-cyclohexylene carbonate) using zinc complex catalyst. *Chin. J. Polym. Sci.* **2011**, *29*, 552–559. [CrossRef]
15. Chen, S.Y.; Xiao, M.; Wang, S.J.; Han, D.M.; Meng, Y.Z. Novel Ternary Block Copolymerization of Carbon Dioxide with Cyclohexene Oxide and Propylene Oxide Using Zinc Complex Catalyst. *J. Poly. Res.* **2012**, *19*. [CrossRef]
16. Bassilakis, R.; Carangelo, R.M.; Wojtowicz, M.A. TG-FTIR analysis of biomass pyrolysis. *Fuel* **2001**, *80*, 1765–1786. [CrossRef]

17. Jiao, L.; Xiao, H.; Wang, Q.; Sun, J. Thermal degradation characteristics of rigid polyurethane foam and the volatile products analysis with TG-FTIR-MS. *Polym. Degrad. STable* **2013**, *98*, 2687–2696. [CrossRef]

18. Bruno, S.S.; Ana Paula, D.M.; Ana Maria, R.F.T. TG-FTIR coupling to monitor the pyrolysis products from agricultural residues. *J. Therm. Anal. Calorim.* **2009**, *97*, 637–642.

19. Rio, J.C.D.; Gutierrez, A.; Hernando, M.; Landin, P.; Romero, J.; Martinez, A.T. Determining the influence of eucalypt lignin composition in paper pulp yield using Py-GC/MS. *J. Anal. Appl. Pyrolysis.* **2005**, *74*, 110–115.

20. Zhu, P.; Sui, S.; Wang, B.; Sun, K.; Sun, G. A study of pyrolysis and pyrolysis products of flame-retardant cotton fabrics by DSC, TGA, and PY-GC/MS. *J. Anal. Appl. Pyrolysis.* **2004**, *71*, 645–655. [CrossRef]

21. Lu, X.Q.; Hanna, J.V.; Johnson, W.D. Source indicators of humic substances: An elemental composition, solid state 13 C CP/MAS NMR and Py-GC/MS study. *Appl. Geochem.* **2000**, *15*, 1019–1033. [CrossRef]

22. Tsuge, S.; Ohtani, H. Structural characterization of polymeric materials by PyrolysisdGC/MS. *Polym. Degrad. STable* **1997**, *58*, 109–130. [CrossRef]

23. Gu, X.L.; Ma, X.; Li, L.X.; Liu, C.; Cheng, K.H.; Li, Z.Z. Pyrolysis of poplar wood sawdust by TG-FTIR and Py-GC/MS. *J. Anal. Appl. Pyrolysis.* **2013**, *102*, 16–23. [CrossRef]

24. Huang, G.; Zou, Y.; Xiao, M.; Wang, S.; Luo, W.; Han, D.; Meng, Y. Thermal degradation of poly (lactide-copropylene carbonate) measured by TG/FTIR and Py-GC/MS. *Polym. Degrad. STable* **2015**, *117*, 16–21. [CrossRef]

25. Luo, W.; Xiao, M.; Wang, S.; Ren, S.; Meng, Y. Thermal degradation behavior of Copoly (propylene carbonateε-caprolactone) investigated using TG/FTIR and Py-GC/MS methodologies. *Polymer Testing.* **2017**, *58*, 13–20. [CrossRef]

26. Kissinger, H.E. Variation of peak temperature with heating rate in differential thermal analysis. *J. Res. Natl. Bur. Stand.* **1956**, *57*, 217–221. [CrossRef]

27. Kissinger, H.E. Reaction kinetics in differential thermal analysis. *Anal. Chem.* **1957**, *29*, 1702–1706. [CrossRef]

28. Akahira, T.; Sunose, T. Method of determining activation deterioration constant of electrical insulating materials. *Res. Rep. Chiba Inst. Technol.* **1971**, *16*, 22–31.

29. Coats, A.W.; Redfern, J.P. Kinetic parameters from thermogravimetric data. *Nature* **1964**, *201*, 68–69. [CrossRef]

30. Yuan, J.J.; Tu, J.L.; Xu, Y.J.; Qin, F.G.F.; Li, B.; Wang, C.Z. Thermal stability and products chemical analysis of olive leaf extract after enzymolysis based on TG-FTIR and Py-GC-MS. *J. Therm. Anal. Calor.* **2018**, *132*, 1729–1740. [CrossRef]

31. Vyazovkin, S.; Burnham, A.K.; Criado, J.M.; Pérez-Maqueda, L.A.; Popescu, C.; Sbirrazzuoli, N. ICTAC Kinetics Committee recommendations for performing kinetic computations on thermal analysis data. *Thermochim. Acta.* **2011**, *520*, 1–19. [CrossRef]

32. Liavitskaya, T.; Birx, L.; Vyazovkin, S. Thermal stability of Malonic Acid Dissolved in Pomy(vinylpyrrolidone) and Other Polymeric Matrices. *Ind. Eng. Chem. Res.* **2018**, *57*, 5228–5233. [CrossRef]

33. Osman, Y.B.; Liavitslaya, T.; Vyazovkin, S. Polyvinylpyrrolidone affects thermal stability of drugs in solid dispersions. *Int. J. Pharm.* **2018**, *551*, 111–120. [CrossRef] [PubMed]

34. Liu, M.; Teng, C.T.; Win, K.Y.; Chen, Y.; Zhang, X.; Yang, D.; Li, Z.; Ye, E. Polymeric Encapsulation of Turmeric Extract for Bioimaging and Antimicrobial Applications. *Macromol. Rapid Commun.* **2018**. [CrossRef] [PubMed]

35. Luinstra, G. Poly (propylene carbonate), old copolymers of propylene oxideand carbon dioxide with new interests: Catalysis and material properties. *Polym. Rev.* **2008**, *48*, 192–219. [CrossRef]

36. Chisholm, M.H.; Navarro-Llobet, D.; Zhou, Z. Poly (propylene carbonate). 1. More about poly (propylene carbonate) formed from the copolymerization of propylene oxide and carbon dioxide employing a zinc glutarate catalyst. *Macromolecules* **2002**, *35*, 6494–6504. [CrossRef]

37. Barreto, C.; Cannon, W.R.; Shanefield, D.J. Thermal decomposition behavior of poly (propylene carbonate): tailoring the composition and thermal properties of PPC. *Polym. Degrad. Stab.* **2012**, *97*, 893–904. [CrossRef]

38. Zhu, Q.; Meng, Y.; Tjong, S.; Zhao, X.; Chen, Y. Thermally stable and high molecular weight poly(propylene carbonate)s from carbon dioxide and propylene oxide. *Polym. Int.* **2002**, *51*, 1079–1085. [CrossRef]

7

Zwitterionic Nanocellulose-Based Membranes for Organic Dye Removal

Carla Vilela [1,*], Catarina Moreirinha [1], Adelaide Almeida [2], Armando J. D. Silvestre [1] and Carmen S. R. Freire [1]

[1] Department of Chemistry, CICECO – Aveiro Institute of Materials, University of Aveiro, 3810-193 Aveiro, Portugal; catarina.fm@ua.pt (C.M.); armsil@ua.pt (A.J.D.S.); cfreire@ua.pt (C.S.R.F.)

[2] Department of Biology and CESAM, University of Aveiro, 3810-193 Aveiro, Portugal; aalmeida@ua.pt

* Correspondence: cvilela@ua.pt

Abstract: The development of efficient and environmentally-friendly nanomaterials to remove contaminants and pollutants (including harmful organic dyes) ravaging water sources is of major importance. Herein, zwitterionic nanocomposite membranes consisting of cross-linked poly(2-methacryloyloxyethyl phosphorylcholine) (PMPC) and bacterial nanocellulose (BNC) were prepared and tested as tools for water remediation. These nanocomposite membranes fabricated via the one-pot polymerization of the zwitterionic monomer, 2-methacryloyloxyethyl phosphorylcholine, within the BNC three-dimensional porous network, exhibit thermal stability up to 250 °C, good mechanical performance (Young's modulus ≥ 430 MPa) and high water-uptake capacity (627%–912%) in different pH media. Moreover, these zwitterionic membranes reduced the bacterial concentration of both gram-positive (*Staphylococcus aureus*) and gram-negative (*Escherichia coli*) pathogenic bacteria with maxima of 4.3– and 1.8–log CFU reduction, respectively, which might be a major advantage in reducing or avoiding bacterial growth in contaminated water. The removal of two water-soluble model dyes, namely methylene blue (MB, cationic) and methyl orange (MO, anionic), from water was also assessed and the results demonstrated that both dyes were successfully removed under the studied conditions, reaching a maximum of ionic dye adsorption of *ca.* 4.4–4.5 mg g^{-1}. This combination of properties provides these PMPC/BNC nanocomposites with potential for application as antibacterial bio-based adsorbent membranes for water remediation of anionic and cationic dyes.

Keywords: bacterial nanocellulose; poly(2-methacryloyloxyethyl phosphorylcholine); zwitterionic nanocomposites; dye removal; water remediation; antibacterial activity

1. Introduction

The need for water remediation systems designed to eliminate contaminants and pollutants ravaging water sources is a global problem and, thus, is part of the goals of the 2030 Agenda for Sustainable Development [1]. Nevertheless, the struggle to remove heavy metal ions, pesticides and other dissolved organic pollutants is a difficult war and some of the efforts of researchers to accomplish such a target are directed towards the development of environmentally friendly porous nanomaterials [2,3]. In fact, big bets are being placed on systems derived from naturally occurring polymers, such as polysaccharides [4,5]. Within the portfolio of commended biopolymers, cellulose and its nanoscale forms, namely cellulose nanocrystals (CNCs), cellulose nanofibrils (CNFs) and bacterial nanocellulose (BNC), show tremendous potential for environmental and water remediation as recently reviewed [6–8]. The high interest in the latter nanocellulose, *viz.* the exopolysaccharide BNC biosynthesized by some non-pathogenic bacterial strains [9], lies in its 3D structure with an ultrafine network of physically entangled cellulose nanofibers, which is responsible for its *in-situ* moldability, shape retention, inherent biodegradability, high water-holding capacity and porous structure [10,11].

BNC, its derivatives and composites [12] have already been used in the manufacture of water remediation systems for the removal of dyes [13,14], oil [15] and heavy metals [16,17]. To the best of our knowledge, the partnership between BNC and a zwitterionic polymer has never been explored for the fabrication of nanocomposites, aiming at the simultaneous removal of anionic and cationic organic dyes. Under these premises, 2-methacryloyloxyethyl phosphorylcholine (MPC) was selected as a non-toxic polymerizable monomer due to its methacrylic functional group and zwitterionic phosphorylcholine moiety, consisting of a phosphate anion and a trimethylammonium cation [18], which are prone to establish electrostatic interactions with positively and negatively charged molecules. Furthermore, the unique hydration state [18], antimicrobial, bioinert and antifouling properties of MPC polymer [19–21] will be a major asset in reducing or avoiding bacterial growth in contaminated water.

The present study contemplates the fabrication of nanocomposite membranes consisting of cross-linked poly(2-methacryloyloxyethyl phosphorylcholine) (PMPC) and BNC via the one-pot polymerization of the corresponding non-toxic zwitterionic monomer (*i.e.* MPC) within the BNC three-dimensional porous network. The structure, morphology, thermal stability, mechanical properties, antibacterial activity towards *Staphylococcus aureus* and *Escherichia coli*, water-uptake capacity and removal of cationic and anionic organic dyes of the resulting nanocomposites were assessed.

2. Materials and Methods

2.1. Chemicals, materials and microorganisms

2-Methacryloyloxyethyl phosphorylcholine (MPC, 97%), 2,2′-azobis(2-methylpropionamidine) dihydrochloride (AAPH, 97%), N,N′-methylenebis(acrylamide) (MBA, 99%), methylene blue (MB, dye content $\geq$ 82%), methyl orange (MO, dye content 85 %) and paraffin oil (puriss., 0.827–0.890 g mL^{-1} at 20 °C) were purchased from Sigma-Aldrich (Sintra, Portugal) and used as received. Ultrapure water (Type 1, 18.2 MX·cm at 25 °C) was obtained from a Simplicity® Water Purification System (Merck, Darmstadt, Germany). Other chemicals and solvents were of laboratory grade.

Bacterial nanocellulose (BNC) wet membranes were biosynthesized in our laboratory using the *Gluconacetobacter sacchari* bacterial strain [22]. *Staphylococcus aureus* (ATCC 6538) and *Escherichia coli* (ATCC 25922) was provided by DSMZ – Deutsche Sammlung von Mikroorganismen und Zellkulturen GmbH (German Collection of Microorganisms and Cell Cultures).

2.2. Preparation of PMPC/BNC nanocomposites

Wet BNC membranes (diameter: *ca.* 70 mm) with 40% water content were placed in stoppered glass-reactors and purged with nitrogen. Simultaneously, aqueous solutions of monomer (MPC, 1:3 and 1:5 BNC/MPC weight fraction), cross-linker (MBA, 5.0% w/w relative to monomer), and radical initiator (AAPH, 1.0% w/w relative to monomer) were prepared and transferred to the glass-reactors containing the drained BNC membranes. After the complete incorporation of the corresponding solution into the BNC membrane, during 1 h in ice, the reaction mixtures were heated in an oil bath at 70 °C and left to react for 6 h. Afterwards, the nanocomposites were repeatedly washed with water, oven dried at 40 °C for 12 h, and kept in a desiccator until further use. All experiments were made in triplicate and samples of cross-linked PMPC were also prepared in the absence of BNC for comparison.

2.3. Characterization methods

2.3.1. Thickness

A hand-held digital micrometer (Mitutoyo Corporation, Tokyo, Japan) with an accuracy of 0.001 mm was used to measure the thickness of the membranes. All measurements were randomly performed at different sites of the membranes.

2.3.2. Ultraviolet-visible spectroscopy (UV–vis)

The transmittance spectra of the samples were acquired with a Shimadzu UV-1800 UV-Vis spectrophotometer (Shimadzu Corp., Kyoto, Japan) equipped with a quartz window plate with 10 mm diameter, bearing the holder in the vertical position. Spectra were recorded at room temperature (RT) in steps of 1 nm in the range 250–700 nm.

2.3.3. Attenuated total reflection-Fourier transform Infrared (ATR-FTIR)

ATR-FTIR spectra were recorded with a Perkin-Elmer FT-IR System Spectrum BX spectrophotometer (Perkin-Elmer Inc., Massachusetts, USA) equipped with a single horizontal Golden Gate ATR cell, over the range of $600–4000$ cm^{-1} at a resolution of 4 cm^{-1} over 32 scans.

2.3.4. Solid-state carbon cross-polarization/magic-angle-spinning nuclear magnetic resonance (^{13}C CP/MAS NMR)

^{13}C CP/MAS NMR spectra were collected on a Bruker Avance III 400 spectrometer (Bruker Corporation, Massachusetts, USA) operating at a B0 field of 9.4 T using 9 kHz MAS with proton 90° pulse of 3 μs, time between scans of 3 s, and a contact time of 2000 μs. ^{13}C chemical shifts were referenced to glycine (C=O at δ 176 ppm).

2.3.5. X-ray diffraction (XRD)

XRD was performed on a Phillips X'pert MPD diffractometer (PANalytical, Eindhoven, Netherlands) using Cu Kα radiation ($\lambda = 1.541$ Å) with a scan rate of 0.05° s^{-1}. The XRD patterns were collected in reflection mode with the membranes placed on a Si wafer (negligible background signal) for mechanical support and thus avoid sample bending.

2.3.6. Scanning electron microscopy (SEM) coupled with energy dispersive X-ray spectroscopy (EDS)

SEM images of the cross-section of the membranes were obtained with a HR-SEM-SE SU-70 Hitachi microscope (Hitachi High-Technologies Corporation, Tokyo, Japan) operating at 4 kV. The microscope was equipped with an EDS Bruker QUANTAX 400 detector for elemental analysis. The samples were fractured in liquid nitrogen, placed on a steel plate and coated with a carbon film prior to analysis.

2.3.7. Thermogravimetric analysis (TGA)

TGA was carried out with a SETSYS Setaram TGA analyzer (SETARAM Instrumentation, Lyon France) equipped with a platinum cell. The samples were heated from RT to 800 °C at a constant rate of 10 °C min^{-1} under a nitrogen atmosphere (200 mL min^{-1}).

2.3.8. Tensile tests

Tensile tests were performed on a uniaxial Instron 5564 testing machine (Instron Corporation, Maryland, USA) in the traction mode at a cross-head velocity of 10 mm min^{-1} using a 500 N static load cell. The specimens were rectangular strips (50×10 mm^2) previously dried at 40 °C and equilibrated at RT in a 50% relative humidity (RH) atmosphere prior to testing. All measurements were performed on five replicates and the results were expressed as the average value.

2.3.9. Water-uptake capacity

The water-uptake of the nanocomposites under different pH conditions was determined via immersion of dry specimens with 10×10 mm^2 in aqueous solutions of 0.01 M HCl (pH 2.1), phosphate buffer saline (pH 7.4) and 0.01 M NaOH (pH 12) at RT for 48 h. After removing the specimens out of the respective medium, the wet surfaces were dried in filter paper, and the wet weight (W_w) was measured. The water-uptake is calculated by the equation: $W_{uptake}(\%) = (W_w - W_0) \times W_0^{-1} \times 100$, where W_0 is the initial weight of the dry membrane.

2.4. In vitro antibacterial activity

The antibacterial activity of the nanocomposite membranes was tested against *S. aureus* and *E. coli*. The bacterial pre-inoculum cultures were grown for 24 h in tryptic soy broth (TSB) growth medium at 37 °C under shaking at 120 rpm. Before the assay, the density of the bacterial culture was adjusted to 0.5 McFarland in phosphate buffered saline (PBS) solution (pH 7.4) to obtain a bacterial concentration of 10^8 to 10^9 colony forming units *per* mL (CFU mL^{-1}). Each membrane sample (50 × 50 mm^2) was placed in contact with 5 mL of bacterial suspension. A bacteria cell suspension was tested as the control and BNC was tested as a blank reference. All samples were incubated at 37 °C under horizontal shaking at 120 rpm. At 24 h contact time, aliquots (100 μL) of each sample and controls were collected and the bacteria cell concentration (CFU mL^{-1}) was determined by plating serial dilutions on tryptic soy agar (TSA) medium. The plates were incubated at 37 °C for 24 h. The CFU were determined on the most appropriate dilution on the agar plates. Three independent experiments were carried out and, for each, two replicates were plated. The bacteria reduction of the samples was calculated as follows: $log\ reduction = log\ \text{CFU}_{\text{control}} - log\ \text{CFU}_{\text{membrane}}$.

2.5. Dye removal capacity

The dye removal capacity of the PMPC/BNC nanocomposite membranes was evaluated by immersing dry samples (20 × 20 mm^2) in 25 mL of methyl blue (MB) and methyl orange (MO) aqueous solutions (25 mg L^{-1}, pH 5.7), and stirred (200 rpm) for 12 h at RT. Then, the membranes were removed from the solution and the residual concentration of dye determined by UV–vis spectroscopy (Shimadzu UV-1800 UV-Vis spectrophotometer, Kyoto, Japan) at 655 nm for MB and 463 nm for MO. Linear calibration curves for each dye in the range 0.4–3.1 μg mL^{-1} were obtained: $y = 0.1919x + 0.0014$ ($R^2 = 0.9991$) for MB and $y = 0.0881x - 0.0046$ ($R^2 = 0.9992$) for MO. The dye removal amount (mg g^{-1}) was calculated by: $q = (C_i - C_t) \times W^{-1} \times V$, where C_i is the initial dye concentration (mg L^{-1}), C_t is the dye concentration at time t (h), W is the weight (g) of the membrane and V is the volume (L) of the dye solution.

Additionally, a dry sample of PMPC/BNC_2 nanocomposite (20 × 20 mm^2) was immersed in 25 mL of paraffin oil containing 1 mL of MB and MO aqueous solutions (25 mg L^{-1}).

2.6. Statistical analysis

Statistical significance was determined using an analysis of variance (ANOVA) and Tukey's test (OriginPro, version 9.0.0, OriginLab Corporation, Northampton, MA, USA). Statistical significance was established at $p < 0.05$.

3. Results and Discussion

3.1. PMPC/BNC nanocomposites: preparation and characterization

The one-pot *in-situ* free radical polymerization of MPC, inside the swollen BNC network and using MBA as cross-linker, was used to produce two PMPC/BNC nanocomposites (Figure 1) with distinct compositions (Table 1). The cross-linker was chosen based on former studies [23,24] and utilized with the goal of preserving the water-soluble zwitterionic homopolymer inside the BNC porous network during washing and utilization. The resulting nanocomposites contain 21 ± 3 wt.% and 46 ± 13 wt.% of BNC ($W_{\text{BNC}}/W_{\text{total}}$), and concomitantly 79 ± 3 wt.% and 54 ± 13 wt.% of PMPC ($W_{\text{PMPC}}/W_{\text{total}}$), which correspond to nanocomposite materials containing 479 ± 118 mg and 859 ± 90 mg of PMPC *per* cm^3 of membrane, respectively, as listed in Table 1. The thickness of the membranes increased from 92 ± 21 μm for neat BNC to 133 ± 65 μm for PMPC/BNC_1 and 226 ± 35 μm for PMPC/BNC_2 (Table 1) due to the inclusion of the cross-linked PMPC into the three-dimensional structure of BNC. The membranes are macroscopically homogeneous with no discernible irregularities on either side of the materials surfaces, indicating a good dispersion of the cross-linked PMPC polymer inside the BNC network. After the incorporation of PMPC into the BNC network, the transparency of the nanocomposites significantly increased, as displayed in Figure 1B and confirmed by transmittance values in the visible range (400–700 nm) of 58.1–65.6% for PMPC/BNC_1 and 60.5–68.2% for PMPC/BNC_2 (Figure 1C). In the ultraviolet region (200–400 nm), the transmittance remained below 5% until 265 nm for PMPC/BNC_1 and 250 nm for PMPC/BNC_2, and then steadily increased to 58.0% and 60.5% at 400 nm for PMPC/BNC_1 and PMPC/BNC_2, respectively. Furthermore, PMPC/BNC_2 presents higher values of transmittance and concomitantly lower absorbance values, which points to a transmittance augment with higher content of cross-linked PMPC (Figure 1C). An analogous trend was observed for other BNC-based nanocomposites containing for example polycaprolactone [25], poly(methacroylcholine chloride) [23] and polyaniline [26].

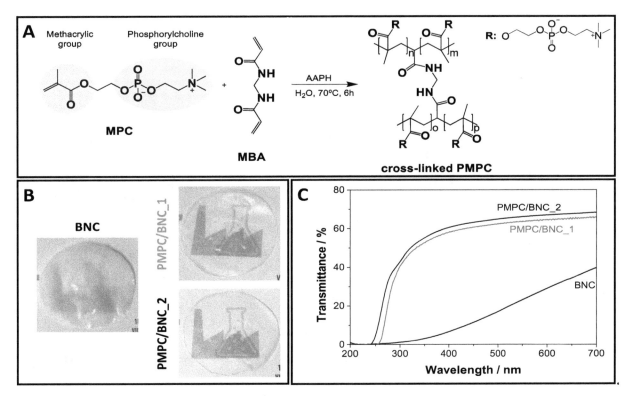

Figure 1. **(A)** Radical polymerization of MPC in the presence of MBA as cross-linker, yielding cross-linked PMPC, **(B)** photographs of neat BNC and nanocomposites PMPC/BNC_1 and PMPC/BNC_2, and **(C)** the corresponding UV-visible transmission spectra.

Table 1. List of the studied membranes with the respective weight compositions and thicknesses.

Samples	Composition [a]			Thickness/µm
	W_{BNC}/W_{total}	W_{PMPC}/W_{total}	W_{PMPC}/V_{total} (mg cm^{-3}) [b]	
BNC	1.0	–	–	92 ± 21
PMPC/BNC_1	0.46 ± 0.13	0.54 ± 0.13	479 ± 118	133 ± 65
PMPC/BNC_2	0.21 ± 0.03	0.79 ± 0.03	859 ± 90	226 ± 35

[a] The composition was calculated by considering the weight of the nanocomposite membrane (W_{total}), BNC (W_{BNC}) and PMPC cross-linked polymer ($W_{PMPC} = W_{total} - W_{BNC}$); [b] Ratio between the mass of the cross-linked PMPC (W_{PMPC}) and the volume of the nanocomposite membrane (V_{total}); all values are the mean of at least three replicates with the respective standard deviations.

The infrared spectra of neat BNC, cross-linked PMPC, and nanocomposites PMPC/BNC_1 and PMPC/BNC_2 are shown in Figure 2. The ATR-FTIR spectra of the PMPC/BNC membranes present the absorption bands of cellulose at 3340 cm^{-1} (O–H stretching), 2900 cm^{-1} (C–H stretching), 1310 cm^{-1} (O–H in plane bending) and 1030 cm^{-1} (C–O stretching) [27], jointly with those of the cross-linked PMPC at 1715 cm^{-1} (C=O stretching), 1479 cm^{-1} (N$^+$(CH$_3$)$_3$ bending), 1228 cm^{-1} (P=O stretching),

1056 cm^{-1} (P–O–C stretching) and 953 cm^{-1} (N$^+$(CH$_3$)$_3$ stretching) [28,29]. The presence of these absorption bands and the absence of one at about 1640 cm^{-1} corresponding to the C=C stretching of the methacrylic group of the starting monomer corroborated the occurrence of the *in-situ* free radical polymerization of MPC inside the BNC network. Furthermore, the relative intensity of the bands assigned to the cross-linked PMPC is in accordance with the W_{PMPC}/W_{total} ratio measured for each nanocomposite (Table 1).

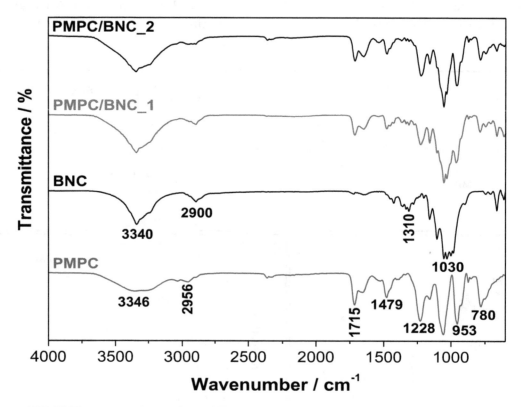

Figure 2. ATR-FTIR spectra of cross-linked PMPC, neat BNC, and nanocomposites PMPC/BNC_1 and PMPC/BNC_2.

The solid-state ^{13}C CP/MAS NMR spectra (Figure 3) of the membranes show the typical carbon resonances of cellulose at δ 65.2 ppm (C6), 71.6–74.5 ppm (C2,3,5), 88.9 ppm (C4) and 105.1 ppm (C1) [27], in combination with those of cross-linked PMPC at δ 18.4 ppm (CH$_3$ of polymer backbone), 44.9 ppm (quaternary C of polymer backbone), 54.3 ppm (CH$_2$ of polymer backbone and N$^+$(CH$_3$)$_3$), 59.7 ppm (OCH$_2$CH$_2$N$^+$(CH$_3$)$_3$), 65.5 ppm (OCH$_2$CH$_2$O and CH$_2$CH$_2$N$^+$(CH$_3$)$_3$) and 176.6 ppm (C=O). In addition, the truancy of the resonances allocated to the C=C double bond of the methacrylic group of the monomer [30] and cross-linker, supports their complete consumption during the polymerization and/or removal during the washing steps, as previously established by ATR-FTIR analysis.

The XRD patterns of the nanocomposites were compared with those of the individual components, namely cross-linked PMPC and BNC, to obtain an indication of the nanomaterials' crystallinity. Figure 4 shows the amorphous character of the cross-linked PMPC with a very broad band centered at $2\theta \approx 18°$, and the crystalline nature of BNC with a diffraction pattern characteristic of cellulose I (native cellulose) composed of highly-ordered and least-ordered regions. The nanocomposites display a diffractogram with three peaks corresponding to the (101) plane at $2\theta \approx 14.7°$, (10$\bar{1}$) plane at $2\theta \approx 16.8°$ and (002) plane at $2\theta \approx 22.8°$ [27], which are representative of the cellulosic substrate. The addition of the cross-linked PMPC is evident through the reduction of the peaks of the (101) and (10$\bar{1}$) planes, most likely linked to the augment of disordered cellulose domains due to the presence of the amorphous polymer. A comparable trend was reported for other BNC-based nanocomposites containing for instance poly(bis[2-(methacryloyloxy)ethyl] phosphate) [31] and poly(4-styrene sulfonic acid) [32].

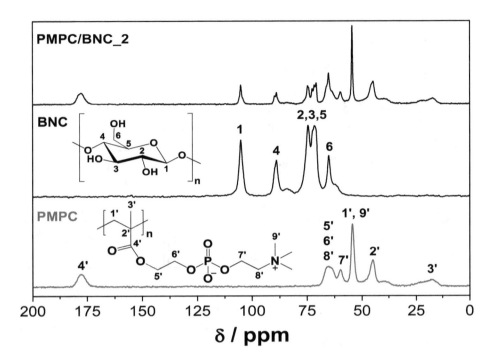

Figure 3. ^{13}C CP/MAS NMR spectra of cross-linked PMPC, neat BNC and nanocomposite PMPC/BNC_2.

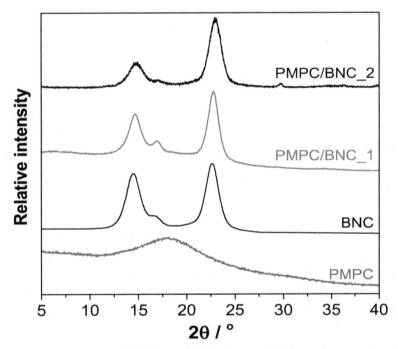

Figure 4. X-ray diffractograms of cross-linked PMPC, neat BNC, and nanocomposites PMPC/BNC_1 and PMPC/BNC_2.

SEM micrographs of the cross-section of neat BNC and nanocomposites PMPC/BNC_1 and PMPC/BNC_2 are compiled in Figure 5A. It is clearly visible that the lamellar microstructure of neat BNC disappeared in the nanocomposites due to the filling of the lamellar spaces by the cross-linked PMPC, particularly in the case of PMPC/BNC_2 with 859 ± 90 mg of PMPC *per* cm^3 of membrane. The SEM/EDS analysis reiterates the presence of PMPC and BNC through the detection of carbon (C), nitrogen (N), oxygen (O) and phosphorous (P) peaks at 0.27, 0.39, 0.51 and 2.01 keV, respectively, as illustrated in Figure 5B for PMPC/BNC_2. Moreover, the SEM/EDS mapping (Figure 5C) confirmed the uniform distribution of nitrogen and phosphorous of the cross-linked polymer within the BNC nanofibrous network, since both elements are only present in the zwitterionic PMPC.

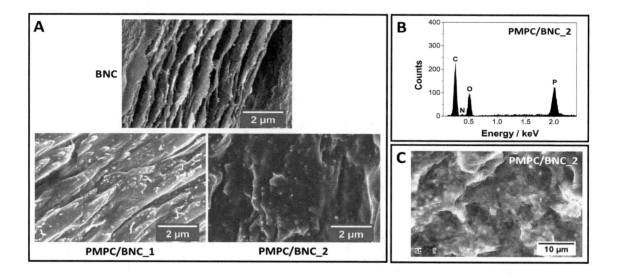

Figure 5. (**A**) SEM micrographs of the cross-section of neat BNC, and nanocomposites PMPC/BNC_1 and PMPC/BNC_2; EDS spectrum (**B**) and micrograph (**C**) for nitrogen and phosphorous elemental mapping of nanocomposite PMPC/BNC_2.

3.2. Thermal stability

TGA analysis was used to study the thermal stability of the PMPC/BNC nanocomposite membranes, as well as of their individual components. Figure 6A shows the thermograms of cross-linked PMPC and neat BNC, while Figure 6B presents the thermograms of PMPC/BNC_1 and PMPC/BNC_2 nanocomposites. The thermal degradation profile of the cross-linked PMPC is characterized by two consecutive steps (apart from the dehydration at about 100 °C with a loss of *ca.* 7.5%) with maximum decomposition temperatures of 284 °C (loss of *ca.* 21%) and 385 °C (loss of *ca.* 28%, Figure 6A) corresponding to the pyrolysis of the PMPC skeleton. The thermogram of BNC shows a single weight-loss step with initial and maximum decomposition temperatures of 290 °C and 342.5 °C (loss of *ca.* 68%, Figure 6A), respectively, allocated to the pyrolysis of the cellulose skeleton [33,34]. PMPC left a residue at 800 °C corresponding to about 34% of the initial mass, whereas BNC only left a residue of *ca.* 18% at the end of the analysis.

The thermal degradation profiles of both nanocomposites (Figure 6B) follow a double weight-loss step, aside from the water loss below 100 °C (loss of *ca.* 10%). PMPB/BNC_1 has maximum decomposition temperatures at 290 °C (loss of *ca.* 24%) and 382 °C (loss of *ca.* 17%) with a final residue of 35%, while for PMPC/BNC_2 the maximum occurs at 288 °C (loss of *ca.* 19%) and 383 °C (loss of *ca.* 20%) with a residue of 32% at the end of the analysis (800 °C). This two-step pathway is associated first with the simultaneous pyrolysis of cellulose and cross-linked PMPC, and the last stage with only the zwitterionic polymer. This pattern points to the reduction of the thermal stability of the nanocomposites when compared to neat BNC, as already described for other BNC-based nanocomposites containing polymers with lower thermal stability [35]. Even so, the two PMPC/BNC membranes exhibit good thermal stability up to 250 °C (Figure 6B).

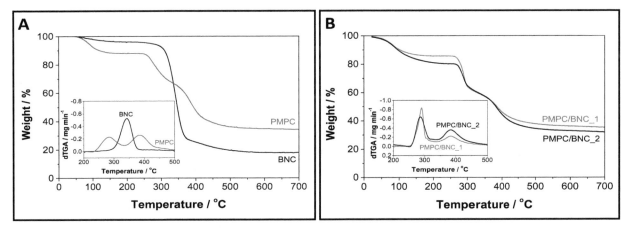

Figure 6. Thermograms of (**A**) cross-linked PMPC and neat BNC, and (**B**) membranes PMPC/BNC_1 and PMPC/BNC_2 under nitrogen atmosphere. The inset curves correspond to the derivative.

3.3. Mechanical properties

Figure 7 compiles the tensile tests data, namely in terms of Young's modulus, tensile strength and elongation at break, determined from the stress-strain curves. Although the tensile tests were not performed for the cross-linked PMPC due to its lack of film-forming aptitude, the cooperative effect between PMPC and BNC relies on the mechanical properties of both nanocomposites. Overall, the Young's modulus and tensile strength of the two membranes increased with the increasing content of the cellulosic substrate (Figure 7A,B), on account of its good mechanical performance, namely Young's modulus of 6.6 ± 1.8 GPa and tensile strength of 221 ± 48 MPa. In fact, the former parameter increased from 430 ± 150 MPa for PMPC/BNC_2 with 21 wt.% of BNC to 3.3 ± 0.8 GPa for PMPC/BNC_1 with 46 wt.% of BNC (Figure 7A), whereas the tensile strength increased from 18 ± 4 MPa for PMPC/BNC_2 to 69 ± 15 MPa for PMPC/BNC_1 (Figure 7B). In contrast, the elongation at break decreased with the increasing content of BNC from $6.0 \pm 1.4\%$ for PMPC/BNC_2 to $2.8 \pm 0.3\%$ for PMPC/BNC_1, as shown in Figure 7C. This means that the nanocomposites are more pliable than the stiff BNC nanofibers with an elongation at break of $4.7 \pm 1.0\%$.

The dependence of the membranes' mechanical performance on the amount of BNC is in tune with earlier studies of other BNC-based nanocomposites with polymers of low mechanical properties [14,35,36]. For example, Zhijiang et al. [14] prepared a chitosan/BNC-based hydrogel composite for dye removal, whose Young's modulus increased from 96.5 MP for pure chitosan (dry state) to 244 MPa after the incorporation of BNC nanofibers grafted with carbon nanotubes into the chitosan hydrogel. The same behavior was obtained for the tensile strength and elongation at break [14].

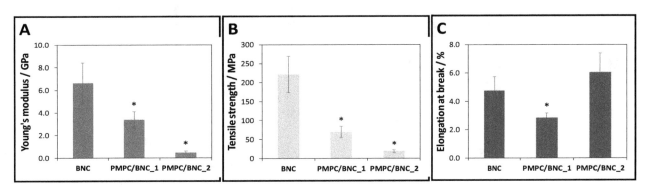

Figure 7. (**A**) Young's modulus, (**B**) tensile strength and (**C**) elongation at break of neat BNC and PMPC/BNC nanocomposites; the error bars correspond to the standard deviations; the asterisk (*) denotes statistically significant differences with respect to the neat BNC ($p < 0.05$).

3.4. In vitro antibacterial activity

Materials with antibacterial activity are relevant for application in multiple fields [37,38] since they can inhibit the growth and simultaneously kill pathogenic bacteria that are harmful to human health [39]. The MPC polymer is known for having antimicrobial and antifouling properties [19–21], which can be a major benefit in reducing/avoiding bacterial growth in contaminated water. This hypothesis was validated by assessing the growth inhibition of gram-positive (*S. aureus*) and gram-negative (*E. coli*) bacteria. *E. coli* was selected for being frequently present in contaminated water, which is a strong indication of recent sewage or fecal contamination. *S. aureus* is not so frequently present in contaminated waters; however, different strains have already been detected in urban wastewater, namely the methicillin-resistant *S. aureus* ST398 [40].

Figure 8 outlines the antibacterial activity of PMPC/BNC nanocomposites and of the neat BNC membrane for comparison purposes. The inoculation of both bacteria in culture media without any sample was used as an experimental control. The neat BNC membrane, along with the experimental control, do not affect the bacterial viability of both *S. aureus* (Figure 8A) and *E. coli* (Figure 8B). This was expected given that BNC is reported not to inhibit the growth of *S. aureus* [41,42], *E. coli* [36,41,42], and other microorganisms such as *Pseudomonas aeruginosa*, *Bacillus subtilis* [42] and *Candida albicans* [43]. In fact, BNC can even be used as a substrate for microbial cell culture [44].

The bacterial killing of *S. aureus* by the two PMPC/BNC nanocomposite membranes is markedly concentration-dependent, as portrayed in Figure 8A. The PMPC/BNC_1 nanocomposite with 54 wt.% of cross-linked PMPC originated a significant reduction ($p < 0.05$) of bacterial concentration relatively to the control, causing a maximum of 2.5–log CFU reduction after 24 h of incubation. The PMPC/BNC_2 with 79 wt.% of cross-linked PMPC reached a higher bacterial inactivation of 4.3–log CFU reduction after 24 h, which indicates that this membrane can be considered an effective antibacterial because according to the American Society of Microbiology (ASM), every new approach has to prove an efficacy of 3–$\log_{10}$ reduction of CFU before being considered antimicrobial or antibacterial [43]. This antibacterial activity is mainly attributed to the trimethylammonium cation that is known for imparting antimicrobial properties [45]. When comparing the activity of the PMPB/BNC membranes with literature, Bertal et al. [46] verified that the triblock copolymer containing PMPC originated an inhibitory zone up to six times greater than the corresponding control against *S. aureus* and a reduction of bacterial growth by 45% compared with the experiments carried out in the absence of PMPC-based copolymer. The authors also claimed that the addition of the copolymer to a 3D-skin model infected with *S. aureus* reduced bacterial recovery by 38% compared to that of controls over 24–48 h [46].

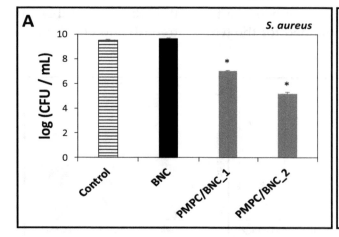 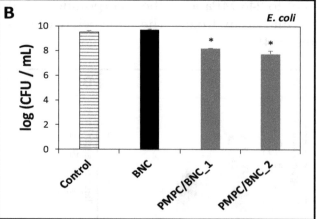

Figure 8. Effect of BNC, PMPC/BNC_1 and PMPC/BNC_2 on the bacterial killing (CFU) of (**A**) *S. aureus* and (**B**) *E. coli* after 24 h of exposure; error bars represent the standard deviation (three independent experiments); the asterisk (*) denotes statistically significant differences with respect to the control and neat BNC ($p < 0.05$).

Regarding the *E. coli* bacteria (Figure 8B), the picture is quite different and both nanocomposites exhibit a lower reduction with 1.3– and 1.8–log CFU reduction for PMPC/BNC_1 and PMPC/BNC_2, respectively. A similar behavior was reported by Fuchs et al. [47] that witnessed no antibacterial activity towards *E. coli* for one MPC copolymer. In fact, this could be expected given that *E. coli* is a gram-negative bacterium whose killing mechanism is more difficult to prevent due to the low permeability of their membranes as discussed previously in detail [48,49].

3.5. Water-uptake and dye removal capacity

Table 2 presents the water-uptake values for BNC and the two PMPC/BNC nanocomposite membranes after immersion in aqueous solutions of 0.01 M HCl (pH 2.1), phosphate buffer saline (pH 7.4) and 0.01 M NaOH (pH 12.0) for 48 h at RT. Overall, the water-uptake vividly increased with the increasing content of cross-linked PMPC. At pH 7.4, it increased from 101 ± 12% for neat BNC up to 639 ± 23% for PMPC/BNC_1 (54 wt.% of PMPC) and 899 ± 44% for PMPC/BNC_2 (79 wt.% of PMPC) (Table 2). In acidic aqueous solutions, PMPC/BNC_1 can absorb 6.3 ± 0.4 g of water *per* g of membrane, while for PMPC/BNC_2 the value is 9.1 ± 0.2 g of water *per* g of membrane. At pH 12, PMPC/BNC_1 absorbs 6.4 ± 0.4 g of water *per* g of membrane, whereas for PMPC/BNC_2 the water-uptake is 9.1 ± 0.3 g of water *per* g of membrane.

Table 2. Water-uptake (water-uptake) of neat BNC and the two PMPC/BNC nanocomposite membranes at different pH media for 48 h at RT.

Membranes	Water-Uptake / %		
	pH 2.1 [a]	pH 7.4 [b]	pH 12.0 [c]
BNC	109 ± 14	101 ± 12	103 ± 10
PMPC/BNC_1	627 ± 38	639 ± 23	640 ± 42
PMPC/BNC_2	911 ± 26	899 ± 44	912 ± 27

[a] Measured after immersion in 0.01 M of HCl aqueous solution; [b] Measured after immersion in phosphate buffer solution; [c] Measured after immersion in 0.01 M of NaOH aqueous solution. All values are the mean of three replicates with the respective standard deviations.

The larger water-uptake of the nanocomposites is correlated with the hydrophilic nature of the phosphorylcholine moiety of the cross-linked PMPC. Additionally, water-uptake is not pH-dependent since there are no significant differences (the means difference is not significant at $\alpha = 0.05$) for the individual membranes under the distinct conditions of acidity or basicity. This can be explained by the unique hydration state of the PMPC chains, where the phosphorylcholine moieties have a hydrophobic hydration layer that do not disturb the hydrogen bonding between the water molecules, as discussed by Ishihara et al. [18]. This is an important characteristic in the water remediation context given that contaminated water can have different pH values. Furthermore, the higher water-uptake of PMPC/BNC_2 is an indication of a higher removal capacity of water-soluble dyes. After 48 h of immersion in aqueous solutions of different pH values, the two nanocomposites were oven dried (at 40 °C) and the final weights demonstrated that the polymer loss ranges between 1%–2%, which emphasizes the insignificant leaching of the cross-linked PMPC from the BNC network.

The removal of two model ionic organic dyes, namely methylene blue (MB) and methyl orange (MO), from water samples at room temperature after 12 h was assessed as a proof-of-concept. While MB is a heterocyclic cationic aromatic compound that is used either as a dye or a drug with for example antimalarial, antidepressant and anxiolytic activity [50], MO is a heterocyclic anionic aromatic compound that is widely used in the textile, pharmaceutical and food industries, and also as an acid-base indicator. Both azo dyes are potentially toxic towards humans and the environment [51].

Figure 9A shows that the PMPC/BNC membranes can indeed retain the model water pollutants as confirmed by the different color of the nanomaterials. This is further corroborated by the data shown in Figure 9B where the dye removal capacity is plotted for each membrane. The pure BNC can remove 0.55 ± 0.12 mg of MB and 0.50 ± 0.06 mg of MO *per* g of membrane. These low removal values were expected, given the lack of binding sites in pure BNC for both cationic and anionic organic dyes. Furthermore, these values are comparable with the dyeability reported by Shim and Kim [52] in their study about the coloration of BNC fabrics with different dyes using *in situ* and *ex situ* methods.

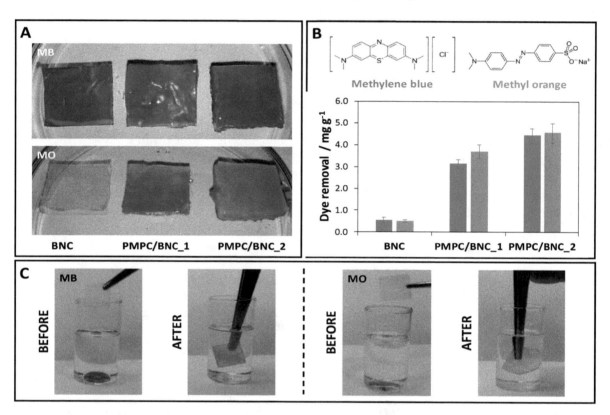

Figure 9. Photographs (**A**) and (**B**) dye removal capacity of BNC, PMPC/BNC_1 and PMPC/BNC_2 after 12 h of immersion in the dye aqueous solution, and (**C**) photographs of the MB and MO aqueous solutions removal from paraffin oil by PMPC/BNC_2 nanocomposite.

Concerning the nanocomposites, PMPC/BNC_1 can remove 3.14 ± 0.19 mg g^{-1} of MB and 3.32 ± 0.31 mg g^{-1} of MO, whereas PMPC/BNC_2 has a removal capacity of 4.44 ± 0.32 mg g^{-1} for MB and 4.56 ± 0.43 mg g^{-1} for MO. Comparing with pure BNC, the dye removal capacity of PMPC/BNC_1 is 5.7 and 7.4 times higher for MB and MO, respectively, while PMPC/BNC_2 removes 8.1 and 9.1 times more MB and MO, respectively, than pure BNC. The higher dye removal capacity of PMPC/BNC_2 is consistent with its higher PMPC content (Table 1). Moreover, the two nanocomposites can remove both cationic and anionic dyes due to the zwitterionic nature of the cross-linked PMPC which can establish electrostatic interactions with either MB or MO model dyes. Worth mentioning is the fact that the PMPC/BNC nanocomposites can easily and quickly remove both MB and MO (25 mg mL^{-1}) from the bottom of a paraffin oil container without the removal of any oil, as exemplified for PMPC/BNC_2 in Figure 9C. This is a good indication of the lack of affinity of the nanocomposites towards the hydrophobic oil and affinity for water or aqueous solutions. A similar behavior was observed for MB (aqueous solution, 100 mg L^{-1}) removal from silicone oil by sulfated-cellulose nanofibrils aerogels [53].

When compared with literature, the dye removal capacity of the PMPC/BNC nanocomposites is lower than that achieved for example with highly carboxylated (COO$^-$) nanocrystalline cellulose with a maximum removal capacity of 101 mg g^{-1} for MB [54], or with the amino-functionalized cellulose nanofibrils-based aerogels with 266 mg g^{-1} for MO [55]. These higher removal capacities are most likely associated with the simultaneous high content of surface binding sites and specific surface area in the first case [54], and the aerogel structure in the second case, which translates into materials with very high porosity and low density [55]. Still, the dye removal values of the PMPC/BNC membranes prepared in the present study are comparable for instance with those achieved with the sulfated-cellulose nanofibrils aerogels that removed *ca.* 5 mg g^{-1} of MB at an adsorbent dosage of 16 mg mL^{-1} [53].

Hence, the adsorbent nanocomposites developed in the present work present a customizable combination of properties, namely antibacterial activity, water-uptake and dye removal capacity, that depend on the amount of the individual components (*i.e.* PMPC and BNC), and that reveal their potential application in the context of water remediation.

4. Conclusions

The combination of the zwitterionic poly(2-methacryloyloxyethyl phosphorylcholine) and the hydrophilic bacterial nanocellulose yielded nanocomposite membranes that are proficient in adsorbing anionic and cationic organic dyes. The optically transparent nanocomposites display high water-uptake capacity in different pH media, thermal stability up to 250 °C, and good mechanical properties (Young's modulus ≥ 430 MPa). Moreover, these zwitterionic membranes inhibited the growth of both Gram-positive (*S. aureus*) and Gram-negative (*E. coli*) pathogenic bacteria with maxima of 4.3– and 1.8–log CFU reduction, respectively, for the nanocomposite composed of 79 wt.% of cross-linked PMPC. Furthermore, their dye removal capacity was demonstrated by a dye adsorption amount of 4.44 ± 0.32 mg g^{-1} of MB and 4.56 ± 0.43 mg g^{-1} of MO for the membrane with the higher content of zwitterionic polymer (*i.e.* 79 wt.% of PMPC). The successful fabrication of these zwitterionic PMPC/BNC nanocomposite membranes with antibacterial activity opens novel avenues for the generation of bio-based adsorbents to address the complex issue of water remediation of anionic and cationic dyes.

Author Contributions: C.V. designed and performed the experiments, analyzed the data and wrote the paper; C.M. and A.A. carried out the antibacterial assays and analyzed the corresponding data; A.J.D.S. and C.S.R.F. contributed to the structural, morphological, thermal and mechanical data interpretation; all authors participated in the critical revision of the paper.

References

1. United Nations Transforming Our World: The 2030 Agenda for Sustainable Development. Available online: https://sustainabledevelopment.un.org/post2015/transformingourworld (accessed on 2 March 2019).
2. Li, R.; Zhang, L.; Wang, P. Rational design of nanomaterials for water treatment. *Nanoscale* **2015**, *7*, 17167–17194. [CrossRef] [PubMed]
3. Khan, S.T.; Malik, A. Engineered nanomaterials for water decontamination and purification: From lab to products. *J. Hazard. Mater.* **2019**, *363*, 295–308. [CrossRef] [PubMed]
4. Vilela, C.; Pinto, R.J.B.; Pinto, S.; Marques, P.A.A.P.; Silvestre, A.J.D.; Freire, C.S.R. *Polysaccharide Based Hybrid Materials*, 1st ed.; Springer: Berlin, Germany, 2018; ISBN 978-3-030-00346-3.
5. Corsi, I.; Fiorati, A.; Grassi, G.; Bartolozzi, I.; Daddi, T.; Melone, L.; Punta, C.; Corsi, I.; Fiorati, A.; Grassi, G.; et al. Environmentally sustainable and ecosafe polysaccharide-based materials for water nano-treatment: An eco-design study. *Materials* **2018**, *11*, 1228. [CrossRef]
6. Mahfoudhi, N.; Boufi, S. Nanocellulose as a novel nanostructured adsorbent for environmental remediation: A review. *Cellulose* **2017**, *24*, 1171–1197. [CrossRef]
7. Voisin, H.; Bergström, L.; Liu, P.; Mathew, A. Nanocellulose-based materials for water purification. *Nanomaterials* **2017**, *7*, 57. [CrossRef] [PubMed]
8. Wang, D. A critical review of cellulose-based nanomaterials for water purification in industrial processes. *Cellulose* **2019**, *26*, 687–701. [CrossRef]
9. Jacek, P.; Dourado, F.; Gama, M.; Bielecki, S. Molecular aspects of bacterial nanocellulose biosynthesis. *Microb. Biotechnol.* **2019**. [CrossRef] [PubMed]
10. Figueiredo, A.R.P.; Vilela, C.; Neto, C.P.; Silvestre, A.J.D.; Freire, C.S.R. Bacterial cellulose-based nanocomposites: Roadmap for innovative materials. In *Nanocellulose Polymer Composites*; Thakur, V.K., Ed.; Scrivener Publishing LLC: Salem, MA, USA, 2015; pp. 17–64.
11. Vilela, C.; Pinto, R.J.B.; Figueiredo, A.R.P.; Neto, C.P.; Silvestre, A.J.D.; Freire, C.S.R. Development and applications of cellulose nanofibers based polymer composites. In *Advanced Composite Materials: Properties and Applications*; Bafekrpour, E., Ed.; De Gruyter Open: Berlin, Germany, 2017; pp. 1–65.
12. Torres, F.G.; Arroyo, J.J.; Troncoso, O.P. Bacterial cellulose nanocomposites: An all-nano type of material. *Mater. Sci. Eng. C* **2019**, *98*, 1277–1293. [CrossRef]

13. Chen, S.; Huang, Y. Bacterial cellulose nanofibers decorated with phthalocyanine: Preparation, characterization and dye removal performance. *Mater. Lett.* **2015**, *142*, 235–237. [CrossRef]

14. Zhijiang, C.; Ping, X.; Cong, Z.; Tingting, Z.; Jie, G.; Kongyin, Z. Preparation and characterization of a bi-layered nano-filtration membrane from a chitosan hydrogel and bacterial cellulose nanofiber for dye removal. *Cellulose* **2018**, *25*, 5123–5137. [CrossRef]

15. Sai, H.; Fu, R.; Xing, L.; Xiang, J.; Li, Z.; Li, F.; Zhang, T. Surface modification of bacterial cellulose aerogels' web-like skeleton for oil/water separation. *ACS Appl. Mater. Interfaces* **2015**, *7*, 7373–7381. [CrossRef]

16. Lu, M.; Guan, X.-H.; Xu, X.-H.; Wei, D.-Z. Characteristic and mechanism of Cr(VI) adsorption by ammonium sulfamate-bacterial cellulose in aqueous solutions. *Chin. Chem. Lett.* **2013**, *24*, 253–256. [CrossRef]

17. Huang, X.; Zhan, X.; Wen, C.; Xu, F.; Luo, L. Amino-functionalized magnetic bacterial cellulose/activated carbon composite for Pb2+ and methyl orange sorption from aqueous solution. *J. Mater. Sci. Technol.* **2018**, *34*, 855–863. [CrossRef]

18. Ishihara, K.; Mu, M.; Konno, T.; Inoue, Y.; Fukazawa, K. The unique hydration state of poly(2-methacryloyloxyethyl phosphorylcholine). *J. Biomater. Sci. Polym. Ed.* **2017**, *28*, 884–899. [CrossRef]

19. Goda, T.; Ishihara, K.; Miyahara, Y. Critical update on 2-methacryloyloxyethyl phosphorylcholine (MPC) polymer science. *J. Appl. Polym. Sci.* **2015**, *132*, 41766. [CrossRef]

20. Fujiwara, N.; Yumoto, H.; Miyamoto, K.; Hirota, K.; Nakae, H.; Tanaka, S.; Murakami, K.; Kudo, Y.; Ozaki, K.; Miyake, Y. 2-Methacryloyloxyethyl phosphorylcholine (MPC)-polymer suppresses an increase of oral bacteria: A single-blind, crossover clinical trial. *Clin. Oral Investig.* **2019**, *23*, 739–746. [CrossRef] [PubMed]

21. Kwon, J.-S.; Lee, M.-J.; Kim, J.-Y.; Kim, D.; Ryu, J.-H.; Jang, S.; Kim, K.-M.; Hwang, C.-J.; Choi, S.-H. Novel anti-biofouling bioactive calcium silicate-based cement containing 2-methacryloyloxyethyl phosphorylcholine. *PLoS ONE* **2019**, *14*, e0211007. [CrossRef] [PubMed]

22. Trovatti, E.; Serafim, L.S.; Freire, C.S.R.; Silvestre, A.J.D.; Neto, C.P. Gluconacetobacter sacchari: An efficient bacterial cellulose cell-factory. *Carbohydr. Polym.* **2011**, *86*, 1417–1420. [CrossRef]

23. Vilela, C.; Sousa, N.; Pinto, R.J.B.; Silvestre, A.J.D.; Figueiredo, F.M.L.; Freire, C.S.R. Exploiting poly(ionic liquids) and nanocellulose for the development of bio-based anion-exchange membranes. *Biomass Bioenergy* **2017**, *100*, 116–125. [CrossRef]

24. Saïdi, L.; Vilela, C.; Oliveira, H.; Silvestre, A.J.D.; Freire, C.S.R. Poly(N-methacryloyl glycine)/nanocellulose composites as pH-sensitive systems for controlled release of diclofenac. *Carbohydr. Polym.* **2017**, *169*, 357–365. [CrossRef]

25. Barud, H.S.; Ribeiro, S.J.L.; Carone, C.L.P.; Ligabue, R.; Einloft, S.; Queiroz, P.V.S.; Borges, A.P.B.; Jahno, V.D. Optically transparent membrane based on bacterial cellulose/polycaprolactone. *Polímeros* **2013**, *23*, 135–138. [CrossRef]

26. Pleumphon, C.; Thiangtham, S.; Pechyen, C.; Manuspiya, H.; Ummartyotin, S. Development of conductive bacterial cellulose composites: An approach to bio-based substrates for solar cells. *J. Biobased Mater. Bioenergy* **2017**, *11*, 321–329. [CrossRef]

27. Foster, E.J.; Moon, R.J.; Agarwal, U.P.; Bortner, M.J.; Bras, J.; Camarero-Espinosa, S.; Chan, K.J.; Clift, M.J.D.; Cranston, E.D.; Eichhorn, S.J.; et al. Current characterization methods for cellulose nanomaterials. *Chem. Soc. Rev.* **2018**, *47*, 2609–2679. [CrossRef] [PubMed]

28. Bellamy, L.J. *The Infrared Spectra of Complex Molecules*, 3rd ed.; Chapman and Hall, Ltd.: London, UK, 1975; ISBN 0412138506.

29. Xu, S.; Ye, Z.; Wu, P. Biomimetic controlling of CaCO3 and BaCO3 superstructures by zwitterionic polymer. *ACS Sustain. Chem. Eng.* **2015**, *3*, 1810–1818. [CrossRef]

30. Ishihara, K.; Ueda, T.; Nakabayashi, N. Preparation of phospholipid polymers and their properties as polymer hydrogel membranes. *Polym. J.* **1990**, *22*, 355–360. [CrossRef]

31. Vilela, C.; Martins, A.P.C.; Sousa, N.; Silvestre, A.J.D.; Figueiredo, F.M.L.; Freire, C.S.R. Poly(bis[2-(methacryloyloxy)ethyl] phosphate)/bacterial cellulose nanocomposites: Preparation, characterization and application as polymer electrolyte membranes. *Appl. Sci.* **2018**, *8*, 1145. [CrossRef]

32. Gadim, T.D.O.; Figueiredo, A.G.P.R.; Rosero-Navarro, N.C.; Vilela, C.; Gamelas, J.A.F.; Barros-Timmons, A.; Neto, C.P.; Silvestre, A.J.D.; Freire, C.S.R.; Figueiredo, F.M.L. Nanostructured bacterial cellulose-poly(4-styrene sulfonic acid) composite membranes with high storage modulus and protonic conductivity. *ACS Appl. Mater. Interfaces* **2014**, *6*, 7864–7875. [CrossRef]

33. Wang, S.; Liu, Q.; Luo, Z.; Wen, L.; Cen, K. Mechanism study on cellulose pyrolysis using thermogravimetric analysis coupled with infrared spectroscopy. *Front. Energy Power Eng. China* **2007**, *1*, 413–419. [CrossRef]

34. Faria, M.; Vilela, C.; Mohammadkazemi, F.; Silvestre, A.J.D.; Freire, C.S.R.; Cordeiro, N. Poly(glycidyl methacrylate)/ bacterial cellulose nanocomposites: Preparation, characterization and post-modification. *Int. J. Biol. Macromol.* **2019**, *127*, 618–627. [CrossRef]

35. Vilela, C.; Gadim, T.D.O.; Silvestre, A.J.D.; Freire, C.S.R.; Figueiredo, F.M.L. Nanocellulose/ poly(methacryloyloxyethyl phosphate) composites as proton separator materials. *Cellulose* **2016**, *23*, 3677–3689. [CrossRef]

36. Figueiredo, A.R.P.; Figueiredo, A.G.P.R.; Silva, N.H.C.S.; Barros-Timmons, A.; Almeida, A.; Silvestre, A.J.D.; Freire, C.S.R. Antimicrobial bacterial cellulose nanocomposites prepared by in situ polymerization of 2-aminoethyl methacrylate. *Carbohydr. Polym.* **2015**, *123*, 443–453. [CrossRef]

37. Vilela, C.; Kurek, M.; Hayouka, Z.; Röcker, B.; Yildirim, S.; Antunes, M.D.C.; Nilsen-Nygaard, J.; Pettersen, M.K.; Freire, C.S.R. A concise guide to active agents for active food packaging. *Trends Food Sci. Technol.* **2018**, *80*, 212–222. [CrossRef]

38. Bui, V.; Park, D.; Lee, Y.-C. Chitosan combined with ZnO, TiO$_2$ and Ag nanoparticles for antimicrobial wound healing applications: A mini review of the research trends. *Polymers* **2017**, *9*, 21. [CrossRef]

39. Konai, M.M.; Bhattacharjee, B.; Ghosh, S.; Haldar, J. Recent progress in polymer research to tackle infections and antimicrobial resistance. *Biomacromolecules* **2018**, *19*, 1888–1917. [CrossRef]

40. Gómez, P.; Lozano, C.; Benito, D.; Estepa, V.; Tenorio, C.; Zarazaga, M.; Torres, C. Characterization of staphylococci in urban wastewater treatment plants in Spain, with detection of methicillin resistant Staphylococcus aureus ST398. *Environ. Pollut.* **2016**, *212*, 71–76. [CrossRef]

41. Padrão, J.; Gonçalves, S.; Silva, J.P.; Sencadas, V.; Lanceros-Méndez, S.; Pinheiro, A.C.; Vicente, A.A.; Rodrigues, L.R.; Dourado, F. Bacterial cellulose-lactoferrin as an antimicrobial edible packaging. *Food Hydrocoll.* **2016**, *58*, 126–140. [CrossRef]

42. Shao, W.; Wang, S.; Wu, J.; Huang, M.; Liu, H.; Min, H. Synthesis and antimicrobial activity of copper nanoparticle loaded regenerated bacterial cellulose membranes. *RSC Adv.* **2016**, *6*, 65879–65884. [CrossRef]

43. Vilela, C.; Oliveira, H.; Almeida, A.; Silvestre, A.J.D.; Freire, C.S.R. Nanocellulose-based antifungal nanocomposites against the polymorphic fungus Candida albicans. *Carbohydr. Polym.* **2019**. [CrossRef]

44. Yin, N.; Santos, T.M.A.; Auer, G.K.; Crooks, J.A.; Oliver, P.M.; Weibel, D.B. Bacterial cellulose as a substrate for microbial cell culture. *Appl. Environ. Microbiol.* **2014**, *80*, 1926–1932. [CrossRef]

45. Li, J.; Cha, R.; Mou, K.; Zhao, X.; Long, K.; Luo, H.; Zhou, F.; Jiang, X. Nanocellulose-based antibacterial materials. *Adv. Healthc. Mater.* **2018**, *7*, 1800334. [CrossRef]

46. Bertal, K.; Shepherd, J.; Douglas, C.W.I.; Madsen, J.; Morse, A.; Edmondson, S.; Armes, S.P.; Lewis, A.; MacNeil, S. Antimicrobial activity of novel biocompatible wound dressings based on triblock copolymer hydrogels. *J. Mater. Sci.* **2009**, *44*, 6233–6246. [CrossRef]

47. Fuchs, A.V.; Ritz, S.; Pütz, S.; Mailänder, V.; Landfester, K.; Ziener, U. Bioinspired phosphorylcholine containing polymer films with silver nanoparticles combining antifouling and antibacterial properties. *Biomater. Sci.* **2013**, *1*, 470–477. [CrossRef]

48. Band, V.I.; Weiss, D.S. Mechanisms of antimicrobial peptide resistance in gram-negative bacteria. *Antibiotics* **2014**, *4*, 18–41. [CrossRef] [PubMed]

49. Masi, M.; Réfregiers, M.; Pos, K.M.; Pagès, J.M. Mechanisms of envelope permeability and antibiotic influx and efflux in Gram-negative bacteria. *Nat. Microbiol.* **2017**, *2*, 17001. [CrossRef]

50. Delport, A.; Harvey, B.H.; Petzer, A.; Petzer, J.P.; Brain Dis, M. Methylene blue and its analogues as antidepressant compounds. *Metab. Brain Dis.* **2017**, *32*, 1357–1382. [CrossRef] [PubMed]

51. Forgacs, E.; Cserháti, T.; Oros, G. Removal of synthetic dyes from wastewaters: A review. *Environ. Int.* **2004**, *30*, 953–971. [CrossRef] [PubMed]

52. Shim, E.; Kim, H.R. Coloration of bacterial cellulose using in situ and ex situ methods. *Text. Res. J.* **2018**, *89*, 1297–1310. [CrossRef]

53. Wang, D.C.; Yu, H.Y.; Fan, X.; Gu, J.; Ye, S.; Yao, J.; Ni, Q.Q. High aspect ratio carboxylated cellulose nanofibers cross-linked to robust aerogels for superabsorption-flocculants: Paving way from nanoscale to macroscale. *ACS Appl. Mater. Interfaces* **2018**, *10*, 20755–20766. [CrossRef]

54. He, X.; Male, K.B.; Nesterenko, P.N.; Brabazon, D.; Paull, B.; Luong, J.H.T. Adsorption and desorption of methylene blue on porous carbon monoliths and nanocrystalline cellulose. *ACS Appl. Mater. Interfaces* **2013**, *5*, 8796–8804. [CrossRef]

55. Tang, J.; Song, Y.; Zhao, F.; Spinney, S.; Bernardes, J.S.; Tam, K.C. Compressible cellulose nanofibril (CNF) based aerogels produced via a bio-inspired strategy for heavy metal ion and dye removal. *Carbohydr. Polym.* **2019**, *208*, 404–412. [CrossRef]

8

Degradation Studies Realized on Natural Rubber and Plasticized Potato Starch Based Eco-Composites Obtained by Peroxide Cross-Linking

Elena Manaila [1], Maria Daniela Stelescu [2] and Gabriela Craciun [1],*

[1]   Electron Accelerators Laboratory, National Institute for Laser, Plasma and Radiation Physics, 409 Atomistilor Street, 077125 Magurele, Romania; elena.manaila@inflpr.ro

[2]   National R&D Institute for Textile and Leather—Leather and Footwear Research Institute, 93 Ion Minulescu Street, 031215 Bucharest, Romania; dmstelescu@yahoo.com

*   Correspondence: gabriela.craciun@inflpr.ro

Abstract: The obtaining and characterization of some environmental-friendly composites that are based on natural rubber and plasticized starch, as filler, are presented. These were obtained by peroxide cross-linking in the presence of a polyfunctional monomer used here as cross-linking co-agent, trimethylolpropane trimethacrylate. The influence of plasticized starch amount on the composites physical and mechanical characteristics, gel fraction and cross-link density, water uptake, structure and morphology before and after accelerated (thermal) degradation, and natural (for one year in temperate climate) ageing, was studied. Differences of two orders of magnitude between the degradation/aging methods were registered in the case of some mechanical characteristics, by increasing the plasticized starch amount. The cross-link density, water uptake and mass loss were also significant affected by the plasticized starch amount increasing and exposing for one year to natural ageing in temperate climate. Based on the results of Fourier Transform Infrared Spectroscopy (FTIR) and cross-link density measurements, reaction mechanisms attributed to degradation induced by accelerated and natural ageing were done. SEM micrographs have confirmed in addition that by incorporating a quantity of hydrophilic starch amount over 20 phr and by exposing the composites to natural ageing, and then degradability can be enhanced by comparing with thermal degradation.

Keywords: natural rubber; plasticized starch; polyfunctional monomers; physical and mechanical properties; cross-link density; water uptake

1. Introduction

Natural rubber (NR) is renewable, non-toxic, has excellent physical properties and due to its low price is the most used elastomer worldwide in industry or in a variety of applications in which the final products are in contact with food or potable water [1]. The most common physical–chemical treatment of rubber is curing (cross-linking, vulcanization) by sulphur, peroxides, ultraviolet light, electron beam, and microwave irradiation, but sulphur and peroxide curing systems still remain the most desirable. The application of sulphur systems leads to the formation of sulphidic cross-links between elastomer chains [2]. In peroxide curing, high thermal stabile C–C bonds are formed. Therefore, peroxide cured elastomers exhibit high-temperature ageing resistance and low compression that is set at high temperatures. There are some disadvantages when compared peroxide to sulfur cured systems, such as low scarce safety and worse dynamic and elastic properties of vulcanizates [3]. But, the cross-linking with peroxides can be effectively improved by the use of co-agents [4–6], because they are able to boost peroxide efficiency by suppressing side reactions to a large extent, like chain scission and disproportionation [3,7], or by the formation of co-agents bridges between polymer chains as extra cross-links [8,9]. Trimethylolpropane trimethacrylate (TMPT), is a polyfunctional monomer that is used as co-agent [10,11] in order to improve the cross-linking process because is able to increase the rate state of cure and, as a consequence, to improve the physical properties of the processed material [8,12]. Anyway, the NR is used in form of mixtures, which generally contain active fillers, plasticizers, cross-linking agents, and other ingredients that give different characteristics to the final product [12,13]. New environmental-friendly elastomeric materials can be obtained by the use of natural fillers in NR and other rubber blends instead of hazardous active fillers, such as silica or carbon black, which are very well known for the harmful effects on human health [14]. So, it is preferable to replace the classics with other compatible types of filler that should maintain, or even improve, the mechanical and usable properties of NR or other rubber products. Because of the increased interests in replacement of non-renewable rubber materials with some based on components originated from natural resources, the use of natural fillers is considered as being a promising solution [15,16]. Even though their nature is known to be hydrophilic, starches are considered among the most promising available natural biopolymers and may successful candidate for the development of novel composite materials based on NR [17,18]. These are cheap, abundant available, biodegradable, recyclable, renewable, and present thermoplastic behavior [19–21]. Thermoplastic properties of starch are obtained by the disruption and plasticization of native starch with plasticizers agents (glycerol, water, and other polyols) [21–23].

The goal of the paper is to present the obtaining and characterization of a new environmental-friendly elastomeric composite that is based on NR and plasticized starch (PS). The cross-linking method was by the use of peroxide in the presence of the trimethylolpropane trimethacrylate (TMPT). The elastomeric composite behavior under accelerated (thermal) aging and natural aging in temperate climate for one year was investigated, because it is very well known that starch-based plastics have some drawbacks, including limited long term stability caused by water absorption, ageing caused degradation, poor mechanical properties and bad processability. Also, the influence of PS amount on the physical and mechanical properties, cross-linking density rate and behavior in aqueous environment, before and after aging was studied. The novelty of the present study consists in the replacing of conventional fillers (silica, carbon black) with an environmental friendly filler (plasticized starch) in order to obtain composites having improved cross-linking density and mechanical properties, and over these, a high degradation degree in natural environment.

2. Results and Discussion

2.1. Physical and Chemical Characteristics

The results of mechanical tests that have been made on unfilled NR and NR filled with PS composites (NR–PS), before and after aging are summarized in Figure 1. An improvement of mechanical properties before aging, excepting 100% Modulus (Figure 1c), due to the plasticized starch introduction, can be observed.

2.1.1. Mechanical Properties of Unfilled and Filled NR before Aging

As it can be seen from Figure 1a, hardness has slowly increased with the PS amount in blend increasing. This is due the reinforcement effect of PS, which incorporated in the NR matrix has conducted to a reduction of plasticity and flexibility of rubber chains and the composite become more rigid [14,24]. Irrespective of the PS amount added, the increase in hardness did not exceed 12.5% for unfilled NR. The results that are presented in Figure 1b show that the unfilled NR has exhibited a higher elastic response when compared to that of NR-PS samples. The composite having 10 phr of PS has a better elastic behavior than unfilled NR. The higher dynamic stiffness of the samples containing over 20 phr of PS can be the reason of the elasticity decreases [25]. The same results were obtained for 100% Modulus Figure 1c. The results presented in Figure 1d show that the tensile strength, which strongly depends on effective and uniform stress distribution, increases as the amount of PS increase, up to 20 phr. As the PS amount still increases, the tensile strength started to decrease by about

45%, because the PS tend to agglomerate, leading to insufficient wetting of filler and bad interface with the matrix. So, the PS acts as flows limiting the tensile strength. Also, another reason may be connected with the presence of some voids trapped in the composite during processing, so a specific attention has to be accorded to the preparation technique when is dealing with high filler loading in order to assure its well dispersion [26,27].

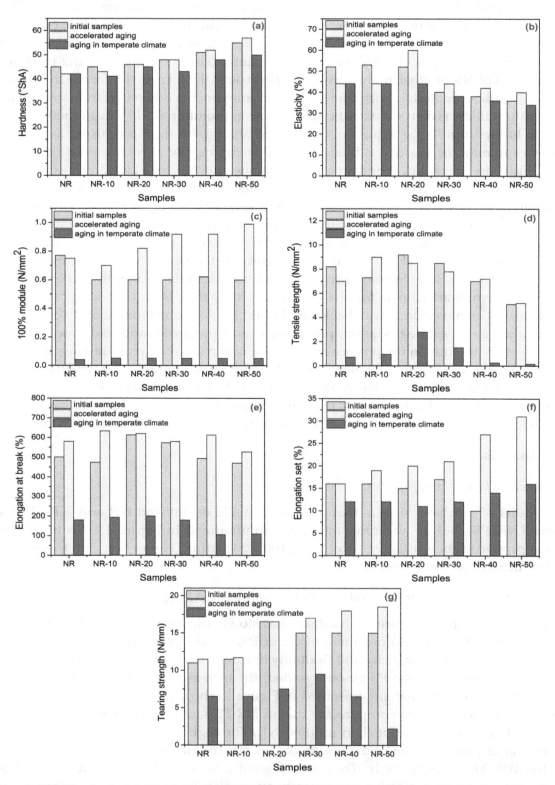

Figure 1. The Hardness (**a**), Elasticity (**b**), 100% Modulus (**c**), Tensile strength (**d**), Elongation at break (**e**), Elongation set, and (**f**) Tearing strength (**g**) variations as a function of PS amount and aging method.

As it can be seen from Figure 1e, elongation at break has the same behavior as tensile strength on the entire PS concentration range. The decrease highlighted over 20 phr of PS, is the consequence of the appeared restriction in the movement of molecular chains, which has lead on a negative effect upon the sample ductility [14,28,29]. Figure 1f shows that the elongation set has decreased with the PS amount in blend increasing, a fact that indicates also an increase in cross-link density. The decrease in residual elongation shows that the sample is vulcanized and thus returns to its original shape easily [14,24]. Figure 1g shows also the same variation trend of tearing strength as tensile strength and elongation at break. So, we can conclude that the PS has a reinforcing effect on NR when it is loaded up to 20 phr.

2.1.2. Mechanical Properties of Unfilled and Filled NR after Aging

Mechanical Properties of Unfilled and Filled NR after Accelerated (Thermal) Ageing

A specimen from each sample was subjected to thermal ageing in an air-circulating oven at 70 °C for 168 h, in order to evaluate the rubber compound properties before and after aging. As it can be seen from Figure 1a–g and Table 1, after the thermal aging, mechanical properties are modified for all the tested composites.

Table 1. Percentage modifications of the mechanical properties of unfilled and filled natural rubber (NR) after thermal aging.

Mechanical Property (%)	Sample Type (PS Loading) (phr of PS at 100 phr of NR)					
	NR	NR-10	NR-20	NR-30	NR-40	NR-50
Hardness	−6.67	−4.44	0	0	+1.96	+3.64
Elasticity	−15.38	−16.98	+15.38	+10.00	+10.53	+11.11
100% Modulus	−2.60	+16.67	+36.67	+53.33	+53.33	+65.00
Tensile strength	−14.63	+23.29	−7.61	−8.24	+2.86	+1.96
Elongation at break	+16.33	+33.83	+1.14	+1.22	+24.34	+12.13
Elongation set	0.00	+18.73	+33.33	+23.53	+170.00	+210.00
Tearing strength	+4.55	+1.74	0	+13.33	+20.00	+23.33

The resistance of rubber based composites to thermal aging is considered as being an essential requirement for better service performance. The increasing of elasticity, 100% modulus, tensile strength, elongation at break, tensile set, and tearing strength of NR-PS composites is due to the presence of peroxide free radicals that were not involved in the cross-linking reactions and that lead to the formation of few new cross-links during thermal aging. The phenomenon is well known as post-curing during aging [30,31]. On the other hand, elasticity, 100% modulus and tensile strength of the unfilled NR have diminished after thermal aging due to the post-curing during aging by which excessive cross-links were formed [30,32]. It seems that the PS presence, even in the small amount of 10 phr, has delayed the formation of excessive cross-links in NR-PS as compared with the unfilled NR.

Mechanical Properties of Unfilled and Filled NR after 1 Year of Natural Ageing in Temperate Climate

Another specimen from each sample was subjected to natural ageing for one year in temperate climate between March 2017 and March 2018, in Bucharest, Romania. In order to determine the effect of outdoor exposure, samples were suspended in vertical position on a special dryer. The influence of the natural environment (heat, cold, frost, sunlight, oxygen, moisture, precipitations) upon the degradation was evaluated by comparing the samples mechanical properties before and after aging processes.

From the results presented in Figure 1a–g and Table 2, it can be seen the important changes suffered by all mechanical properties of unfilled and filled NR composites, after one year of natural ageing in temperate climate staying. The negative modifications of mechanical properties clearly show the sample degradation during the natural process. So, 100% modulus and tensile strength present an important falling-off that indicates structural changes in rubber chains due to the cross-links dissociation.

Table 2. Percentage modifications of the mechanical properties of unfilled and filled NR after one year of natural ageing in temperate climate.

Mechanical Property (%)	Sample Type (PS Loading) (phr of PS at 100 phr of NR)					
	NR	NR-10	NR-20	NR-30	NR-40	NR-50
Hardness	−6.67	−8.89	−2.17	−10.42	−5.88	−9.09
Elasticity	−15.38	−16.98	−15.38	−5.00	−5.26	−5.56
100% Modulus	−94.81	−91.67	−91.67	−91.67	−91.67	−91.67
Tensile strength	−91.64	−86.99	−69.57	−82.35	−96.29	−96.86
Elongation at break	−64.00	−59.20	−67.37	−68.59	−78.50	−76.60
Elongation set	−25.00	−25.00	−26.67	−29.41	+40.00	+60.00
Tearing strength	−40.91	−43.48	−54.55	−36.67	−56.67	−85.33

As a consequence, the composites lose the elastic properties and the ability to act as an effective matrix material to transmit stress [30,33]. An important decrease of elongation at break was also observed. This is as a consequence of both cross-linking and scission reactions, which negative modify the elastic nature of rubber chains [30,34,35].

2.1.3. Gel Fraction and Cross-Link Density

Figure 2a,b presents the results regarding the gel fraction and cross-link density investigations before and after aging. In Table 3, the percentage changes of these parameters for unfilled and filled NR after one year of natural ageing in temperate climate are presented.

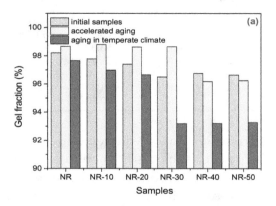 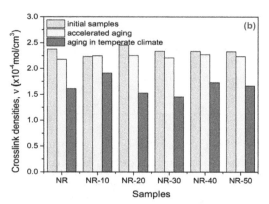

Figure 2. Gel Fraction (**a**) and Cross-link density (**b**) variations as a function of plasticized starch (PS) amount and aging method.

Table 3. Percentage modifications of gel fraction and cross-link densities of unfilled and filled NR after 1 year of natural ageing in temperate climate.

Sample Type (PS Loading)	Gel Fraction		Cross-Link Density	
	Thermal Aging	Natural Aging for 1 Year	Thermal Aging	Natural Aging for 1 Year
NR	+0.47	−0.56	−8.28	−32.13
NR-10	+1.05	−0.83	+0.71	−14.44
NR-20	+1.26	−0.78	−8.07	−37.81
NR-30	+2.24	−3.42	−5.47	−37.83
NR-40	−0.57	−3.55	−2.60	−26.03
NR-50	−0.41	−3.48	−3.99	−28.46

As it can be seen from Figure 2a,b, before being subjected to aging tests, the introduction of PS has induced a decrease of samples gel fraction and a not so significant modification of cross-link density. Continuing the loading with PS over 20 phr, the cross-link density remains on a path. The slight variations of cross-link density may be explained by the changing in phase structure character of NR in which the PS was introduced. But, for most applications, the cross-link density should not be so high and it must be sufficient to give the rubber mechanical integrity so that it can bear loads and present deformation recovery. A high cross-link density immobilized the polymer chains, fact that lead to a hard and brittle rubber [36,37].

Figure 2a,b and Table 3 show that after the accelerated (thermal) aging, an increase in the gel fraction is observed for both unfilled and filled NR composites up to the loading with 30 phr. Over this, the gel fraction has decreases and remains on a path. The cross-link density was more affected by the thermal aging. But, after one year of natural degradation in temperate climate, all the samples gel fractions and cross-link densities have strongly decreased. NR samples have showed a reduction of 32.13%, while NR-PS samples of 37.81% for the loading of 20 phr and 37.83% for 30 phr, respectively. The results are in agreement with those that were obtained in mechanical tests and prove a clear degradation of the composites.

2.2. Structural and Morphological Characteristics

2.2.1. Spectral Characterization by Fourier Transform Infrared Spectroscopy Analysis

In order to investigate structural modification of filled NR as compared with unfilled NR before and after aging, the spectral characterization by Fourier Transform Infrared Spectroscopy (FTIR) in the range of 600–4000 cm^{-1}, was done. The results are presented in Figure 3a–c.

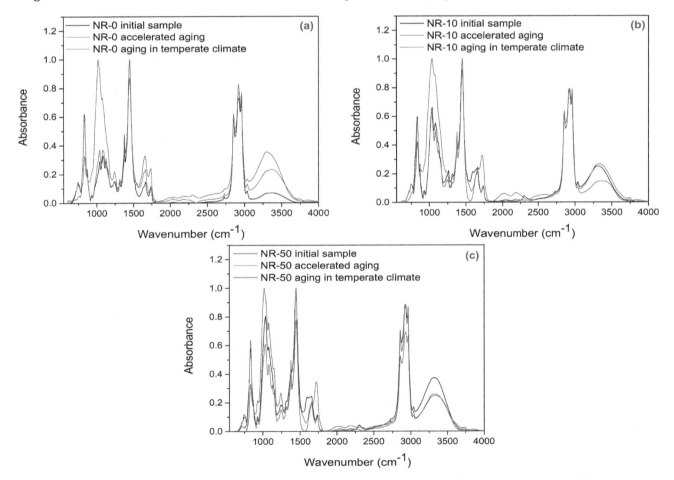

Figure 3. Infrared spectra in the range of 600–4000 cm^{-1} for samples unfilled (**a**), filled with 10 phr PS (**b**) and with 50 phr PS (**c**).

The assignments of the main bands in the NR and NR-PS before and after aging are summarized in Table 4. A comparative analysis between the FTIR spectra of NR (Figure 3a) and NR-PS (Figure 3b,c) highlights notable differences (Table 4).

Table 4. Characteristic infrared bands observed in NR and NR-PS spectra.

Wave Number (cm^{-1})	Assignment
740–760	C–O–C ring vibration from starch or deformation vibration of R_2C=CH–R groups from NR
833	=CH out-of-plane bending vibration from NR rubber
870	C(1)–H(α) bending vibration from starch
930–925	skeletal mode vibrations of α-(1-4) glycosidic linkage (C–O–C) from starch
1034–1038	C–O stretching vibration in C–O–H and C–O–C in the anhydrous glucose ring from starch
1080–1086	C–O–C stretching vibration that indicate the grafting of PS on NR
1125–1126	C–O stretching of C–O–C (from starch) or of alcohols >HC–OH resulted from the degradation
1240–1260	carbonyl ((>C=O) and hydroxyl (–OH) compound resulted from the degradation
1310–1315	bending vibration of C–H and C–O groups of aromatic rings (starch)
1370–1380	–CH$_3$ asymmetric deformation of NR
1440–1450	–CH$_2$– deformation vibration from NR or –CH$_2$– symmetric bending vibration from starch
1655–1665	–C=C– stretching vibration in the NR structure or may be due to absorbed water or carboxylate or conjugated ketone (>C=O) resulted from the degradation
1710–1740	the fatty acid ester groups existing in NR or carbonyl group (>C=O) from ketone (R_2C=O) or aldehyde (RCOH) resulted from the oxidative degradation
2852–2854	–CH$_2$– symmetric stretching vibration of NR
2919–2927	–CH$_2$– asymmetric stretching vibration of NR
2958–2960	–CH$_3$ asymmetric stretching vibration of NR
3030–3040	=CH– stretching vibration of –CH=CH$_2$ group from NR
3300–3380	N–H stretching vibration of amide groups from the existing proteins in NR or from OH-stretching vibration (–OH as a result of the degradation by oxidation)

Thus, the band that appeared at 1081 cm^{-1} (C–O–C) indicates the grafting of PS on NR [38]. Also, the band at 930–925 cm^{-1} can be attributed to the skeletal mode vibrations of α-(1-4) glycosidic linkage (C–O–C) and the one between 1100 cm^{-1} and 1030 cm^{-1} is characteristic of the anhydrous glucose ring C–O stretch [39,40]. The absorption bands at 3380 cm^{-1} that appear in NR-0 before aging treatment (Figure 3a) were identified to the proteins from NR [41]. After degradation, the spectra indicate that the intensity of the broad bend near 3380 cm^{-1} has increases. The broad band at 3380 cm^{-1} may be due formation of hydroxyl group (–OH) as a result of the degradation by oxidation [38,42]. It can be seen that after 1 year of natural ageing in temperate climate, all these bands have increased in intensity. The decreasing of *cis*-1,4 double bonds number in the polyisoprene chain at 833 cm^{-1}, the formation of hydroxyl group at 3380 cm^{-1}, the appearance of ketone and aldehyde groups between 1736–1722 cm^{-1}, and the increasing of glycosidic linkage at 930–925 cm^{-1} may be interpreted as consequences of the degradation induced by the accelerated and natural aging processes to which samples have been subjected. It can be seen that after natural ageing, all these bands have increased in intensity. All of these, correlated with the decreasing of *cis*-1,4 double bonds number in the polyisoprene chain at 870–830 cm^{-1}, the formation of hydroxyl group at 3380 cm^{-1}, the appearance of ketone and aldehyde groups between 1736–1722 cm^{-1} and the increasing of glycosidic linkage at 930–925 cm^{-1} may be interpreted as consequences of the degradation. Also, the formation of carbonyl or hydroxyl bonds (>C=O and –OH) and carboxylate or conjugated ketone (RCOOH and R_2C=O) is demonstrated by the occurrence of the absorption bands between 1260–1400 cm^{-1} and 1550–1690 cm^{-1}, respectively [43]. The cross-linking degree and molecular masses decreasing are due to the cleavage of the macromolecules, as demonstrated by the appearance of the absorption bands at 1370–1380 cm^{-1}, 2850–2880 cm^{-1}, and 2950–2980 cm^{-1} that correspond to –CH$_3$ groups. The slight increasing of the absorption bands at 1640–1660 cm^{-1} and 800–900 cm^{-1} due to the number of double bonds increasing, as well as the modification of absorption bands at 800–900 cm^{-1}, 1650–1680 cm^{-1}, and 3010–3040 cm^{-1} that correspond to the changes in the degree of substitution of carbon atoms of the double bond are connected with the degradation process [43].

2.2.2. Mechanisms of Natural Degradation Reactions for NR and NR-PS

In temperate climate, the main ageing processes in NR are the oxidative and thermal-oxidative degradation (photo-degradation) and UV/ozone degradation [36,44].

The initial step of oxidative and thermal-oxidative degradation consists in free-radicals formation on the NR chain by hydrogen abstraction.

The propagation of oxidative degradation takes place in several stages as it can be seen from Figure 4. The propagation first step is the reaction of a free radical with an oxygen molecule (O$_2$) to form a peroxy radical (NR–O–O•), which then abstracts a hydrogen atom from another polymer chain to form a hydroperoxide (NR–O–O–H). The hydroperoxide splits then into two new free radicals, (NR–O• and •OH) that abstract another hydrogen and from other polymer chains.

The termination of the reaction is achieved by the recombination of two radicals or by disproportionation/hydrogen abstraction, as it can be seen from Figure 5.

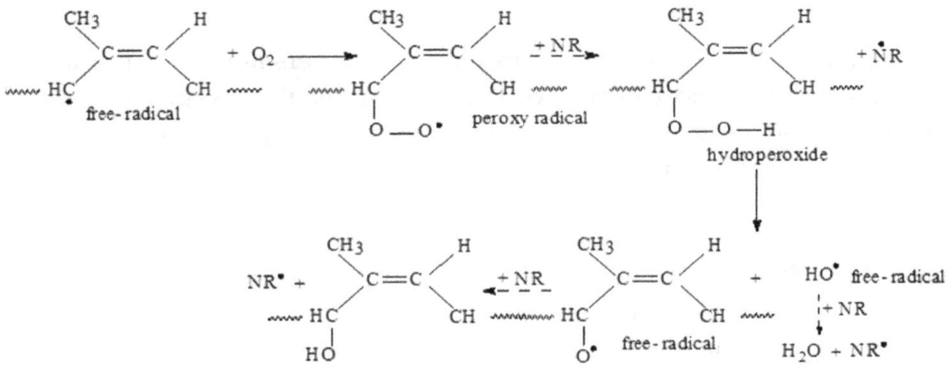

Figure 4. Propagation step of oxidative degradation.

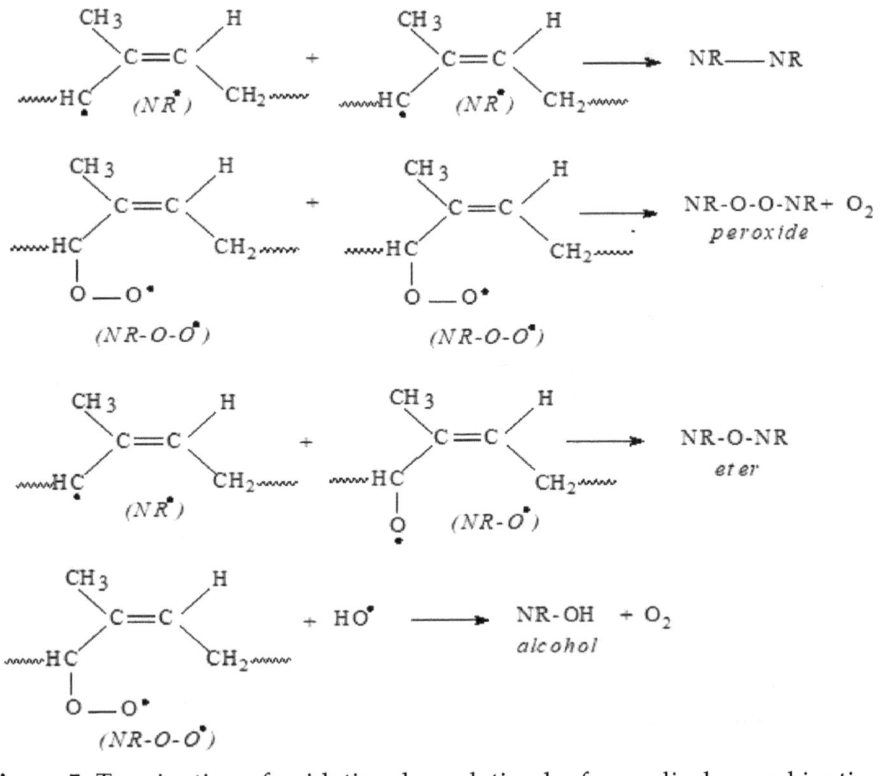

Figure 5. Termination of oxidative degradation by free radical recombination.

The results of these reactions are the polymer enbrittlement and cracking. On the other hand, termination by chain scission, as presented in Figure 6, results in the decrease of the molecular weight leading to softening of the polymer and reduction of the mechanical properties [44].

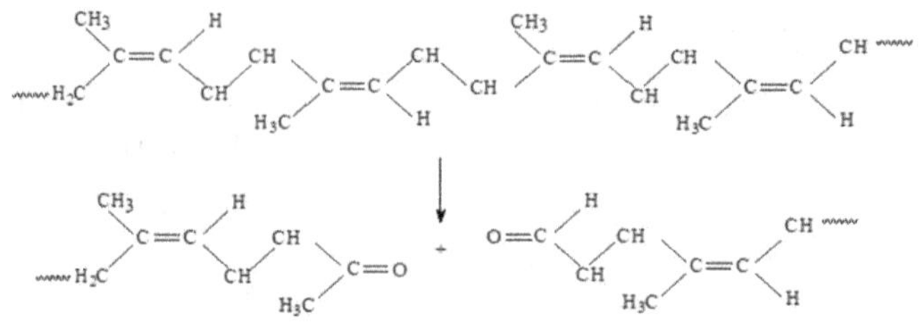

Figure 6. Termination of oxidative degradation by chain scission.

In photo-degradation, if free radicals are directly produced by UV radiation, then all of the subsequent reactions are similar to those of thermal-oxidative degradation, including chain scission, cross-linking, and secondary oxidation. Photo-oxidation can also occur by breaking the bonds that are created between NR and PS (Figure 7). Starch may form other radicals by the cleavage of a glycosidic bond and from β-fragmentation of an oxygen-centred radical resulting from cleavage of a glycosidic bond [45].

Sunlight and ozone rapidly attack the unprotected polymers and can significantly reduce them service life [44]. Ozone, being more corrosive then the molecular oxygen will attack the polymer directly at the carbon-carbon double bond. As it can be seen from Figure 8, the ozone molecule attaches itself to the double bond creating a C_2O_3 ring. The chain scission results in products containing carbonyl groups.

Ozone degradation of vulcanized NR exhibits a very specific cracking, through which it is formed a hard surface layer [46]. The reaction with ozone, leads to chain scission of the polymer chain and the formation of polymeric peroxides that can also increase the rate of oxidative aging [36,47]. As those processes are related to substantial modifications of the macromolecular backbone, substantial damage in mechanical properties are expected, even at low rate of conversion (less than 0.1%) [36,37]. This was observed for all samples aged for one year in temperate climates, for both mechanical properties and cross-linking density.

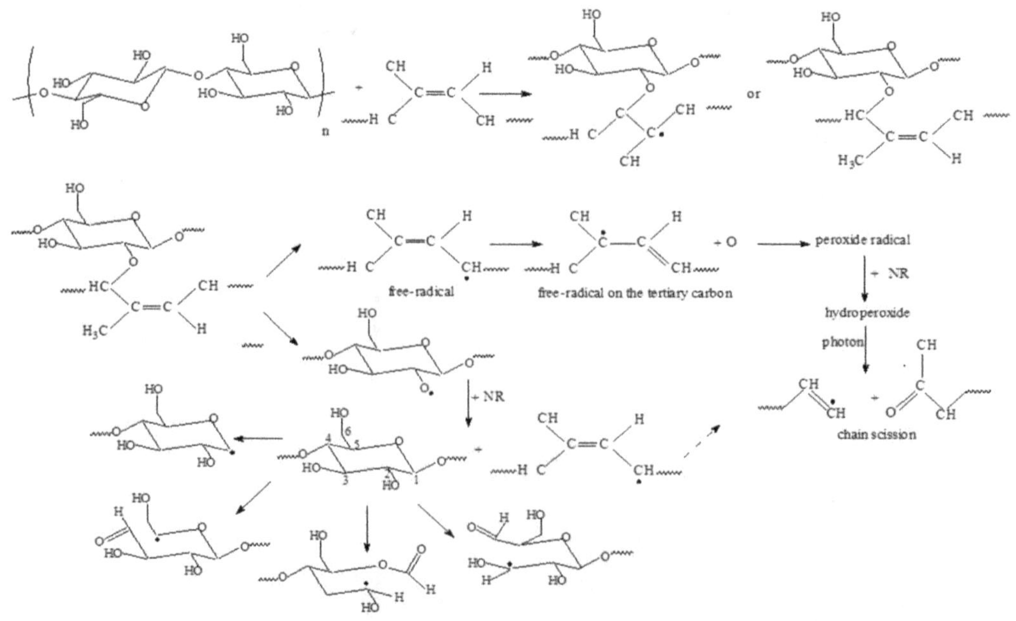

Figure 7. The photo-oxidation reaction on NR—starch bond.

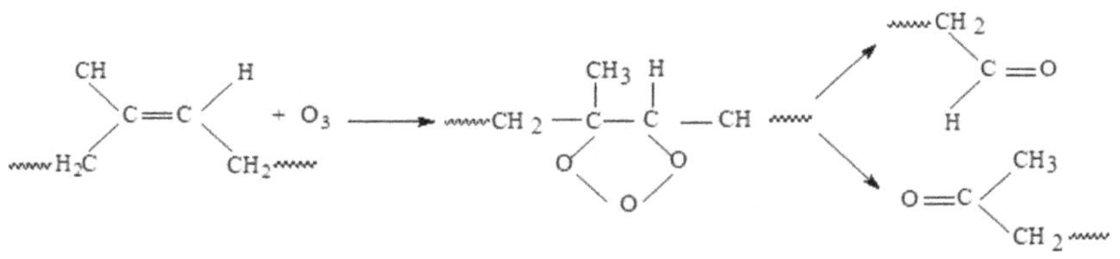

Figure 8. The ozone attack on NR chain.

2.2.3. SEM Analysis

The morphological changes of unfilled and filled NR samples before and after one year of natural ageing in temperate climate were evaluated by surfaces SEM analysis. Before and after ageing, samples were immersed in toluene, in order to remove any split fragments or un-reacted materials. The results are presented in Figures 9 and 10.

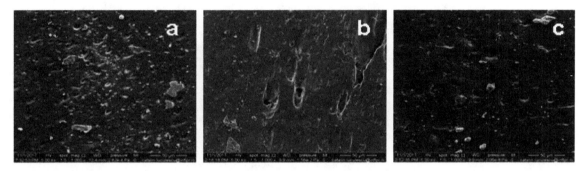

Figure 9. SEM micrographs of (**a**) NR, (**b**) NR-10, and (**c**) NR-50 samples, before one year of natural ageing in temperate climate, at magnification of 1000.

From Figure 9, it can be seen that, before aging, the NR and NR-PS samples surface looks smooth presenting only with small imperfections that can be caused by the presence of some impurities remained even after the immersion in toluene.

Figure 10. SEM micrographs of (**a**) NR, (**b**) NR-10, and (**c**) NR-50 samples, after one year of natural ageing in temperate climate, at magnification of 1000.

From Figure 10, the NR and NR-PS samples degradation after 1 year of natural ageing in temperate climate can be observed. The surfaces of all samples are rough, extensive surface cracks are formed, leading the appearance similar to a mosaic pattern. As the PS amount has increased, the roughness of surface sample also increased and more micro cracks are formed. The surface roughness can be attributed to the decreasing of elasticity, 100% modulus and tensile strength, as the filler loading was increased, but also to the dissociation of existing cross-links or to the structural changes in natural rubber chains [48]. In addition to the oxidative and ozone degradation, the outdoor thermal stress and the dust presence must be taken into account for the contribution to the cracks formation on the composite surface [48–50].

2.3. Water Uptake and Mass Loss

Water uptake tests were done before and after ageing in order to demonstrate the contribution of the strongly hydrophilic polar groups that appeared in the FTIR spectra of the composite, to the natural degradation. The results are presented in Figure 11a–d. As it can be seen, the water absorption of NR-PS composites is strongly dependent on the PS amount, increasing with the PS content due to the hydrophilic nature of starch and the greater interfacial area between the starch and the NR matrix.

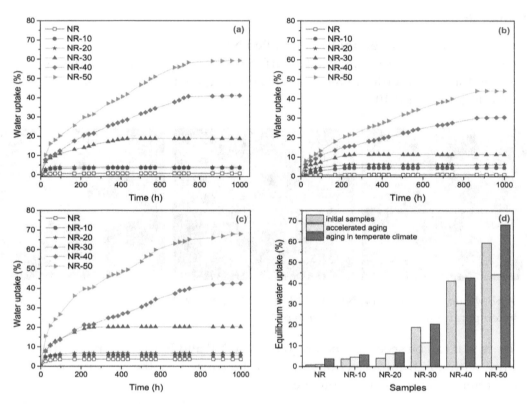

Figure 11. Water uptake (**a**) before aging, (**b**) after the accelerated ageing, (**c**) after one year of natural ageing in temperate climate, and (**d**) at equilibrium before and after ageing.

Before ageing, the smallest absorptions were obtained for PS loadings up to 20 phr. Over this, the water uptake percentages were of 41.19% and 59.31% for 40 phr and 50 phr, respectively (Figure 11a). The post-curing phenomenon, after the thermal aging, has lead to the decrease of water uptake (Figure 11b). When the composites have been exposed to high temperature for longer period (7 days at 70 °C), the oxygen molecules from the air have been diffused into the surface, but some of them were immediately consumed by oxidation reactions to produce cross-linking, scission of the rubber chains or cross-links, and combination with molecular chains of rubber [51]. So, we can conclude that the water uptake is related with the cross-link densities of the composites. After 1 year of natural ageing in temperate climate, the water uptake percentages are increased, as it can be seen from Figure 11c,d. The phenomenon theory is talking about the difficulty of solvent molecule to penetrate the carbon linkages (C–C) because of the strong bonding and high rigidity [51,52]. But, in our study, the results show that the water molecules have easy penetrated the NR, the water uptake before and after one year of natural ageing in temperate climate being of 0.71% and 3.68%, respectively. In the same, the NR-PS composites have presented a great increase in water uptake percentages as comparing with the samples before and after thermal ageing, this indicating the decreasing of cross-link density and increasing of chain scissions. The disruption of C–C bonds in NR leads mainly to chain scission, while the disruption of C–O bonds from NR-PS causes the scission of grafting bonds [53]. Since the C–C and C–O bonding energies are comparable (346 vs. 358 kJ = mol), both phenomena can equally occur in degradation. Also, the appearance of some changes in the composites structure supported by FTIR results (decreasing of *cis*-1,4 double bonds number in the polyisoprene chain, the formation of hydroxyl group, the appearance of ketone and aldehyde groups and the increasing of glycosidic linkage) that can be interpreted as being consequences of the natural degradation, can also explain the increases in water absorption. From Figure 11d, clearly the results demonstrate the superiority of the natural ageing over thermal ageing. Also, it can be seen that high PS loadings create conditions for stronger degradation. Before and after immersion in water, all of the samples were weighed in order to determine the mass loss. The results are presented in Table 5.

Table 5. Mass lose before and after thermal and natural ageing.

Samples Codes	Mass Loss (%)		
	Before Ageing	After Accelerated Ageing	After 1 Year of Natural Ageing
NR	0.256	0.263	0.282
NR-10	0.263	0.286	0.293
NR-20	0.309	0.409	0.417
NR-30	1.059	0.963	1.095
NR-40	2.083	1.031	2.619
NR-50	2.971	1.126	5.960

From Table 5 it can be seen that NR does not presents significant mass losses before and after accelerated and natural ageing, higher being however after one year in temperate climate (0.026%). The filled NR has shows small mass differences, but increased with the PS amount, before and after ageing, the highest being of 2.99% for the case of NR-50 aged for one year in temperate climate. It should be noted that the mass losses for all of the samples loaded with PS over 30%, were smaller after thermal ageing than before. These results are the consequence of the post-curing effect and are in perfect agreement with those obtained after the mechanical properties evaluation. The results presented in Table 5 for NR and NR-PS after one year of natural ageing in temperate climate, show that higher PS content increases mass loss and enhances the degradation kinetics due to the starch hydrophilic nature, which leads to the moisture retaining. The higher is the starch content in the polymer, the higher is the moisture content that renders faster degradation. This can be correlated with the polymer sample's gross morphology, which was observed to be physically changed, the surface being roughened over the degradation period as it can be seen in SEM micrographs from Figure 10 [54].

3. Materials and Methods

3.1. Materials and Samples Preparation

The raw materials that were used in the experiments were as follows: (a) Natural rubber (NR) for pharmaceutical use, Crep from Sangtvon Rubber Ltd. (Nakhon Si Thammarat, Thailand), in the form of white rubber sheets (Mooney viscosity of 67.64 ML1+4 at 100 °C, volatile materials content of 0.5%, nitrogen content of 0.45%, percentage of ash of 0.25%, impurities content of 0.026%); (b) Soluble potato starch produced by Lach-Ner Ltd. (Neratovice, Czech Republic), water insoluble substances 0.28%; loss on drying 16.9%, easily biodegradable: BOD_5 −0.6 g/g and COD −1.2 mg/g); (c) Glycerine from SC Chimreactiv SRL (Bucharest, Romania) (free acidity 0.02%, density 1.26 g/cm^3, purity 99.5%); (d) IPPD antioxidant (4010 NA) *N*-isopropyl-*N*-phenyl-phenylene diamine from Dalian Richon Chem Co. Ltd. (Dalian, China), 98% purity, molecular mass: 493.6374 g/mol; (e) Peroxide Perkadox, 40 dibenzoyl peroxide, from AkzoNobel Chemicals (Deventer, The Netherlands) (density 160 g/cm^3, 3.8% active oxygen content, 40% peroxide content, pH 7); and, (f) TMPT DL 75 Luvomaxx, trimethylolpropane trimethacrylate as polyfunctional monomer from Lehmann&Voss&Co (Hamburg, Germany) (22% ash, pH 9.2, density 1.36 g/cm^3, 75 ± 3% active ingredient).

Composites that are based on NR and PS have been obtained according with the recipes that are presented in Table 6. Mixtures were cross-linked with peroxide in the presence of TMPT; a polyfunctional monomer was used here as curing co-agent.

Table 6. The recipes used for composites obtaining.

Ingredients (phr)	Mixtures Codes					
	NR	NR-10	NR-20	NR-30	NR-40	NR-50
Natural rubber (NR)	100	100	100	100	100	100
Starch	0	10	20	30	40	50
Glycerine	0	6	12	18	24	30
Peroxyde	8	8	8	8	8	8
TMPT	3	3	3	3	3	3
Antioxidant	1	1	1	1	1	1

PS was obtained by mixing at 70 °C, starch (50%), water (20%), and glycerine (30%) for 15 min at 50–100 rpm until the homogeneity was attended. After, the homogeneous mixture has been left to rest for 1 h, then being introduced in the oven at 80 °C for 22 h and at 110 °C for another 2 h. Finally, has been left to cool down for at least 16 h in a dry place.

The blends were prepared on an electrically roller mixer. The constituents were added in the following sequences and amounts: NR that has been mixed in the roller mixer for 2 min, PS and glycerine (mixing time between 5 and 30 min), antioxidant (mixing time: 1 min), peroxide, and TMPT (mixing time: 1 min). After all of the ingredients were added, they were homogenized for another 2 min and then removed from the roller mixer in the form of a sheet. The sheets have been cured using moulds and vulcanization press in order to obtain rubber plates with the sizes of 150 × 150 × 2 mm^3 being required for die punching test specimens. The compression temperature in the moulding machine was kept constant at 160 °C, for 20 min at a pressure of 300 kN. Cooling time was 10 min at 25 °C and 300 kN.

One specimen from each composite that have been obtained as above was subjected to accelerated ageing (thermal ageing) in an air-circulating oven at 70 °C for 168 h and another one to natural ageing for one year in temperate climate. Physical and mechanical properties, cross-linking density rate, behavior in aqueous environment, and structural and morphological investigations were done before and after ageing.

3.2. Laboratory Tests

3.2.1. Mechanical Characteristics Determining

Hardness, elasticity, 100% Modulus, tensile strength, elongation, and tearing strength were measured. Hardness was measured according to ISO 7619-1/2011 on 6 mm thick samples, while using a hardness tester. Elasticity (the rebound resilience) was evaluated according to ISO 4662/2009 also on 6 mm thick samples, while using the Schob test machine. Tensile strength and tearing strength tests were carried out with a Schopper strength tester at testing speed of 460 mm/min, using dumbbell shaped specimens according to ISO 37/2012, and angular test pieces (Type II) according to EN 12771/2003, respectively.

3.2.2. Sol-Gel Analysis

The sol-gel analysis, were performed on the cross-linked composites in order to determine the mass fraction of insoluble NR resulting from the network-forming cross-linking process. Samples having known mass were swollen in toluene for 72 h in order to remove any split fragments and un-reacted materials, and they were dried in air for six days and then in a laboratory oven at 80 °C for 12 h. Finally, samples were re-weighed and the gel fraction was calculated, as follows:

$$Gel_{fraction} = \frac{m_s}{m_i} \times 100 \tag{1}$$

where m_s and m_i are the mass of the dried sample after extraction and the initial mass of the sample, respectively [55,56].

3.2.3. Cross-Link Density Determining

The samples cross-link density was determined on the basis of equilibrium solvent-swelling measurements in toluene at 23–25 °C, by application of the modified Flory-Rehner equation for tetra functional networks. Samples having thicknesses of 2 mm were initially weighed (m_i) and immersed in toluene for 72 h. The swollen samples were dried to remove the solvent excess and weighed (m_g) being covered, in order to avoid toluene evaporation during weighing. Traces of solvent and other small molecules were eliminated by drying in air for six days and then in an oven at 80 °C for 12 h. Finally, the samples were weighed for the last time (m_s), and volume fractions of polymer in the samples at equilibrium swelling v_{2m} were determined from swelling ratio G, as follows:

$$v_{2m} = \frac{1}{1 + G} \tag{2}$$

$$G = \frac{m_g - m_s}{m_s} \times \frac{\rho_e}{\rho_s} \tag{3}$$

where, ρ_e and ρ_s are the densities of samples and solvent (0.866 g/cm^3 for toluene), respectively.

Densities were determined by hydrostatic weighing method, according to SR ISO 2781/2010. The cross-linking densities v, were determined from measurements in a solvent, while using the Flory–Rehner relationship:

$$v = -\frac{Ln(1 - v_{2m}) + v_{2m} + \chi_{12}v_{2m}^2}{V_1 \left(v_{2m}^{1/3} - \frac{v_{2m}}{2} \right)} \tag{4}$$

where, V_1 is the molar volume of solvent (106.5 cm^3/mol for toluene), v_{2m} is the volume fraction of polymer in the sample at the equilibrium swelling, and χ_{12} is the Flory-Huggins polymer-solvent interaction term (the value of and χ_{12} is 0.393 for toluene) [55,56].

3.2.4. Structural and Morphological Measurements

Structural changes of NR and NR-PS before and after ageing, were highlighted by FTIR measurements that have been done while using the TENSOR 27 (Bruker, Bremen, Germany) FTIR spectrophotometer by ATR measurement method. The spectra were obtained from 30 scans mediation, realized in absorption in the range of 4000–600 cm^{-1}, with a resolution of 4 cm^{-1}.

Morphological measurements of NR and NR-PS, also before and after ageing, were done on the samples surfaces while using the scanning electron microscope FEI/Phillips (FEI Company, Hillsboro, OR, USA). For this, the samples were placed on an aluminum mount, sputtered with gold palladium, and then scanned at an accelerating voltage of 30 kV.

3.2.5. Mechanisms of Degradation Reactions

Based on the results that were obtained by FTIR measurements, reaction mechanisms attributed to degradation induced by natural ageing were done. Generally, degradation can be induced by heat (thermal degradation), oxygen (oxidative and thermal-oxidative degradation), light (photo-degradation), and weathering (generally UV/ozone degradation) [44,45].

3.2.6. Water Uptake Tests

The water uptake tests were done in accordance with SR EN ISO 20344/2004 in order to study the water absorption on NR and NR-PS before and after ageing. For this, the samples were dried in a laboratory oven at 80 °C for 3 h and then were cooled at room temperature in desiccators before weighing. Water absorption tests were conducted by immersing the samples in distilled water in beaker and then maintaining at room temperature (25 ± 2 °C). After immersion, the samples were taken out from the water at periodic intervals and the wet surfaces were quickly dried while using a clean dry cloth or tissue paper before weighing. Absorption was calculated from the weight difference. The percentage of samples weight gaining was measured at different intervals of time. The water uptake was calculated, as follows:

$$\text{Water uptake (\%)} = \frac{m_t - m_i}{m_i} \times 100 \tag{5}$$

where, m_t is the weight of the sample immersed in water at time t and m_i is the initial weight of the oven-dried specimen.

Note: For mechanical, sol-gel analysis, cross-link determining, rubber-filler interaction, and water uptake tests before and after ageing, five samples were taken in work and the results are the averages of these five measurements.

4. Conclusions

A new elastomeric composite that is based on natural rubber and plasticized potato starch was obtained by the peroxide cross-linking method in the presence of the cross-linking co-agent trimethylolpropane trimethacrylate. The composite behavior before and after accelerated (thermal) aging and natural degradation in temperate climate for 1 year was investigated in terms of physical and mechanical characteristics. So, an improvement of mechanical properties, excepting 100% Modulus, due to the plasticized starch introduction, was observed before ageing. Also, by the increasing of the plasticized starch amount in the composite, excepting hardness, all other mechanical characteristics have started to decrease around the loading of 20 phr. Thus, we can say that the plasticized starch loading up to 20 phr has a reinforcing effect on natural rubber. By comparing the mechanical properties of the samples after ageing, we have observed the appearance of the post-curing during aging effect, reflected in minor modifications after the accelerated thermal ageing and notable negative modifications due to the natural process. These results are in accordance with those that were obtained by studying the gel fraction and cross-link density. Structural investigations through the FTIR technique before and after thermal and natural ageing were done. The intensity increasing of some bands corresponding to the formation of polar groups, such as carbonyl and hydroxyl between 1400–1260 cm^{-1}, carboxylate or conjugated ketone between 1690–1550 cm^{-1}, and aldehyde groups between 1736–1722 cm^{-1} were attributed to the degradation, sustaining the process efficiency, and are in a perfect accordance with the termination phase of reaction mechanisms that were achieved. Morphological investigations through the SEM technique have showed that, after aging, the plasticized starch amount increasing has been reflected in both surface roughness and more micro cracks appearance also. The water uptake and mass loss tests that have been done before and after ageing also have demonstrated that the natural degradation is favored by the addition of plasticized starch in the composite due to the appearance of some strongly hydrophilic groups from natural rubber, plasticized starch, and composite also. In the process of replacing conventional fillers, as silica and carbon black, with some natural fillers in order to obtain composites that are highly degradable in natural environment, starch can be considered as being a solution.

Author Contributions: The authors contribution was as fallows: conceptualization and methodology, M.E. and S.M.D.; Investigations, M.E., C.G. and S.M.D.; Writing-Original Draft Preparation, M.E.; Writing-Review & Editing, C.G.

Acknowledgments: The authors want to thank to Dumitru Marius Grivei for the given support for SEM investigations.

References

1. Craciun, G.; Manaila, E.; Stelescu, M.D. New Elastomeric Materials Based on Natural Rubber Obtained by Electron Beam Irradiation for Food and Pharmaceutical Use. *Materials* **2016**, *9*, 999. [CrossRef] [PubMed]

2. Kruželák, J.; Sýkora, R.; Hudec, I. Sulphur and peroxide vulcanisation of rubber compounds-overview. *Chem. Pap.* **2016**, *70*, 1533–1555. [CrossRef]

3. Kruželák, J.; Sýkora, R.; Hudec, I. Peroxide vulcanization of natural rubber. Part II: Effect of peroxides and co-agents. *J. Polym. Eng.* **2015**, *35*, 21–29. [CrossRef]

4. Henning, S.K.; Boye, W.M. Fundamentals of Curing Elastomers with Peroxides and Coagents II: Understanding the Relationship Between Coagent and Elastomer. *Rubber World* **2009**, *240*, 31–39.

5. Rajan, R.; Varghese, S.; George, K.E. Role of coagents in peroxide vulcanizatin of natural rubber. *Rubber Chem. Technol.* **2013**, *86*, 488–502. [CrossRef]

6. Vieira, E.R.; Mantovani, J.D.; de Camargo Forte, M.M. Comparison between peroxide/coagent cross-linking systems and sulfur for producing tire treads from elastomeric compounds. *J. Elastom. Plast.* **2013**, *47*, 347–359. [CrossRef]

7. Bucsi, A.; Szocs, F. Kinetics of radical generation in PVC with dibenzoyl peroxide utilizing high-pressure technique. *Macromol. Chem. Phys.* **2000**, *201*, 435–438. [CrossRef]

8. Alvarez Grima, M.M. Novel Co-Agents for Improved Properties in Peroxide Cure of Saturated Elastomers. Ph.D. Thesis, University of Twente, Enschede, The Netherlands, 16 February 2007.

9. Drobny, J.G. *Ionizing Radiation and Polymers: Principles, Technology and Applications*, 1st ed.; Elsevier Health Sciences; William Andrew: Norwich, NY, USA, 2012; pp. 88–91. ISBN 978-1-4557-7881-.

10. Dikland, H.G.; Ruardy, T.; Van der Does, L.; Bantjes, A. New coagents in peroxide vulcanization of EPM. *Rubber Chem. Technol.* **1993**, *66*, 693–711. [CrossRef]

11. Thitithammawong, A.; Uthaipan, N.; Rungvichaniwat, A. The effect of the ratios of sulfur to peroxide in mixed vulcanization systems on the properties of dynamic vulcanized natural rubber and polypropylene blends. *Songklanakarin J. Sci. Technol.* **2012**, *34*, 653–662. Available online: http://rdo.psu.ac.th/sjstweb/journal/34-6/0597-0721-34-6-653-662.pdf (accessed on 1 August 2018).

12. Manaila, E.; Craciun, G.; Stelescu, M.D.; Ighigeanu, D.; Ficai, M. Radiation vulcanization of natural rubber with polyfunctional monomers. *Polym. Bull.* **2014**, *71*, 57–82. [CrossRef]

13. Stelescu, M.D.; Manaila, E.; Craciun, G.; Dumitrascu, M. New Green Polymeric Composites Based on Hemp and Natural Rubber Processed by Electron Beam Irradiation. *Sci. World J.* **2014**, 684047. [CrossRef] [PubMed]

14. Manaila, E.; Stelescu, M.D.; Craciun, G.; Ighigeanu, D. Wood Sawdust/Natural Rubber Ecocomposites Cross-Linked by Electron Beam Irradiation. *Materials* **2016**, *9*, 503. [CrossRef] [PubMed]

15. Datta, J. Effect of Starch Fillers on the Dynamic Mechanical Properties of Rubber Biocomposite Materials. *Polym. Compos.* **2015**, *23*, 109–112. Available online: http://www.polymerjournals.com/pdfdownload/1189816.pdf (accessed on 26 June 2018). [CrossRef]

16. Datta, J.; Rohn, M. Structure, thermal stability and mechanical properties of polyurethanes, based on glycolysate from polyurethane foam waste prepared with use of 1,6-hexanediol as a glycol. *Polimery* **2008**, *53*, 871–875.

17. Liu, C.; Shao, Y.; Jia, D. Chemically modified starch reinforced natural rubber composites. *Polimer* **2008**, *49*, 2176–2181. [CrossRef]

18. Mente, P.; Motaung, T.E.; Hlangothi, S.P. Natural Rubber and Reclaimed Rubber Composites—A Systematic Review. *Polym. Sci.* **2016**, *2*, 1–19. [CrossRef]

19. Lomeli Ramírez, M.G.; Satyanarayana, K.G.; Iwakiri, S.; Bolzon de Muniz, G.; Tanobe, V.; Flores-Sahagun, T.S. Study of the properties of biocomposites. Part, I. Cassava starch-green coir fibers from Brazil. *Carbohydr. Polym.* **2011**, *86*, 1712–1722. [CrossRef]

20. Mali, S.; Grossmann, M.V.E.; García, M.A.; Martino, M.N.; Zaritzky, N.E. Antiplasticizing effect of glycerol and sorbitol on the properties of cassava starch Films. *Braz. J. Food Technol.* **2008**, *11*, 194–200. Available online: http://bj.ital.sp.gov.br/artigos/html/busca/pdf/v11n3a6107a.pdf (accessed on 26 June 2018).

21. Gaspar, M.; Benko, Z.; Dogossy, G.; Reczey, K.; Czigany, T. Reducing water absorption in compostable starch-based plastics. *Polym. Degrad. Stab.* **2005**, *90*, 563–569. [CrossRef]

22. Mathew, A.P.; Dufresne, A. Plasticized waxy maize starch: Effects of polyols and relative humidity on material properties. *Biomacromolecules* **2002**, *3*, 1101–1108. [CrossRef] [PubMed]

23. Van der Burg, M.C.; Van der Woude, M.E.; Janssen, L.P.B.M. The influence of plasticizer on extruded thermoplastics starch. *J. Vinyl Addit. Technol.* **1996**, *2*, 170–174. [CrossRef]

24. Ahmed, K.; Nizami, S.S.; Raza, N.Z.; Mahmood, K. Effect of micro-sized marble sludge on physical properties of natural rubber composites. *Chem. Ind. Chem. Eng. Q.* **2013**, *19*, 281–293. [CrossRef]

25. Thongsang, S.; Sombatsompop, N. Dynamic Rebound Behavior of Silica/Natural Rubber Composites: Fly Ash Particles and Precipitated Silica. *J. Macromol. Sci. Part B* **2007**, *46*, 825–840. [CrossRef]

26. Chen, R.S.; Ahmad, S.; Ab Ghani, M.H.; Salleh, M.N. Optimization of High Filler Loading on Tensile Properties of Recycled HDPE/PET Blends Filled with Rice Husk. *AIP Conf. Proc.* **2014**, *1614*, 46–51. [CrossRef]

27. Nourbakhsh, A.; Baghlani, F.F.; Ashori, A. Nano-SiO$_2$ filled rice husk/polypropylene composites: Physico-mechanical properties. *Ind. Crop. Prod.* **2011**, *33*, 183–187. [CrossRef]

28. Ahmed, K. Hybrid composites prepared from Industrial waste: Mechanical and swelling behavior. *J. Adv. Res.* **2015**, *6*, 225–232. [CrossRef] [PubMed]

29. Kukle, S.; Gravitis, J.; Putnina, A.; Stikute, A. The effect of steam explosion treatment on technical hemp fibres. In Proceedings of the 8th International Scientific and Practical Conference, Rezekne, Latvia, 20–22 June 2011; pp. 230–237.

30. Hanafi, I.; Muniandy, K.; Othman, N. Fatigue life, morphological studies, and thermal aging of rattan powder-filled natural rubber composites as a function of filler loading and a silane coupling agent. *BioResources* **2012**, *7*, 841–858. Available online: http://ojs.cnr.ncsu.edu/index.php/BioRes/article/view/BioRes_07_1_0841_Ismail_MO_Fatigue_Morpholog_Thermal_Aging_Rattan_Composite/1352 (accessed on 25 June 2018).

31. Abdul Kader, M.; Bhowmick, A.K. Acrylic rubber–fluorocarbon rubber miscible blends: Effect of curatives and fillers on cure, mechanical, aging, and swelling properties. *J. Appl. Polym. Sci.* **2003**, *89*, 1442–1452. [CrossRef]

32. Rattansom, N.; Prasertsri, S. Relationship among Mechanical properties, heat ageing resistance, cut growth behaviour and morphology in natural rubber: Partial replacement of clay with various type of carbon black at similar hardness level. *Polym. Test.* **2009**, *28*, 270–276. [CrossRef]

33. Ismail, H.; Ishiaku, U.S.; Azhar, A.A.; Mohd Ishak, Z.A. A comparative study of the effect of thermo-oxidative aging on the physical properties of rice husk ash and commercial fillers in epoxidized natural rubber compounds. *J. Elastom. Plast.* **1997**, *29*, 270–289. [CrossRef]

34. Azura, A.R.; Ghazali, S.; Mariatti, M. Effects of the filler loading and aging time on the mechanical and electrical conductivity properties of carbon black filled natural rubber. *J. Appl. Polym. Sci.* **2008**, *110*, 747–752. [CrossRef]

35. Khanlari, S.; Kokabi, M. Thermal stability, aging properties, and flame resistance of NR-based nanocomposite. *J. Appl. Polym. Sci.* **2010**, *119*, 855–862. [CrossRef]

36. Martins, A.F.; Visconte, L.L.Y.; Schuster, R.H.; Boller, F.; Nunes, H.C.R.; Nunes, R.C.R. Ageing Effect on Dynamic and Mechanical Properties of NR/Cel II Nanocomposites. *Kautsch. Gummi Kunstst.* **2004**, *57*, 446–451.

37. Somers, A.E.; Bastow, T.J.; Burgar, M.I.; Forsyth, M.; Hill, A.J. Quantifying rubber degradation using NMR. *Polym. Degradr. Stab.* **2000**, *70*, 31–37, . [CrossRef]

38. Riyajan, S.-A.; Sasithornsonti, Y.; Phinyocheep, P. Green natural rubber-g-modified starch for controlling urea release. *Carbohydr. Polym.* **2012**, *89*, 251–258. [CrossRef] [PubMed]

39. Fang, J.M.; Fowler, P.A.; Tomkinson, J.; Hill, C.A.S. The preparation and characterization of a series of chemically modified potato starches. *Carbohydr. Polym.* **2002**, *47*, 245–252. [CrossRef]

40. Mu, T.-H.; Zhang, M.; Raad, L.; Sun, H.-N.; Wang, C. Effect of α-Amylase Degradation on Physicochemical Properties of Pre-High Hydrostatic Pressure-Treated Potato Starch. *PLoS ONE* **2015**, *10*, e0143620. [CrossRef] [PubMed]

41. Eng, A.H.; Tanaka, Y.; Gan, S.N. FTIR studies on amino groups in purified Hevea rubber. *J. Nat. Rubber Res.* **1992**, *7*, 152–155.

42. Kim, I.-S.; Lee, B.-W.; Sohn, K.-S.; Yoon, J.; Lee, J.-H. Characterization of the UV Oxidation of Raw Natural Rubber Thin Film Using Image and FT-IR Analysis. *Elastom. Compos.* **2016**, *51*, 1–9. [CrossRef]

43. Coates, J. Interpretation of Infrared Spectra, A Practical Approach. In *Encyclopedia of Analytical Chemistry*, 1st ed.; Meyers, R.A., Ed.; John Wiley & Sons Ltd.: Chichester, UK, 2000; pp. 10815–10837. Available online: https://pdfs.semanticscholar.org/9203/59b562f615d9f68f8e57b6b6b505aa213174.pdf (accessed on 20 June 2018).

44. Polymer Properties Database. Available online: http://polymerdatabase.com/polymer%20chemistry/Thermal%20Degradation.html (accessed on 12 June 2018).

45. Alberti, A.; Bertini, S.; Gastaldi, G.; Iannaccone, N.; Macciantelli, D.; Torri, G.; Vismara, E. Electron beam irradiated textile cellulose fibres: ESR studies and derivatisation with glycidyl methacrylate (GMA). *Eur. Polym. J.* **2005**, *41*, 1787–1797. [CrossRef]

46. Connors, S.A. Chemical and Physical Characterization of the Degradation of Vulcanized Natural Rubber in the Museum Environment. Master's Thesis, Queen's University, Kingston, ON, Canada, 1998.

47. Palmas, P.; Le Campion, L.; Bourgeisat, C.; Martel, L. Curing and thermal ageing of elastomers as studied by H-1 Broadband and C-13 high-resolution solid-state NMR. *Polymer* **2001**, *42*, 7675–7683. [CrossRef]

48. Muniandy, K.; Hanafi, I.; Othman, N. Studies on natural weathering of rattan powder filled natural rubber composites. *BioResources* **2012**, *7*, 3999–4011. Available online: http://ojs.cnr.ncsu.edu/index.php/BioRes/article/view/BioRes_07_3_3999_Muniandy_Natural_Weathering_Rattan_Powder_Rubber/1653 (accessed on 20 May 2018).

49. Datta, R.N. Rubber-curing systems. In *Current Topics in Elastomers Research*; Bhowmick, A.K., Ed.; CRC Press: Boca Raton, FL, USA, 2008; ISBN 9781420007183.

50. Noriman, N.Z.; Ismail, H. The effects of electron beam irradiation on the thermal properties, fatigue life and natural weathering of styrene butadiene rubber/recycledacrylonitrile–butadiene rubber blends. *Mater. Des.* **2011**, *32*, 3336–3346. [CrossRef]

51. Rohana Yahya, Y.S.; Azura, A.R.; Ahmad, Z. Effect of Curing Systems on Thermal Degradation Behaviour of Natural Rubber (SMR CV 60). *J. Phys. Sci.* **2011**, *22*, 1–14. Available online: http://web.usm.my/jps/22-2-11/22.2.1.pdf (accessed on 26 June 2018).

52. Azura, A.R.; Muhr, A.H.; Thomas, A.G. Diffusion and reactions of oxygen during ageing for conventionally cured natural rubber vulcanisate. *Polym. Plast. Technol. Eng.* **2006**, *45*, 893–896. [CrossRef]

53. Pimolsiriphol, V.; Saeoui, P.; Sirisinha, C. Relationship among Thermal Ageing Degradation, Dynamic Properties, Cure Systems, and Antioxidants in Natural Rubber Vulcanisates. *Polym. Plast. Technol. Eng.* **2007**, *46*, 113–121. [CrossRef]

54. Hoque, M.E.; Ye, T.J.; Yong, L.C.; Dahlan, K.Z.M. Sago Starch-Mixed Low-Density Polyethylene Biodegradable Polymer: Synthesis and Characterization. *J. Mater.* **2013**, *365380*. [CrossRef]

55. Arroyo, M.; Lopez-Manchado, M.A.; Herrero, B. Organo-montmorillonite as substitute of carbon black in natural rubber compounds. *Polymer* **2003**, *44*, 2447–2453. [CrossRef]

56. Chenal, J.M.; Chazeau, L.; Guy, L.; Bomal, Y.; Gauthier, C. Molecular weight between physical entanglements in natural rubber: A critical parameter during strain-induced crystallization. *Polymer* **2007**, *48*, 1042–1046. [CrossRef]

9

Biodegradable and Toughened Composite of Poly(Propylene Carbonate)/Thermoplastic Polyurethane (PPC/TPU): Effect of Hydrogen Bonding

Dongmei Han [1,2], Guiji Chen [1,3], Min Xiao [1], Shuanjin Wang [1,*], Shou Chen [4], Xiaohua Peng [4] and Yuezhong Meng [1,2,*]

[1] The Key Laboratory of Low-carbon Chemistry & Energy Conservation of Guangdong Province/State Key Laboratory of Optoelectronic Materials and Technologies, Sun Yat-Sen University, Guangzhou 510275, China; handongm@mail.sysu.edu.cn (D.H.); chenguiji@kingfa.com.cn (G.C.); stsxm@mail.sysu.edu.cn (M.X.)

[2] School of Chemical Engineering and Technology, Sun Yat-Sen University, Guangzhou 510275, China

[3] Shanghai Kingfa Science and Technology Development Co., Ltd., Shanghai 201714, China

[4] Shenzhen Beauty Star Co., Ltd., Shenzhen 518112, China; chens@beautystar.cn (S.C.); alice@beautystar.cn (X.P.)

* Correspondence: wangshj@mail.sysu.edu.cn (S.W.); mengyzh@mail.sysu.edu.cn (Y.M.)

Abstract: The blends of Poly(propylene carbonate) (PPC) and polyester-based thermoplastic polyurethane (TPU) were melt compounded in an internal mixer. The compatibility, thermal behaviors, mechanical properties and toughening mechanism of the blends were investigated using Fourier transform infrared spectra (FTIR), tensile tests, impact tests, differential scanning calorimetry (DSC), scanning electron microscopy (SEM) and dynamic mechanical analysis technologies. FTIR and SEM examination reveal strong interfacial adhesion between PPC matrix and suspended TPU particles. Dynamic mechanical analyzer (DMA) characterize the glass transition temperature, secondary motion and low temperature properties. By the incorporation of TPU, the thermal stabilities are greatly enhanced and the mechanical properties are obviously improved for the PPC/TPU blends. Moreover, PPC/TPU blends exhibit a brittle-ductile transition with the addition of 20 wt % TPU. It is considered that the enhanced toughness results in the shear yielding occurred in both PPC matrix and TPU particles of the blends.

Keywords: Poly(propylene carbonate); thermoplastic polyurethane; compatibility; toughness

1. Introduction

Poly(propylene carbonate) (PPC) is a biodegradable aliphatic polycarbonate derived from carbon dioxide and propylene oxide. It has been paid much attention because of its high value-added fixation of CO_2, biodegradability and excellent oxygen barrier performance [1–3]. However, because PPC exhibits amorphous nature and possess low glass-transition temperature (Tg), it becomes brittle at low temperature and quickly loses strength at elevated temperature, which severely limits its wide application [4]. Many efforts have been devoted to overcoming these drawbacks. Chemical methods such as terpolymerization of carbon dioxide and propylene oxide with another monomer—for example, cyclohexene oxide (CHO) [5], maleic anhydride [6], or caprolactone [7]—have been employed. Simultaneously, physical blending of PPC with other polymers remains attractive because it is simple and cost-effective. Several PPC-base alloys have been prepared, starch [8,9], poly(ε-hydroxybutyrate-co-ε-hydroxyvalerate) (PHBV) [10], poly(3-hydroxybutyrate) (PHB) [11], poly(lactic acid) (PLA) [12], poly(butylene succinate) (PBS) [13], poly(ethylene-co-vinyl alcohol) (EVOH) [14] or others have been used to modify PPC. Nevertheless, most of these blends exhibited low miscibility between each component due to relatively weak inter-molecular interactions. For this reason, toughening of PPC is rarely achieved.

Thermoplastic polyurethane (TPU) elastomer, a linear segmented block copolymer consist of alternating soft segments and hard segments, are extensively used due to its superior properties, including high strength, high toughness, abrasion resistance, low temperature flexibility, biocompatibility, durability, biostability and so forth. [15–17]. Several works have been reported that TPU can be used as flexibilizer in many brittle polymers, such as polypropylene [18], poly(lactic acid) [19,20], polyacetal [21], poly(butylene terephthalate) [22] and so on.

Because of the similarly chemical structure, it is believed that PPC and TPU have better miscibility. Therefore, according to the phenomena that PPC becomes brittle at low temperature and loses its strength rapidly at high temperature, thermoplastic polyurethanes (TPU) are used to toughen PPC, in this work. Furthermore, the urethane moiety in hard segment of TPU can form hydrogen bonding with the carbonyl groups in PPC (Scheme 1), which in turn increase the interaction of molecular chain between PPC and TPU. In these connections, PPC was blended with TPU is expected to improve its toughness and thermal stability. These improvements will broaden the practical application of the new biodegradable PPC.

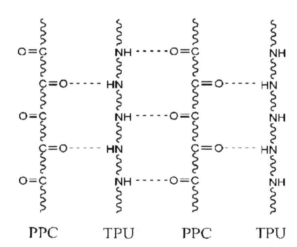

Scheme 1. Schematic illustration of hydrogen-bonding between the molecules of PPC and TPU.

2. Results and Discussion

2.1. FTIR Investigation

It is well known that the interaction between two components of the blend can be identified with FTIR technology. If two polymers are compatible, the absorption band shifts and broadening in the FTIR spectra indicates a chemical interaction between the molecular chains of one polymer and the other [23]. It was reported that there are the hydrogen-bonding interactions in their blends between Poly(propylene carbonate) (PPC) and other polymers, such as starch [8] and poly(ethylene-co-vinyl alcohol) (EVOH) [14]. Similarly, FTIR technique was used to confirm the existence of molecular chain interaction between the PPC and TPU in the blends. FTIR spectra of neat PPC, neat TPU and PPC/TPU blends with different TPU content are shown in Figure 1. The C=O absorption peak at 1749 cm^{-1} and N–H absorption peak at 3339 cm^{-1} were observed in the neat PPC and TPU respectively. In the blends, with the increase of PPC content, the N–H peak shifted towards lower wavenumber in the presence of PPC. It indicates a certain intermolecular hydrogen-bonding interaction between the N–H group of TPU and the carboxyl group of PPC. The peak for C=O group of blends become broader, which due to the shift of C=O groups in pure PPC and TPU is different. The wavenumber of C=O groups in TPU is lower than which in PPC.

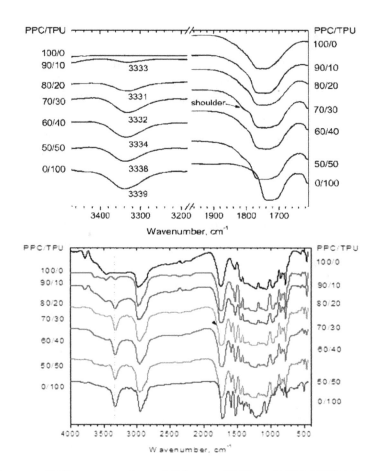

Figure 1. FTIR spectra of neat PPC, neat TPU and PPC/TPU blends.

2.2. Morphology Observation

Figure 2 represents the typical SEM micrographs of cryogenically fractured surfaces of the pure PPC and the PPC/TPU blends with various TPU contents. It can be seen that the pure PPC shows a smooth and uniform surface. For PPC/TPU blends, TPU particles are well dispersed in PPC matrix in fine droplets. The interfaces between PPC and TPU are fuzzy, demonstrating the good interfacial adhesion between suspended TPU particles and PPC matrix. This indicates the improved compatibility between PPC and TPU and which owe to the intermolecular hydrogen-bonding formation. This result is quite different from Huang's work and they reported that PLA is incompatible with TPU, although PLA also has carboxyl groups [24].

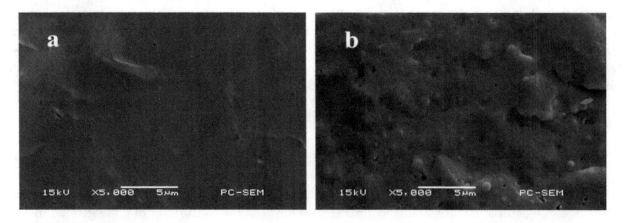

Figure 2. SEM micrographs of PPC and PPC/TPU blends. (**a**) Neat PPC, (**b**) PPC/10% TPU.

2.3. Thermal Behaviors

Figure 3 shows the DSC traces of the neat PPC, neat TPU and PPC/TPU blends recorded on the second heating step. The soft segment of TPU has a glass transition temperature (Tg) of $-23.5\,^{\circ}$C, while there is no obvious glass transition for the hard segment of TPU. The Tg of PPC increases with increasing TPU content. The value reaches 36.1 $^{\circ}$C with the addition of 50 wt % TPU, while that of neat PPC is 31.7 $^{\circ}$C. The increasing of PPC's Tg is presumably due to the hydrogen-bonding as indicated by FTIR investigation. Physical-crosslinking between PPC and TPU molecular chains greatly constrain the molecular movement of PPC matrix. On the contrary, the Tg value for soft segment of TPU decreases with increasing PPC content. It declines to about $-34.2\,^{\circ}$C with the addition of 50 wt % PPC. When PPC component increases to 90 wt %, the Tg value of TPU further decreases to as low as $-45.6\,^{\circ}$C. This is because the strong hydrogen-bonding between carbonyl of soft segment and the urothen groups of the TPU hard segment. The strong hydrogen-bonding decreases in turn constraining of soft segment of TPU, resulting in the decrease of Tg.

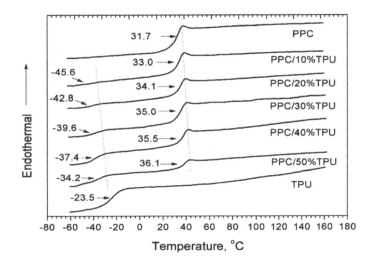

Figure 3. DSC curves of neat PPC, neat TPU and PPC/TPU blends.

The Vicat softening temperatures of the neat PPC, neat TPU and PPC/TPU blends are plotted in Figure 4. The Vicat temperatures shift to higher value with the incorporation of TPU. The value increases about 5 $^{\circ}$C with every 10 wt % TPU addition. The improvement in thermal stability of PPC will certainly broaden the practical application of PPC.

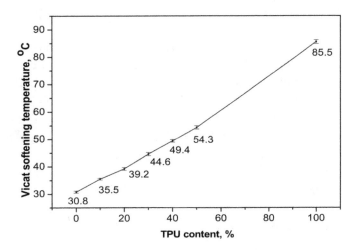

Figure 4. Effect of TPU content on Vicat softening temperatures of PPC/TPU blends.

2.4. Mechanical Properties

The stress-strain curves for the neat PPC, neat TPU and PPC/TPU blends are given in Figure 5. It is apparent that TPU has a low modulus but high strain and stress at break. All PPC/TPU blends exhibit yield characteristics, nevertheless, PPC shows the lowest elongation at break compared with other PPC/TPU blends. The yield strength for each PPC/TPU blend is higher than that of neat PPC, indicating the formation of the hydrogen-bonding between PPC matrix and TPU particles. The addition of TPU significantly changes the tensile behavior of the PPC, especially the enhancement of elongation at break.

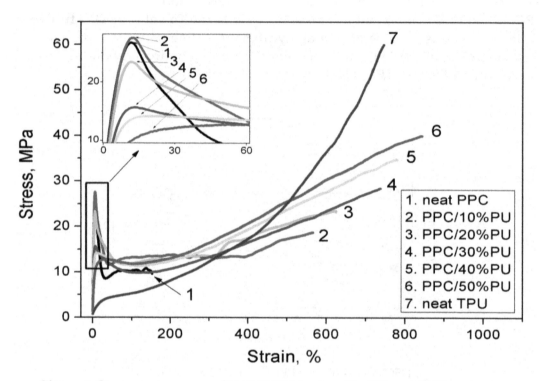

Figure 5. Stress-strain curves of PPC/TPU blends with different TPU contents.

The yield stress, stress at break, elongation at break and tensile energy to break values of neat PPC, neat TPU and PPC/TPU blends are listed in Table 1. It can be seen that the stress at break, elongation at break increase dramatically due to the hydrogen-bonding between PPC and TPU of the blends. Many reports have disclosed that the yield stress of blends decrease with the addition of TPU [20,22,25]. In this work, however, the yield stress of PPC/10%TPU blend is little higher than that of the neat PPC due to the hydrogen-bonding. Because of the lower modulus of TPU, the yield stress of PPC/TPU blends decreases with increasing TPU. Moreover, both elongation at break and tensile energy to break increase sharply with the addition of TPU. It is interesting to note that the PPC/TPU blend with only 10 wt % TPU exhibits a very high elongation at break of 566.2%, together with a little increase of yield stress. The tensile energy increases up to about 5 times for PPC/10%TPU blend and to more than 11 times for PPC/50%TPU blend. PPC/40%TPU and PPC/50%TPU blends even show higher tensile energy and strain to break than that of neat TPU. The results demonstrate that both toughness and tensile strength of PPC can be improved by the simply introduction of TPU.

The charpy impact strength of the neat PPC and PPC/TPU blends as a function of TPU content and is shown in Figure 6. The impact strength of the PPC/TPU blends increases slightly with the addition of 10 wt % TPU. The impact strength increases dramatically with further increasing TPU content. Some of the impact specimens cannot be broken completely during impact testing. The PPC/TPU blends exhibit a brittle-ductile transition behavior when 20 wt % TPU added because the maximum impact strength of our impact tester is 125 kJ/m^2 for standard specimen according to ASTM D256-05. These demonstrate the effective improvement of impact strength of PPC by the incorporation of TPU.

Table 1. Mechanical property of PPC/TPU blends with different TPU contents.

TPU Content, %	Yield Stress, MPa	Stress at Break, MPa	Elongation at Break, %	Tensile Energy to Break, MJ/m^3	Stress at 100%, MPa	Stress at 200%, MPa	Stress at 300%, MPa
0	26.5 ± 1.2	9.60 ± 0.6	180 ± 20	17.1 ± 3	10.3 ± 0.3	-	-
10	27.6 ± 0.9	18.7 ± 0.7	570 ± 30	82.2 ± 9	13.2 ± 0.2	13.8 ± 0.2	13.3 ± 0.3
20	23.5 ± 0.8	23.2 ± 0.4	620 ± 40	103 ± 7	11.9 ± 0.4	12.9 ± 0.3	13.5 ± 0.2
30	18.6 ± 0.5	27.5 ± 0.7	750 ± 50	124 ± 11	9.9 ± 0.2	10.7 ± 0.2	13.3 ± 0.2
40	14.3 ± 0.3	33.5 ± 2	780 ± 30	159 ± 9	11.4 ± 0.4	12.4 ± 0.3	15.2 ± 0.4
50	12.8 ± 0.5	40.0 ± 3	840 ± 50	195 ± 14	11.8 ± 0.2	12.9 ± 0.4	16.4 ± 0.4

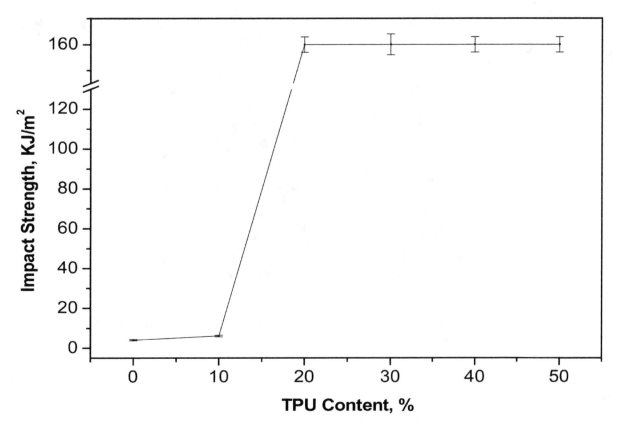

Figure 6. Impact strength of neat PPC and PPC/TPU blends as a function of TPU contents.

2.5. Toughening Mechanism

The results of tensile and impact experiments show that the highly effective toughness of TPU on PPC, especially for the PPC/TPU blends with more than 20 wt % TPU. Figure 7 shows the micrographs of the impact fractured surfaces of PPC/TPU blends. According to neat PPC (Figure 2a), with the addition of 10 wt % TPU, the fractured surface seems very roughness (Figure 7a). From the enlarged image (Figure 7b), the suspended TPU droplets can be clearly seen with observable gaps between PPC matrix and TPU particles. The dispersed TPU particles play a crucial role as stress concentration point within PPC matrix, where the deformation is much easier to happen than other areas within PPC/TPU blends. TPU particles serve as the plasticizing deformation phase because of its high elasticity. The deformation can absorb impact energy, therefore, the impact strength of PPC/10%TPU blend increases about 50 wt % compared with that of neat PPC. The PPC/TPU blends tend to become a tough failure with increasing TPU content. The dispersed TPU particles, as a large number of stress concentration points within PPC matrix, initiate crazing and shear banding, thus, absorb a large number of impact energy and impede crack growth under a strong external shock. PPC/30%TPU blend shows a typical tough fracture (Figure 7c). It can be seen that the cavitation and extensive plastic deformation of both PPC matrix and suspended TPU particles, implying that shear yielding is more predominant than crazing of PPC matrix and TPU particles.

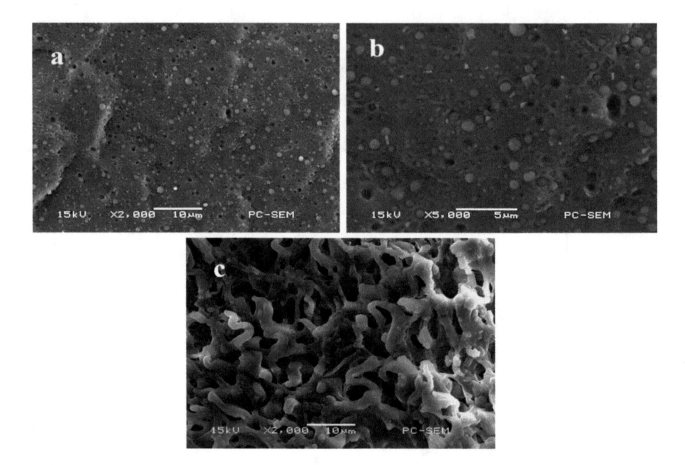

Figure 7. SEM micrographs of the impact fractured surfaces of PPC/TPU blends: (**a**) PPC/10%TPU, (**b**) enlarge view of a, (**c**) PPC/30%TPU.

In order to investigate the toughening mechanism, the different regions on the fractured surface of the impact fractured PPC/30%TPU blend sample is examined under SEM (Figure 8). The sample was not cracked completely and the cryogenically fractured surface along the impact direction (Figure 8a). Figure 8b–d shows different deformation states during impact testing. In the beginning of impact testing, the cracks were easily induced because of the high impact velocity and the stress concentration near the notch. The sample started to fracture immediately without large yielding as depicted in Figure 8b. Many voids around TPU particles were generated and some TPU particles showed small deformation and tended to be pulled out of PPC matrix. Thereafter, the impact energy was dissipated and the impact velocity decreased slightly. As shown in Figure 8c, further deformation of TPU particles can be seen clearly in point B. The PPC matrix around TPU particles deformed much obviously with shear yielding. This process absorbed the most energy of the impact fracture. As the impact velocity decreased sequentially, the sample showed again a tough failure in point C as shown in Figure 8d. The toughing mechanism can be explained as follows. First, TPU particles decrease the yield stress of PPC matrix following to generate shear yielding. Secondly, the PPC matrix deforms before cracking due to the decrease of impact velocity, which is attributed to the impact energy absorption from point A to point B. The micrograph of the freeze-fractured surface of the impact unfractured region is shown in Figure 8e. From this figure, it can be seen that the fractured surface is much smoother than other regions, only a few gaps between PPC matrix and TPU particles can be observed. Because the most of applied impact energy is dissipated before point D, there is not enough impact energy to destroy the blend sample with high TPU contents.

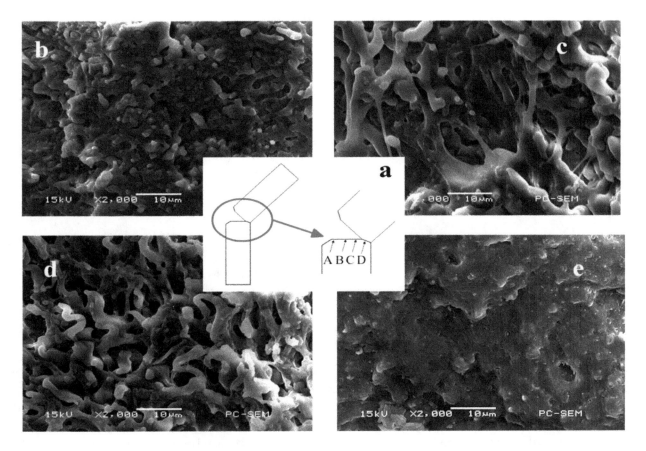

Figure 8. SEM micrographs of different regions on the impact fractured surface of the PPC/30%TPU sample. (**a**) Different regions of the impact fractured surface, (**b–e**) correspond to A–D in a, respectively.

2.6. Dynamic Mechanical Analysis

The dynamic mechanical analysis (DMA) can be used to characterize the secondary motion and properties of polymer molecules at low temperature [26], except for the characterization of the polymerized glass transition temperature.

The plots of storage modulus as a function of temperature for PPC, TPU and PPC/TPU are showed in Figure 9. It can be seen that in addition to a stress relaxation between 10 and 50 °C, pure PPC also has a secondary transition in the low temperature range of −100°C to −50 °C. Similarly, in addition to glass transition between −50 °C and 0 °C, pure TPU has a large secondary transition between −150 °C and −60 °C, which is important to the excellent flexibility and good performance at low temperature. Where When the TPU is in a highly elastic state, the storage modulus of the blends material gradually decreases as the TPU content increases, which indicates that the flexibility of the blends material is improved, which is one of the reasons why stretching and impact toughness is getting better with the increase of the TPU content.

Figure 10a shows the tg δ curves of PPC, TPU and their blends in the higher temperature region. The two peaks in the curve represent the glass transitions of PPC and TPU, respectively. The high temperature is according to the glass transition of PPC. The lower temperature is according to the glass transition of TPU. With the increase of TPU content, the corresponding peak of PPC becomes lower and lower and the corresponding peak of TPU becomes higher and higher. The corresponding value of the peak position is the glass transition temperature of the material and which is listed in Table 2. For PPC, as the content of TPU increases, the glass transition temperature of PPC increases gradually, from 44.7 °C of pure PPC to 53.5 °C when 50% TPU is added and 8.8 °C is increased. For TPU, as the content of PPC increases, the glass transition temperature of TPU gradually decreases, which is consistent with the results obtained by DSC analysis.

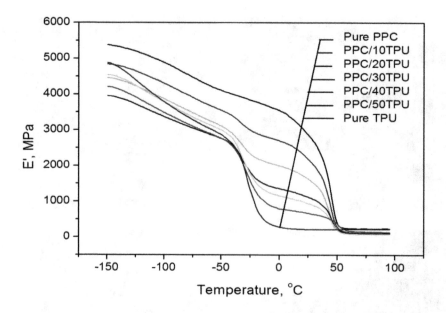

Figure 9. Storage modulus versus temperature for PPC, TPU and PPC/TPU blends.

It could be seen the tg δ curves of PPC, TPU and its blends in the lower temperature region in Figure 10b. It shows that both PPC and TPU have a β-transformation in the low temperature region and with the increase of TPU content, the β-transition peak of PPC in the blends material is becoming smaller, while the β-transition of TPU is more and more obvious. It indicates that as the TPU content increases, the high-temperature β-transition in the blends slowly disappears and replaced by a lower-temperature β-transition. The shift of β to low temperature means that the low temperature toughness of the material is improved. Therefore, the brittle-ductile transition temperature of the alloy material gradually decreases with the increase of the TPU content, showing better flexibility at low temperature.

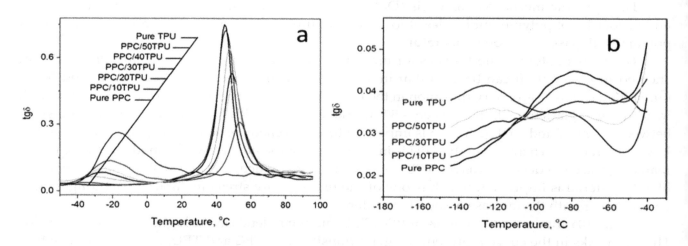

Figure 10. tg δ versus temperature for PPC, TPU and PPC/TPU blends: (**a**) higher temperature, (**b**) lower temperature.

Table 2. Glass transition temperature for PPC, TPU and their blends from dynamic mechanical analysis (DMA).

TPU Content, %	0	10	20	30	40	50	100
Tg (PPC), °C	44.7	45.4	47.5	48.9	50.2	53.5	-
Tg (TPU), °C	-	−35.9	−31.6	−26.3	−24.8	−22.1	−16.8

3. Materials and Methods

3.1. Materials

The high molecular weight Poly(propylene carbonate) (PPC), with a number-average molecular weight (Mn) of 109,000 Da and a polydispersity (PD) of 1.91, was provided by Tian'guan Enterprise Group Co. (Henan, China). Polyester-based TPU (grade S685AL, density = 1.2 g/cm^3), was obtained from Kin Join Co., Ltd. (Taiwan). Both PPC and TPU pellets were dried in a vacuum oven for 24 h at 80 °C before blending.

3.2. Preparation of PPC/TPU Blends

The PPC/TPU blends with weight ratios of 100/0, 90/10, 8/20, 70/30, 60/40, 50/50 and 0/100 were prepared in a Haake Rheomix RT 600 mixer (Haaker, Germany). The mixing was carried out at 160 °C with a rotary speed of 40 rpm for 7 min. The prepared blends were then melt pressed at 170 °C into standard dumbbell tensile bars (ASTM D638) and standard V-shaped notched impact bars (ASTM D256-05).

3.3. Fourier Transform Infrared (FTIR) Spectra

FTIR spectra were recorded on a Perkin-Elmer FTIR-100 spectrometer (Perkin-Elmer, Waltham, MA, USA). Samples were first melt pressed to thin film and then scanned with wavenumbers from 4000 to 400 cm^{-1} with a resolution of 4.0 cm^{-1}.

3.4. Scanning Electron Microscopy (SEM) Observation

The morphologies of the blends were observed using SEM (Jeol JSM-6380, JEOL, Tokyo, Japan). All of the Specimens were fractured in liquid nitrogen; the fracture surface was then coated with a thin layer of gold. The fracture surface after impact tests were also observed by the same SEM apparatus.

3.5. Differential Scanning Calorimetry (DSC) Investigation

The samples were scanned with differential scanning calorimetry (DSC, Netzsch 204, Selb, Bavaria, Germany). The temperature was initially heated from room temperature to 190 °C at a rate of 10 °C/min, maintained at 190 °C for a period of 3 min to eliminate previous thermal history, then cooled down to −100 °C at 10 °C/min, maintained at −100 °C for 3 min. The samples were subsequently scanned back to 190 °C with a heating rate of 10 °C/min. All the scanning processes were under a protective atmosphere of N2.

3.6. Vicat Softening Temperature

The Vicat softening temperatures of the blends were measured by a Vicat tester (New SANS, Shen Zhen, China) at 10 N load and heating rate of 50 °C/h according to ASTM D 1525. Six specimens have been examined for each blend and average value was reported accordingly.

3.7. Static Tensile Properties

The static tensile properties were investigated using a universal mechanical testing machine (New SANS) at 25 °C with a relative humidity of 50 ± 5%. The crosshead speed was set at 50 mm/min. Five specimens of each sample were measured and the average results were recorded. All the samples were conditioned at 25 °C and 50% RH for 24 h before testing.

3.8. Charpy Impact Strength

The Charpy impact strengths of notched specimens were determined using a Charpy impact tester (SANS ZBC-4B, MTSSANS, Eden Prairie, Minnesota, USA) at 25 °C and 50% RH according to ASTM D256-05. The maximum velocity of the pendulum is 2.9 m/s, generating impact energy of 4 J.

Each blend of five specimens was measured and the average results were recorded. All the samples were conditioned for 24 h at 25 °C and 50% RH before testing.

3.9. Dynamic Mechanical Analysis

The dynamic mechanical analysis of the material was carried out by a dynamic mechanical analyzer (DMA, model 242D, Netzsch, Selb, Germany). The test was conducted in a single cantilever vibration mode. The vibration frequency was 3.3 Hz and the amplitude was 100 μm. The size of the specimen is 35 mm × 10 mm × 1 mm and the temperature is set from −150 °C to 150 °C, with a heating rate of 5 °C/min.

4. Conclusions

PPC/TPU blends can be readily prepared by melt blending. Experimental results indicate the existence of intermolecular hydrogen-bonding between the carbonyl groups of PPC and the urethane groups of TPU. The SEM examination and thermal analysis of the PPC/TPU blends show a compatible blend system. The incorporation of TPU can obviously enhance the thermal stability and broaden the application temperature of PPC according to DSC and Vicat softening temperature investigation. Moreover, the toughness of the blends increases dramatically with increasing TPU content. The blend exhibits a brittle-ductile transition temperature at about 25 °C with the addition of 20 wt % TPU. Based on the experiment results, we have proposed a toughening mechanism of TPU for PPC matrix. The shear yielding occurs in both suspended TPU particles and PPC matrix, which accounts for the sharp increase in the mechanical properties of PPC.

Author Contributions: D.H., G.C., Y.M., S.W. and M.X. conceived and designed the experiments; D.H. and G.C. performed the experiments and analyzed the data; S.C. and X.P. contributed analysis tools. D.H. and G.C. wrote this paper.

Acknowledgments: This work was supported by funding from National Natural Science Foundation of China (No. 21376276, 51673131); Guangdong Province Sci & Tech Bureau (Nos. 2017B090901003, 2016B010114004, 2016A050503001); the Special-funded Program on National Key Scientific Instruments and Equipment Development of China (No. 2012YQ230043); and Fundamental Research Funds for the Central Universities (171gjc37).

References

1. Inoue, S.; Tsuruta, T.; Kobayash, M.; Koinuma, H. Reactivities of Some Organozinc Initiators for Copolymerization of Carbon Dioxide and Propylene Oxide. *Die. Makromol. Chem.* **1972**, *155*, 61–73. [CrossRef]
2. Darensbourg, D. Making Plastics from Carbon Dioxide: Salen Metal Complexes as Catalysts for the Production of Polycarbonates from Epoxides and CO_2. *J. Chem. Rev.* **2007**, *107*, 2388–2410. [CrossRef] [PubMed]
3. Luinstra, G.A. Poly(Propylene Carbonate), Old Copolymers of Propylene Oxide and Carbon Dioxide with New Interests: Catalysis and Material Properties. *Polym. Rev.* **2008**, *48*, 192–219. [CrossRef]
4. Inoue, S.; Tsuruta, T. Synthesis and thermal degradation of carbon dioxide-epoxide copolymer. *Appl. Polym. Symp.* **1975**, *26*, 257.
5. Ren, W.M.; Zhang, X.; Liu, Y.; Li, J.F.; Wang, H.; Lu, X.B. Highly Active, Bifunctional Co(III)-Salen Catalyst for Alternating Copolymerization of CO_2 with Cyclohexene Oxide and Terpolymerization with Aliphatic Epoxides. *Macromolecules* **2010**, *43*, 1396–1402. [CrossRef]
6. Song, P.F.; Xiao, M.; Du, F.G.; Wang, S.J.; Gan, L.Q.; Liu, G.Q.; Meng, Y.Z. Synthesis and properties of aliphatic polycarbonates derived from carbon dioxide, propylene oxide and maleic anhydride. *J. Appl. Polym. Sci.* **2008**, *109*, 4121–4129. [CrossRef]
7. Seong, J.E.; Na, S.J.; Cyriac, A.; Kim, B.W.; Lee, B.Y. Terpolymerization of CO_2 with Propylene Oxide and ε-Caprolactone Using Zinc Glutarate Catalyst. *Macromolecules* **2003**, *36*, 8210–8212. [CrossRef]
8. Zeng, S.S.; Wang, S.J.; Xiao, M.; Han, D.M.; Meng, Y.Z. Preparation and properties of biodegradable blend containing poly (propylene carbonate) and starch acetate with different degrees of substitution. *Carbohydr. Polym.* **2011**, *86*, 1260–1265. [CrossRef]

9. Ge, X.C.; Li, X.H.; Zhu, Q.; Li, L.; Meng, Y.Z. Preparation and properties of biodegradable poly(propylene carbonate)/starch composites. *Polym. Eng. Sci.* **2004**, *44*, 2134–2140. [CrossRef]

10. Tao, J.; Song, C.J.; Cao, M.F.; Hu, D.; Liu, L.; Liu, N.; Wang, S.F. Thermal properties and degradability of poly(propylene carbonate)/poly(β-hydroxybutyrate-*co*-β-hydroxyvalerate) (PPC/PHBV) blends. *Polym. Degrad. Stab.* **2009**, *94*, 575–583. [CrossRef]

11. Yang, D.Z.; Hu, P. Miscibility, crystallization, and mechanical properties of poly(3-hydroxybutyrate) and poly(propylene carbonate) biodegradable blends. *J. Appl. Polym. Sci.* **2008**, *109*, 1635–1642. [CrossRef]

12. Ma, X.F.; Yu, J.G.; Wang, N. Compatibility characterization of poly (lactic acid)/poly (propylene carbonate) blends. *J. Polym. Sci. Part B Polym. Phys.* **2006**, *44*, 94–101. [CrossRef]

13. Pang, M.Z.; Qiao, J.J.; Jiao, J.; Wang, S.J.; Xiao, M.; Meng, Y.Z. Miscibility and properties of completely biodegradable blends of poly(propylene carbonate) and poly(butylene succinate). *J. Appl. Polym. Sci.* **2008**, *107*, 2854–2860. [CrossRef]

14. Jiao, J.; Wang, S.J.; Xiao, M.; Xu, Y.; Meng, Y.Z. Processability, property, and morphology of biodegradable blends of poly (propylene carbonate) and poly (ethylene-*co*-vinyl alcohol). *Polym. Eng. Sci.* **2007**, *47*, 174–180. [CrossRef]

15. Simmons, A.; Hyvarinen, J.; Poole-Warren, L. The effect of sterilisation on a poly (dimethylsiloxane)/poly (hexamethylene oxide) mixed macrodiol-based polyurethane elastomer. *Biomaterials* **2006**, *27*, 4484–4497. [CrossRef] [PubMed]

16. Lebedev, E.V.; Ishchenko, S.S.; Denisenko, V.D.; Dupanov, V.O.; Privalko, E.G.; Usenko, A.A.; Privalko, V.P. Physical characterization of polyurethanes reinforced with the in situ-generated silica-polyphosphate nano-phase. *Compos. Sci. Technol.* **2006**, *66*, 3132–3137. [CrossRef]

17. Xian, W.Q.; Song, L.N.; Liu, B.H.; Ding, H.L.; Li, Z.; Cheng, M.P.; Ma, L. Rheological and mechanical properties of thermoplastic polyurethaneelastomer derived from CO_2 copolymer diol. *J. Appl. Polym. Sci.* **2018**, *135*, 45974. [CrossRef]

18. Lu, Q.W.; Macosko, C.W.; Lu, Q.W.; Macosko, C.W. Comparing the compatibility of various functionalized polypropylenes with thermoplastic polyurethane (TPU). *Polymer* **2004**, *45*, 1981–1991. [CrossRef]

19. Feng, F.; Ye, L. Morphologies and mechanical properties of polylactide/thermoplastic polyurethane elastomer blends. *J. Appl. Polym. Sci.* **2011**, *119*, 2778–2783. [CrossRef]

20. Li, Y.J.; Shimizu, H. Toughening of polylactide by melt blending with a biodegradable poly (ether) urethane elastomer. *Macromol. Biosci.* **2007**, *7*, 921–928. [CrossRef] [PubMed]

21. Palanivelu, K.; Balakrishnan, S.; Rengasamy, P. Thermoplastic polyurethane toughened polyacetal blends. *Polym. Test.* **2000**, *19*, 75–83. [CrossRef]

22. Palanivelu, K.; Sivaraman, P.; Reddy, M.D. Studies on thermoplastic polyurethane toughened poly (butylene terephthalate) blends. *Polym. Test.* **2002**, *21*, 345–351. [CrossRef]

23. Peng, S.W.; Wang, X.Y.; Dong, L.S. Special interaction between poly (propylene carbonate) and corn starch. *Polym. Compos.* **2005**, *26*, 37–41. [CrossRef]

24. Han, J.J.; Huang, H.X. Preparation and characterization of biodegradable polylactide/thermoplastic polyurethane elastomer blends. *J. Appl. Polym. Sci.* **2011**, *120*, 3217–3223. [CrossRef]

25. Chang, F.C.; Yang, M.Y. Mechanical fracture behavior of polyacetal and thermoplastic polyurethane elastomer toughened polyacetal. *Polym. Eng. Sci.* **1990**, *30*, 543–552. [CrossRef]

26. Borah, J.S.; Chaki, T.K. Dynamic mechanical, thermal, physico-mechanical and morphological properties of LLDPE/EMA blends. *J. Polym. Res.* **2011**, *18*, 569–578. [CrossRef]

Evaluation of Mechanical and Interfacial Properties of Bio-Composites Based on Poly(Lactic Acid) with Natural Cellulose Fibers

Laura Aliotta [1,2], **Vito Gigante** [1], **Maria Beatrice Coltelli** [1], **Patrizia Cinelli** [1,2,*] **and Andrea Lazzeri** [1,2,*]

[1] Department of Civil and Industrial Engineering, University of Pisa, Via Diotisalvi, 2, 56122 Pisa, Italy; laura.aliotta@dici.unipi.it (L.A.); vito.gigante@dici.unipi.it (V.G.); maria.beatrice.coltelli@unipi.it (M.B.C.)

[2] Interuniversity National Consortium of Materials Science and Technology (INSTM), Via Giusti 9, 50121 Florence, Italy

[*] Correspondence: patrizia.cinelli@unipi.it (P.C.); andrea.lazzeri@unipi.it (A.L.)

Abstract: The circular economy policy and the interest for sustainable material are inducing a constant expansion of the bio-composites market. The opportunity of using natural fibers in bio-based and biodegradable polymeric matrices, derived from industrial and/or agricultural waste, represents a stimulating challenge in the replacement of traditional composites based on fossil sources. The coupling of bioplastics with natural fibers in order to lower costs and promote degradability is one of the primary objectives of research, above all in the packaging and agricultural sectors where large amounts of non-recyclable plastics are generated, inducing a serious problem for plastic disposal and potential accumulation in the environment. Among biopolymers, poly(lactic acid) (PLA) is one of the most used compostable, bio-based polymeric matrices, since it exhibits process ability and mechanical properties compatible with a wide range of applications. In this study, two types of cellulosic fibers were processed with PLA in order to obtain bio-composites with different percentages of microfibers (5%, 10%, 20%). The mechanical properties were evaluated (tensile and impact test), and analytical models were applied in order to estimate the adhesion between matrix and fibers and to predict the material's stiffness. Understanding these properties is of particular importance in order to be able to tune and project the final characteristics of bio-composites.

Keywords: bio-composites; mechanical properties; poly(lactic acid); cellulose fibers

1. Introduction

The increasing environmental awareness coupled with the circular economy policy, supported by new regulations, are driving plastic industries as well as consumers toward the selection of ecologically friendly raw materials for their plastic products. Several products developed for large application fields are based on natural fibers in composites with a polypropylene matrix [1]. These materials are not compostable and are hardly recyclable; thus, work is in progress to investigate new composites with biopolymers as polymeric matrices (bio-composites), offering the advantage of bio-recycling options at the end of their service life through composting or anaerobic digestion. In this contest, bio-based polymers reinforced with natural fibers are beneficial to prepare biodegradable composite materials [2]. The most used biopolymers for this application are poly(lactic acid) (PLA), cellulose esters, polyhydroxyalkanoates (PHAs), and starch-based plastics [3,4].

In applications where biodegradability offers clear advantages for customers and the environment, such as single-use applications (packaging and agriculture), it is expected that the demand for these biopolymers will increase [5–7].

In this context, poly(lactic acid) (PLA) is certainly one of the best candidates, being compostable and produced from renewable resources such as sugar beets or corn starch [8,9]. In addition to its biodegradability and renewability, PLA exhibits at room temperature a Young's modulus of about 3 GPa, a tensile strength between 50 and 70 MPa with an elongation at break of about 4%, and an impact strength close to 2.5 kJ/m^2 [10].

Although PLA is considered a sustainable alternative to traditional petroleum-based plastics, many drawbacks must be overcome in order to enlarge its application field. In particular, PLA has a relatively higher cost (2.5–3.0 Euro/Kg) compared to commodity petro-derived polymers (1 Euro/Kg), it has low flexibility, bad impact resistance, low thermal stability (due to its high glass transition temperature, $T_g \approx 60\ °C$), and low crystallization rates that could limit its applications [11].

Generally, composite materials show enhanced mechanical and physical properties when compared to their individual composite components [12,13]. However, especially when the fibers are very short and randomly oriented, the resulting composite does not necessarily provide enhanced properties. In this case, the benefit of composite production is envisaged in cost savings, lighter weight, and promoted degradability.

The study of the interaction between the fiber and the polymeric matrix in a composite plays an important role because it influences both physical and mechanical properties of the final materials. In particular, the adhesion—that is the ability to transfer stresses across the interface—is often related to a combination of different factors such as the interface thickness, the interphase layer, the adhesion strength, and the surface energy of the fibers [14–16].

Natural fibers have many advantages compared to synthetic ones. They are recyclable, biodegradable, renewable, have relatively high strength and stiffness, and do not cause skin irritation [17]. On the other hand, there are also some disadvantages such as moisture uptake, the presence of color, the presence of odor when heated or burned during processing, quality variations, and low thermal stability. Many investigations have been carried out on the potential of natural fibers as reinforcement for composites, and in several cases the results have shown that natural fiber composites reached a good stiffness, but their final strength was not improved [18,19].

Composite manufacturing industries are looking for plant-based natural fiber reinforcements, such as flax, hemp, jute, sisal, kenaf, and banana as alternative materials to replace synthetic fibers. Lignocellulose fibers have also been considered for replacing glass fibers [20] as lignocellulose fibers are cheaper, lighter than glass fibers, and safer to be handled by workers [21].

Due to their advantages of low cost, biodegradability, large availability, and valuable mechanical and physical properties [22], a wide variety of lignocellulose fibers and natural fillers—coming from agricultural and industrial crops such as corn, wheat, bagasse, orange and apple peel algae, and sea grasses-derived fibers—have been used in the production of composites in various industrial sectors, such as packaging, automotive industry, and building [23–26].

For these reasons, several bio-composites were produced with a polymeric biodegradable matrix such as PLA and natural fibers. Tserki et al. [27] investigated the usefulness of lignocellulose waste flours derived from spruce, olive husks, and paper flours as potential reinforcements for the preparation of cost-effective bio-composites using PLA as the matrix. Petinakis et al. [28] studied the effect of wood flour content on the mechanical properties and fracture behavior of PLA/wood flour composites. In several natural fiber bio-composites with PLA as the matrix, the interfacial adhesion between the polymeric matrix and the fibers was poor [29]. Thus, the incorporation of lignocellulose materials into biodegradable polymer matrices, such as PLA, generally has the effect of improving the mechanical properties, such as tensile modulus, but sometimes the strength and toughness of these bio-composites are not improved.

Although several reviews [30–33] deal with lignocellulose-based composites including preparation methods and properties, most of them do not consider a deep analysis of interfacial adhesion between fiber and matrix or the application of mathematical models to explain them, which are very useful for predicting and tuning the properties of bio-composites. Thus, work remains to be done on the collective analysis of various applications of cellulose-based material. Natural fibers contain large amount of cellulose, hemicelluloses, lignin, and pectin, tending to be polar and hydrophilic, while polymeric materials are generally not polar and exhibit significant hydrophobicity [34]. The weak interfacial bonding between highly polar natural fibers and a non-polar organophilic matrix can lead to the worsening of the final properties of the bio-composites, ultimately hindering their industrial usage. Different strategies have been applied to eliminate this deficiency in compatibility and interfacial bond strength, including the use of surface modification techniques [35].

The hydrophilic nature of natural fibers decreases their adhesion to a hydrophobic matrix and, as a result, it may cause a loss of strength. To prevent this, the fiber surface may be modified in order to promote adhesion. Several methods have been proposed to modify natural fibers' surface, such as graft copolymerization of monomers onto the fiber surface and the use of maleic anhydride copolymers, alkyl succinic anhydride, stearic acid, etc. [36].

In this work, different amounts of two types of short cellulosic fibers (with different aspect ratios) added in a PLA polymeric matrix were investigated to evaluate the final effect on the mechanical properties. Furthermore, in order to have an estimation of the matrix/fiber adhesion, the B parameter calculated from the Pukanszky's model [37] was determined. The increase in stiffness of the final composite was also investigated using different analytical models existing in the literature with the aim to find the best fit with experimental data.

2. Results and Discussion

Results of the thermal gravimetric analysis are reported in Figure 1. From the weight loss peaks in the weight-to-temperature graph, it is evident that the fibers will not degrade during extrusion and injection molding, since the maximum temperature reached during processing is similar to the extrusion temperature, that was equal to 190 °C. This is an advantage of cellulose fibers versus other natural fibers which very often present thermal degradation during processing with negative effects on color and odor of the produced bio-composites.

In the graph, a small weight loss at temperatures lower than 100 °C can be observed and attributed to the loss of the residual moisture trapped in the fibers. The degradation of the fibers occurs at relatively quite high temperatures, beyond 300 °C, well above those reached in the processing of composites. Consequently, we can expect that the fibers inside the composites are stable and are not degraded, as confirmed by the nice white color of the composites and the absence of odor.

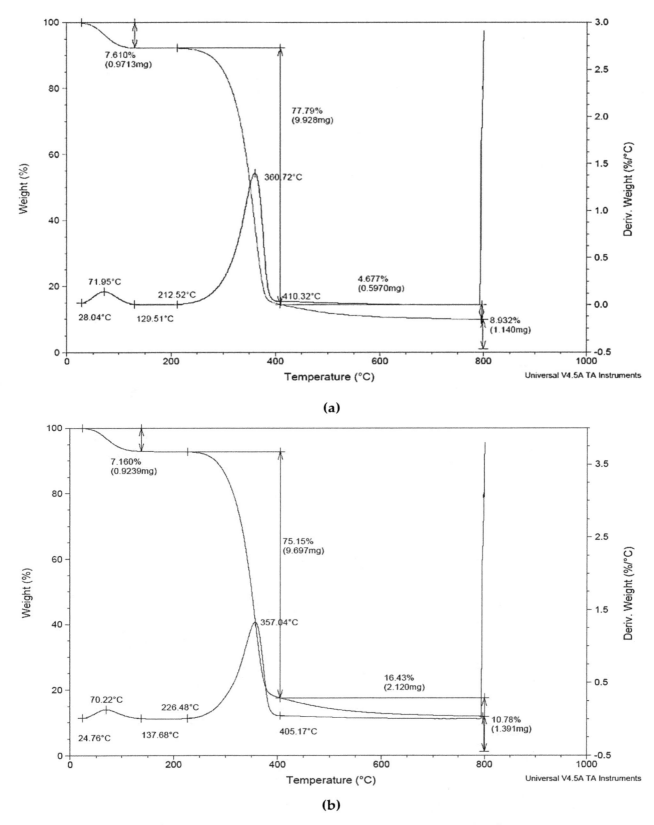

Figure 1. Thermogravimetric analysis (TGA) graphs of: (a) Arbocel® 600BE/PU and (b) Arbocel® BWW40.

From the results of the mechanical tests, it can be observed that, as expected, increasing the fiber content increases the elastic modulus of the composites (Figure 2a), in agreement with the trend normally observed in other studies in which cellulosic fibers were used [38] in polymeric matrices.

This behavior is very common, and the stiffness increment is generally related to the higher rigidity of the reinforcement versus the polymeric matrix However, BWW40 fibers show less marked increments in the Young's modulus compared to 600BE/PU fibers. This is likely due to the fibers' orientation and their higher aspect ratio with respect to the former. The higher aspect ratio for BWW40 can in fact cause twisting phenomena (that in general are encountered for natural fibers [39,40]) that can influence not only the elastic modulus but also the fibers' adhesion.

On the other hand, no significant increase in the final strength and strain at break of the composite are registered (Figure 2b,c). In particular for the composite with Arbocel® BWW 40, the stress at break decreases with the filler content. Even the Charpy impact resistance does not show significant improvements (Figure 2d). From these results, we can suppose that a very little or entirely null stress transfer takes place between the fiber and the matrix, due to a lack of fibers–matrix adhesion.

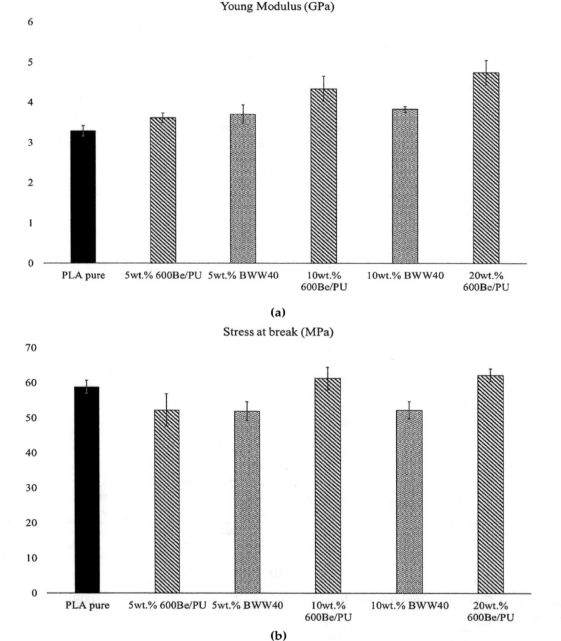

Figure 2. *Cont.*

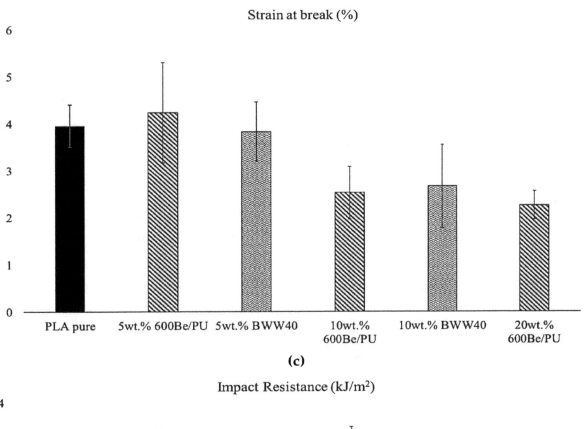

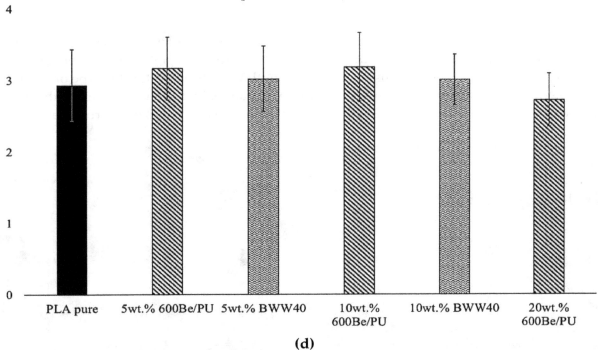

Figure 2. Mechanical properties of poly(lactic acid) (PLA)–Arbocel® composites: (**a**) Young's modulus, (**b**) stress at break, (**c**) strain at break, and (**d**) impact resistance.

In Figure 3, the Pukánszky's plot for the two different types of Arbocel® used is reported. An approximately linear trend of $ln\ \sigma_{red}$ can be extrapolated to calculate the B parameter.

The values obtained for the parameter B (reported in Table 1) confirmed that the adhesion between these cellulosic fibers and the PLA matrix is very low, explaining the results of the mechanical tests in which no significant improvements in strength were observed.

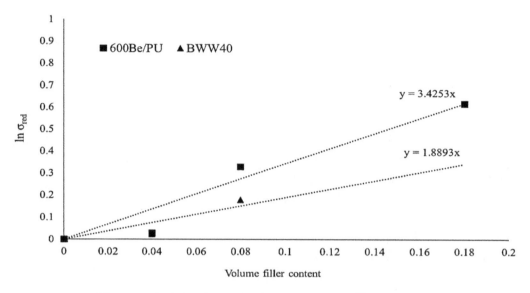

Figure 3. Pukánszky's plot for PLA–Arbocel® composites.

Table 1. B values.

Arbocel® Type	B
600BE/PU	3.42
BWW40	1.89

In particular, for BWW40 fibers, the B parameter is lower than that of 600BE/PU fibers. This means worse adhesion that can also be related to the higher aspect ratio and possible twisting phenomena. It is therefore explained why a moderate loss in tensile strength was observed for this type of composite.

In Figure 4, two SEM images are displayed reporting details of both Arbocel® fibers within the PLA matrix. We can observe a detachment and pull-out of the fibers due to the poor matrix–fibers adhesion.

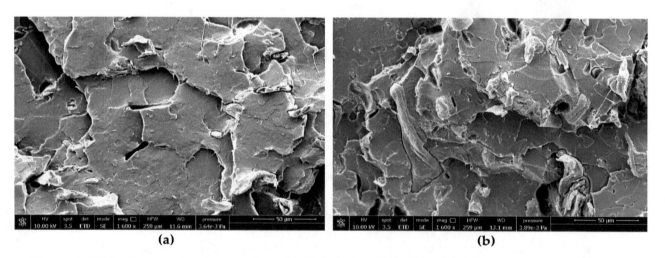

Figure 4. SEM micrographs of composites: (**a**) PLA + 10 wt % Be600/PU, (**b**) PLA + 10 wt % BWW40.

The results obtained are consistent with those in the literature, in which a lack of adhesion was encountered in similar composite materials. The study of interfacial adhesion is, in fact, a well-known problem when natural fibers and synthetic polymers are used [4,41]. A compatibilization is necessary if we want to obtain a composite with tailored mechanical properties and good efficiency in the transferring of the stress from the matrix to the fibers.

Furthermore, in this work, multiple analytical models were applied to investigate which better predicts the experimental data. The results of these analyses are reported in Figure 5.

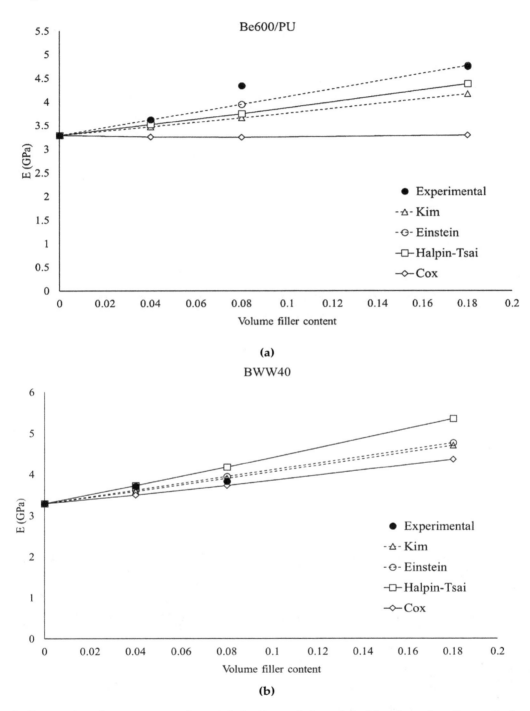

(a)

(b)

Figure 5. Comparison between experimental elastic modulus of the blends and mathematical models for (**a**) Be600/PU–PLA composites and (**b**) BWW40–PLA composites.

It can be observed that the Cox model provides underestimated stiffness predictions. This is due to the fact that the Cox model is referred to as the shear lag theory in which long, straight, and discontinuous fibers completely embedded in a continuous matrix were considered [42]. Consequently, this model is not very accurate when the fibers' aspect ratios are very small [43] and the adhesion is not good. This explains why the prediction is completely inaccurate for Be600/PU fibers, which have a very short aspect ratio (equal to 3), and improves with BWW40 fibers (aspect ratio equal to 10).

The Kim's model derives from the shear lag theory (like the Cox model) but was extended to resolve discrepancies of the Cox model in the case of short fiber-reinforced composites. In this case, the predicted values of the elastic modulus were similar to the experimental ones [43]. Effectively, in our case, this model gives good results for both types of fiber and consequently may be efficiently applied to these systems.

The Einstein's method is very simple and in general is applied for spheres, hence for fillers having a low aspect ratio. This model does not contain information about the geometry of the reinforcement, but the stiffness of the composites, depends only on the filler volume content [44]. Despite its simplicity, this model is able to efficiently estimate the composites' stiffness, probably because of the fibers' low aspect ratio.

The Halpin-Tsai equation also fits the experimental data with good accuracy. This model is often used to predict the elastic modulus of fibers that are randomly oriented on a plane [30]. In our case, the fibers are very short and not aligned. As a consequence, this model that contains an expression for the evaluation of both longitudinal and transversal moduli provides a good, but not perfect fitting.

3. Material and Methods

3.1. Materials and Characterization

The PLA used was 2003D, derived from natural resources and purchased from NatureWorks (Minnetonka, MN, USA) (grade for thermoforming and extrusion processes) [melt flow index (MFI): 6 g/10 min (210 °C, 2.16 kg), nominal average molar mass: 200,000 g/mol]. This type of PLA contains about 3–6% of D-lactic acid units in order to lower the melting point and the crystallization tendency, improving the processing ability. Two different types of commercial cellulosic short fibers, kindly provided by J Rettenmaier Sohne® (Rosenberg, Germany), were used. The trade names and the main properties of these two types of fibers are:

- ARBOCEL® 600BE/PU (mean diameter 20 μm, mean fiber length 60 μm, and consequently, aspect ratio 3, bulk density: 200–260 g/L, fiber density 1.44 g/cm³)
- ARBOCEL® BWW40 (mean diameter 20 μm, mean fiber length 200 μm, and consequently, aspect ratio 10, bulk density: 110–145 g/L, fiber density 1.44 g/cm³)

The morphology of these Arbocel® fibers before processing is shown in Figure 6.

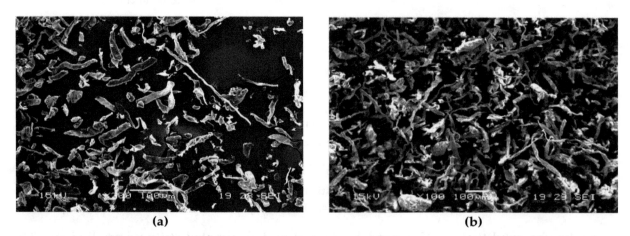

(a)   (b)

Figure 6. Scanning Electron Microscope (SEM) micrograph showing micro-cellulose fibers before processing: (**a**) Arbocel® 600BE7PU and (**b**) Arbocel® BWW40.

An increasing amount of cellulose fibers (at 5, 10, and 20 wt % corresponding to 4, 8, and 18 vol %, respectively) were added to the PLA matrix in order to produce bio-composites.

The materials, dried for at least 24 h in an air-circulated oven, were mixed in the correct quantities and then processed on a Thermo Scientific MiniLab Haake (Vreden. Germany)twin-screw extruder at a screw rate of 110 rpm/min and a cycle time of 60 s. After extrusion, the molten materials were transferred through a preheated cylinder to a Thermo Scientific Haake MiniJet II mini injection molder, for the preparation of the specimens for the Charpy and tensile tests. The Haake MiniJet II was equipped with an internal microprocessor capable of monitoring all the working parameters such as time, temperature, and injection pressure. The operative conditions of extrusion and injection molding are reported in Table 2.

Table 2. Processing conditions of Minilab and Minijet.

Minilab	
Extrusion temperature (°C)	190
Cycle time (s)	60
Screw rate (rpm)	110
Minijet	
Cylinder temperature (°C)	190
Mould temperature (°C)	60
Pressure (bar)	680
Residence time (s)	15

It is important to observe that for the blends containing ARBOCEL® BWW40, it was impossible to produce specimens containing 20 wt % of fibers because the molten material was too viscous, and the specimens were not consistent.

The tensile and impact properties of pure PLA and its composites containing different percentages of Arbocel® fibers were determined.

The tensile tests were carried out at room temperature, at a crosshead speed of 10 mm/min on an Instron universal testing machine 5500R equipped with a 10 kN load cell and interfaced with a computer running MERLIN software (INSTRON version 4.42 S/N–014733H) 24 h after specimen production. At least five specimens (gauge dimensions: $25 \times 5 \times 1.5$ mm) were tested for each sample, and the average values reported.

The impact tests were performed on V-notched specimens (width: 10 mm, length: 80 mm, thickness: 4 mm, V-notch 2 mm) using a 15 J Charpy pendulum of an Instron CEAST 9050. The standard ISO179:2000 was followed. At least 10 specimens for each blend were tested at room temperature.

The fibers and their composites were investigated by Scanning Electron Microscope (SEM) (FEI Quanta 450 FEG).

The thermogravimetric analysis (TGA) of Arbocel® fibers was also performed on a TGA Rheometric Scientific at a scanning velocity of 10 °C/min from room temperature up to 1000 °C, using nitrogen as purge gas.

3.2. Theoretical Analysis

In this work, different analytical models were applied in order to estimate the fiber/matrix adhesion and to predict the elastic modulus of PLA–cellulose composites containing different amount of fibers.

It is well known that the strength of a composite varies on the basis of its fiber content. Adhesion between fibers and polymeric matrix has a very large effect on this property, and in particular, it was demonstrated that the reinforcement characteristics seem to have a larger effect on strength than on stiffness [45]. For rigid fillers and for fibers with a low aspect ratio (as in this study), the reinforcing effect of a filler or a fiber can be expressed quantitatively by the following equation, proposed by Pukánszky [14]:

$$\sigma_c = \sigma_m \frac{1 - \varphi_f}{1 + 2.5\varphi_f} exp\left(B\varphi_f\right) \tag{1}$$

where the terms σ_c and σ_m, in this case, are the tensile stress of the composite and of the matrix, respectively, φ_f is the volumetric filler fraction, while the term $(1 - \varphi_f)/(1 + 2.5\varphi_f)$ indicates the decreasing of the effective load-bearing cross section due to reinforcement introduction. Finally, the term $\exp(B\varphi_f)$ takes into account the filler–matrix interactions, by means of the interaction parameter B [41]. We can write Equation (1) in linear form:

$$ln(\sigma_{red}) = log\frac{\sigma_c\left(1 + 2.5\varphi_f\right)}{\sigma_m\left(1 - \varphi_f\right)} = B\varphi_f \tag{2}$$

Plotting the natural logarithm of Pukánszky's reduced tensile strength (that is adimensional) against volume fraction (in the following graph this will be named Pukánszky's plot) results in a linear correlation in which the linear slope is proportional to the interaction parameter B [37]. In this way, by applying Equation (2), it is possible to calculate the B parameter for the two different types of fibers and consequently obtain a simple estimation of their adhesion to the PLA matrix.

For the prediction of the elastic modulus, the present system, based on a thermoplastic matrix in which random short fibers are dispersed, is not easy to evaluate. In fact, in this case, a great number of geometric, topological, and mechanical parameters are necessary [46]. Theoretical approaches usually attempt to exploit as much readily available information (which in most cases consists of the mechanical properties of matrix and fibers and the reinforcement volume fraction) as possible, while suitable assumptions cover missing data. Referring in particular to the elastic modulus, the existing expressions can be obtained from the elasticity theory—from a sort of mixture rule—or they are simply an attempt to match theoretical curves with experimental data [42–44,47,48]. Some of these analytical models (reported in Table 3) consider in particular the aspect ratio, the packing factor, and the Poisson ratio, in order to better predict the elastic modulus of composites containing increasing amounts of reinforcement.

Table 3. List of the analytical expressions used in this work for the prediction of the composites' Young's modulus.

Model	$E_{composite}$
Einstein	$E_c = E_m\left(1 + 2.5\ \varphi_f\right)$
Kim	$E_c = \varphi_m E_m + \varphi_f E_f \cdot \left\{1 + \left(\sqrt{\frac{E_m}{E_f}} - 1\right) \cdot \frac{\tanh\ (n\cdot a_r)}{(n\cdot a_r)}\right\}$
Cox	$E_c = \varphi_m E_m + \varphi_f E_f \cdot \left(1 - \frac{\tanh\ (n\cdot a_r)}{(n\cdot a_r)}\right)$
Halpin-Tsai	$E_c = \frac{3}{8}E_l + \frac{5}{8}E_t$

In Table 2, E_f and E_m are the elastic modulus of the fibers and matrix, respectively, φf is the fibre volume fraction, a_r is the fibers' aspect ratio. The adimensional parameter n is defined as:

$$\frac{2E_m}{E_f(1+v)ln\left(\frac{P}{\varphi_f}\right)} \tag{3}$$

where v is the Poisson ratio of the matrix (≈ 0.4), and P is the fibers' packing factor with the value $2\pi/\sqrt{3}$.

Furthermore, in the Halpin-Tsai model, the two terms E_l and E_t are, respectively, the longitudinal and the tangential modulus, quantified by the following expressions:

$$E_l = E_m \cdot \frac{1 + 2\cdot a_r \cdot \left(\frac{\frac{E_f}{E_m}-1}{\frac{E_f}{E_m}+2a_r}\right)\cdot\varphi_f}{1 - \left(\frac{\frac{E_f}{E_m}-1}{\frac{E_f}{E_m}+2a_r}\right)\cdot\varphi_f} \tag{4}$$

$$E_t = E_m \cdot \frac{1 + 2 \cdot a_r \cdot \left(\frac{\frac{E_f}{E_m} - 1}{\frac{E_f}{E_m} + 2} \right) \cdot \varphi_f}{1 - \left(\frac{\frac{E_f}{E_m} - 1}{\frac{E_f}{E_m} + 2} \right) \cdot \varphi_f} \tag{5}$$

4. Conclusions

In this study PLA–cellulose Arbocel® fiber composites were produced and studied. Two different types of cellulose fibers having different aspect ratios were used for producing cohesive, white, nice-looking, and odorless bio-based composite materials, whose mechanical and thermal properties still meet the requirements for practical applications, such as in the packaging and agricultural sectors.

The addition of 600BE/PU cellulose fibers up to 20 wt % does not worsen the starting PLA properties (unlike with BWW 40 fibers). Consequently, 600BE/PU fibers can be used without any compatibilization in order to lower the final product cost and at the same time increase the stiffness and promote the biodegradability of the materials [20]. A compatibilization between polymeric matrix and fibers would be necessary (as verified by Pukánszky's B parameter) if there is an interest in obtaining composites with improved tensile and Charpy impact properties with respect to those of the raw PLA-based materials, as these might be required in more demanding sectors such as automotive or electronics.

The stiffness of the composites was predicted by applying and comparing different analytical models. It was observed that a very simple model such as the Einstein's model gives positive results. At the same time, it is not adequate to apply the Cox's model and it is necessary to use an adjustment of it (Kim's model). However, because of the random orientation of the fibers, the Halpin-Tsai model gives a good estimation and is preferable because it is able to provide information not only on the final composite stiffness, but also on the transversal and longitudinal composite stiffness.

Author Contributions: A.L. supervised the study results, and discussion. P.C. co-supervised the study and reviewed the paper. L.A. performed the experimental research and wrote a draft of the paper. V.G. performed the mechanical analysis and collaborated to write the paper, M.B.C. analyzed the data and contributed to the results and discussion.

Acknowledgments: Authors acknowledge J Rettenmaier Sohne, Germany, for cooperation in providing the cellulose fibers.

References

1.   Hornsby, P.; Hinrichsen, E.; Tarverdi, K. Preparation and properties of polypropylene composites reinforced with wheat and flax straw fibres—Part 2: Analysis of composite microstructure and mechanical properties. *J. Mater. Sci.* **1997**, *1*, 1009–1015. [CrossRef]

2.   Liang, J.Z.; Ma, W.Y. Young's modulus and prediction of plastics/elastomer blends. *J. Polym. Eng.* **2012**, *32*, 343–348. [CrossRef]

3.   Bledzki, A.K.; Reihmane, S.; Gassan, J. Properties and modification methods for vegetable fibers for natural fiber composites. *J. Appl. Polym. Sci.* **1996**, *59*, 1329–1336. [CrossRef]

4.   Oksman, K.; Skrifvars, M.; Selin, J.F. Natural fibres as reinforcement in polylactic acid (PLA) composites. *Compos. Sci. Technol.* **2003**, *63*, 1317–1324. [CrossRef]

5.   Râpă, M.; Popa, M.; Cinelli, P.; Lazzeri, A.; Burnichi, R.; Mitelut, A.; Grosu, E. Biodegradable Alternative to Plastics for Agriculture Application. *AcRomanian Biotechnol. Lett.* **2011**, *16*, 59–64.

6.   Garrison, T.F.; Murawski, A.; Quirino, R.L. Bio-Based Polymers with Potential for Biodegradability. *Polymers* **2016**, *8*, 262. [CrossRef]

7.   Künkel, A.; Becker, J.; Borger, L.; Hamprecht, J.; Koltzenburg, S.; Loos, R.; Schick, M.B.; Schlegel, K.; Sinkel, C.; Skupin, G.; Yamamoto, M. Polymers, Biodegradable. *Ullmann's Encycl. Ind. Chem.* **2016**, 1–29. [CrossRef]

8.   Murariu, M.; Dechief, A.-L.; Ramy-Ratiarison, R.; Paint, Y.; Raquez, J.-M.; Dubois, P. Recent advances in production of poly(lactic acid) (PLA) nanocomposites: A versatile method to tune crystallization properties of PLA. *Nanocomposites* **2015**, *1*, 71–82. [CrossRef]

9. Aliotta, L.; Cinelli, P.; Coltelli, M.B.; Righetti, M.C.; Gazzano, M.; Lazzeri, A. Effect of nucleating agents on crystallinity and properties of poly (lactic acid) (PLA). *Eur. Polym. J.* **2017**, *93*, 822–832. [CrossRef]

10. Raquez, J.M.; Habibi, Y.; Murariu, M.; Dubois, P. Polylactide (PLA)-based nanocomposites. *Prog. Polym. Sci.* **2013**, *38*, 1504–1542. [CrossRef]

11. Liu, H.; Zhang, J. Research progress in toughening modification of poly(lactic acid). *J. Polym. Sci. Part B Polym. Phys.* **2011**, *49*, 1051–1083. [CrossRef]

12. Yu, L.; Dean, K.; Li, L. Polymer blends and composites from renewable resources. *Prog. Polym. Sci.* **2006**, *31*, 576–602. [CrossRef]

13. Bajpai, P.K.; Singh, I.; Madaan, J. Development and characterization of PLA-based green composites: A review. *J. Thermoplast. Compos. Mater.* **2012**. [CrossRef]

14. Lazzeri, A.; Phuong, T.V. Dependence of the Pukánszky's Interaction Parameter B on the Interface Shear Strength (IFSS) of Nanofiller- and Short Fiber-Reinforced Polymer Composites. *Composites Science and Technology* **2014**, *93*, 106–113. [CrossRef]

15. Lauke, B. Determination of adhesion strength between a coated particle and polymer matrix. *Compos. Sci. Technol.* **2006**, *66*, 3153–3160. [CrossRef]

16. Zappalorto, M.; Salviato, M.; Quaresimin, M. Influence of the interphase zone on the nanoparticle debonding stress. *Compos. Sci. Technol.* **2011**, *72*, 49–55. [CrossRef]

17. Peijs, T.; Garkhail, S.; Heijenrath, R.; van Den Oever, M.; Bos, H. Thermoplastic composites based on flax fibres and polypropylene: Influence of fibre length and fibre volume fraction on mechanical properties. *Macromol. Symp.* **2011**, *127*, 193–203. [CrossRef]

18. Oksman, K.; Wallström, L.; Berglund, L.A.; Filho, R.D.T. Morphology and mechanical properties of unidirectional sisal- epoxy composites. *J. Appl. Polym. Sci.* **2002**, *84*, 2358–2365. [CrossRef]

19. Mathew, A.P.; Oksman, K.; Sain, M. Mechanical properties of biodegradable composites from poly lactic acid (PLA) and microcrystalline cellulose (MCC). *J. Appl. Polym. Sci.* **2005**, *97*, 2014–2025. [CrossRef]

20. Lawrence, T.; Drzal, A.K.; Mohanty, M.M. Bio-composite materials as alternatives to petroleum-based composites for automotive applications. In Proceedings of the Automotive Composites Conference, Troy, MI, USA, 19–20 September 2001.

21. Jawaid, M.; Abdul Khalil, H.P.S. Cellulosic/synthetic fibre reinforced polymer hybrid composites: A review. *Carbohydr. Polym.* **2011**, *86*, 1–18. [CrossRef]

22. Ferrero, B.; Boronat, T.; Moriana, R.; Fenollar, O.; Balart, R. Green composites based on wheat gluten matrix and posidonia oceanica waste fibers as reinforcements. *Polym. Compos.* **2013**, *34*, 1663–1669. [CrossRef]

23. Seggiani, M.; Cinelli, P.; Verstichel, S.; Puccini, M.; Anguillesi, I.; Lazzeri, A. Development of Fibres-Reinforced Biodegradable Composites. *Chem. Eng. Trans.* **2015**, *43*, 1813–1818. [CrossRef]

24. Chiellini, E.; Cinelli, P.; Imam, S.H.; Mao, L. Composite Films Based on Biorelated Agro-Industrial Waste and Poly(vinyl alcohol). Preparation and Mechanical Properties Characterization. *Biomacromolecules* **2001**, *2*, 1029–1037. [CrossRef] [PubMed]

25. Seggiani, M.; Cinelli, P.; Geicu, M.; Elen, P.M.; Monica, P.; Lazzeri, A. Microbiological valorisation of bio-composites based on polylactic acid and wood fibres. *Chem. Eng. Trans.* **2016**, *49*, 127–132. [CrossRef]

26. Seggiani, M.; Cinelli, P.; Balestri, E.; Mallegni, N.; Stefanelli, E.; Rossi, A.; Id, C.L.; Id, A.L. Novel Sustainable Composites Based on Degradability in Marine Environments. *Materials* **2018**, *11*, 772. [CrossRef] [PubMed]

27. Tserki, V.; Matzinos, P.; Panayiotou, C. Effect of compatibilization on the performance of biodegradable composites using cotton fiber waste as filler. *J. Appl. Polym. Sci.* **2003**, *88*, 1825–1835. [CrossRef]

28. Petinakis, E.; Yu, L.; Edward, G.; Dean, K.; Liu, H.; Scully, A.D. Effect of Matrix–Particle Interfacial Adhesion on the Mechanical Properties of Poly(lactic acid)/Wood-Flour Micro-Composites. *J. Polym. Environ.* **2009**, *17*, 83. [CrossRef]

29. Tao, Y.; Wang, H.; Li, Z.; Li, P.; Shi, S.Q. Development and Application of Wood Flour-Filled Polylactic Acid Composite Filament for 3D Printing. *Materials* **2017**, *10*, 339. [CrossRef]

30. Petinakis, E.; Yu, L.; Simon, G. Natural Fibre Bio-Composites Incorporating Poly (Lactic Acid). In *Fiber Reinforced Polymers—The Technology Applied for Concrete Repair*; Masuelli, M., Ed.; Intechopen: London, UK, 2013; pp. 41–59, ISBN 978-953-51-0938-9. [CrossRef]

31. Gurunathan, T.; Mohanty, S.; Nayak, S.K. A review of the recent developments in biocomposites based on natural fibres and their application perspectives. *Compos. Part A Appl. Sci. Manuf.* **2015**, *77*, 1–25. [CrossRef]

32. Muthuraj, R.; Misra, M.; Mohanty, A.K. *Biocomposites*; Elsevier: Amsterdam, The Netherlands, 2015; ISBN 9781782423737.

33. La Mantia, F.P.; Morreale, M. Green composites: A brief review. *Compos. Part A Appl. Sci. Manuf.* **2011**, *42*, 579–588. [CrossRef]

34. Satyanarayana, K.G.; Arizaga, G.G.C.; Wypych, F. Biodegradable composites based on lignocellulosic fibers—An overview. *Prog. Polym. Sci.* **2009**, *34*, 982–1021. [CrossRef]

35. Huda, M.S.; Drzal, L.T.; Misra, M.; Mohanty, A.K. Wood-fiber-reinforced poly(lactic acid) composites: Evaluation of the physicomechanical and morphological properties. *J. Appl. Polym. Sci.* **2006**, *102*, 4856–4869. [CrossRef]

36. Herrera-Franco, P.J.; Valadez-Gonzalez, A. A study of the mechanical properties of short natural-fiber reinforced composites. *Compos. Part B Eng.* **2005**, *36*, 597–608. [CrossRef]

37. Pukánszky, B. Influence of interface interaction on the ultimate tensile properties of polymer composites. *Composites* **1990**, *21*, 255–262. [CrossRef]

38. Phuong, V.T.; Gigante, V.; Aliotta, L.; Coltelli, M.B.; Cinelli, P.; Lazzeri, A. Reactively extruded ecocomposites based on poly(lactic acid)/bisphenol A polycarbonate blends reinforced with regenerated cellulose microfibers. *Compos. Sci. Technol.* **2017**, *139*, 127–137. [CrossRef]

39. Tucker III, C.L.; Liang, E. Stiffness predictions for unidirectional short-fiber composites: Review and evaluation. *Compos. Sci. Technol.* **1999**, *59*, 655–671. [CrossRef]

40. Gigante, V.; Aliotta, L.; Phuong, T.V.; Coltelli, M.B.; Cinelli, P.; Lazzeri, A. Effects of waviness on fiber-length distribution and interfacial shear strength of natural fibers reinforced composites. *Compos. Sci. Technol* **2017**, *152*, 129–138. [CrossRef]

41. Huber, T.; Mussig, J. Fibre matrix adhesion of natural fibres cotton, flax and hemp in polymeric matrices analyzed with the single fibre fragmentation test. *Compos. Interfaces* **2008**, *15*, 335–349. [CrossRef]

42. Cox, H.L. The elasticity and strength of paper and other fibrous materials. *Br. J. Appl. Phys.* **1952**, *3*, 72–79. [CrossRef]

43. Kim, H.G.; Kwac, L.K. Evaluation of elastic modulus for unidirectionally aligned short fiber composites. *J. Mech. Sci. Technol.* **2009**, *23*, 54–63. [CrossRef]

44. Ahmed, S.; Jones, F.R. A review of particulate reinforcement theories for polymer composites. *J. Mater. Sci.* **1990**, *25*, 4933–4942. [CrossRef]

45. Renner, K.; Kenyó, C.; Móczó, J.; Pukánszky, B. Micromechanical deformation processes in PP/wood composites: Particle characteristics, adhesion, mechanisms. *Compos. Part A Appl. Sci. Manuf.* **2010**, *41*, 1653–1661. [CrossRef]

46. Bourkas, G.; Prassianakis, I.; Kytopoulos, V.; Sideridis, E.; Younis, C. Estimation of Elastic Moduli of Particulate Composites by New Models and Comparison with Moduli Measured by Tension, Dynamic, and Ultrasonic Tests. *Adv. Mater. Sci. Eng.* **2010**, *2010*, 891824. [CrossRef]

47. Halpin, J.C.; Kardos, J.L. The Halpin-Tsai equations: A review. *Polym. Eng. Sci.* **1976**, *16*, 344–352. [CrossRef]

48. Fu, S.Y.; Feng, X.Q.; Lauke, B.; Mai, Y.W. Effects of particle size, particle/matrix interface adhesion and particle loading on mechanical properties of particulate-polymer composites. *Compos. Part B Eng.* **2008**, *39*, 933–961. [CrossRef]

11

Roles of Silk Fibroin on Characteristics of Hyaluronic Acid/Silk Fibroin Hydrogels for Tissue Engineering of Nucleus Pulposus

Tze-Wen Chung [1,2,*], Weng-Pin Chen [3,4,*], Pei-Wen Tai [1,†], Hsin-Yu Lo [1] and Ting-Ya Wu [1]

[1]   Department of Biomedical Engineering, National Yang-Ming University, Taipei 11221, Taiwan; ylno02ekil@yahoo.com.tw (P.-W.T.); catfish19930628@gmail.com (H.-Y.L.); wuyaya0317@gmail.com (T.-Y.W.)

[2]   Center for Advanced Pharmaceutical Science and Drug Delivery, National Yang-Ming University, Taipei 11221, Taiwan

[3]   Department of Mechanical Engineering, National Taipei University of Technology, Taipei 10608, Taiwan

[4]   Additive Manufacturing Center for Mass Customization Production, National Taipei University of Technology, Taipei 10608, Taiwan

*   Correspondence: twchung@ym.edu.tw (T.-W.C.); wpchen@ntut.edu.tw (W.-P.C.)

†   Co-first author.

Abstract: Silk fibroin (SF) and hyaluronic acid (HA) were crosslinked by horseradish peroxidase (HRP)/H_2O_2, and 1,4-Butanediol di-glycidyl ether (BDDE), respectively, to produce HA/SF-IPN (interpenetration network) (HS-IPN) hydrogels. HS-IPN hydrogels consisted of a SF strain with a high content of tyrosine (e.g., strain A) increased viscoelastic modules compared with those with low contents (e.g., strain B and C). Increasing the quantities of SF in HS-IPN hydrogels (e.g., HS7-IPN hydrogels with weight ratio of HA/SF, 5:7) increased viscoelastic modules of the hydrogels. In addition, the mean pores size of scaffolds of the model hydrogels were around 38.96 ± 5.05 μm which was between those of scaffolds H and S hydrogels. Since the viscoelastic modulus of the HS7-IPN hydrogel were similar to those of human nucleus pulposus (NP), it was chosen as the model hydrogel for examining the differentiation of human bone marrow-derived mesenchymal stem cell (hBMSC) to NP. The differentiation of hBMSC induced by transforming growth factor β3 (TGF-β3) in the model hydrogels to NP cells for 7 d significantly enhanced the expressions of glycosaminoglycan (GAG) and collagen type II, and gene expressions of aggrecan and collagen type II while decreased collagen type I compared with those in cultural wells. In summary, the model hydrogels consisted of SF of strain A, and high concentrations of SF showed the highest viscoelastic modulus than those of others produced in this study, and the model hydrogels promoted the differentiation of hBMSC to NP cells.

Keywords: silk fibroin; hyaluronic acid; Tyrosine; viscoelastic modulus of HS-IPN hydrogels; hBMSC differentiations; nucleus pulposus

1. Introduction

Hydrogels can be produced by crosslinking polymers to form interpenetration network (IPN) with varying mechanical properties [1,2] for tissue engineering such as cardiac tissue repairs [3], controlling the fates of stem cells [1], and drug delivery, etc. [4]. Since hydrogels are highly permeable to nutrients and water-soluble metabolites, they can support cell growth and proliferation which are suitable for tissue engineering (TE). Hydrogels for TE usually consist of synthetic polymers, such as polyurethane and polyvinyl alcohol (PVA) [5,6], or natural polymers, such as HA and collagen [7,8].

HA is a natural glycosaminoglycan with carboxylic groups; it is an important component of the extracellular matrices (ECM) in various tissues and play important roles in cell proliferation and migration [9]. For instance, the interactions of HA in a cardiac patch and CD44 of BMSC enhanced cardiac differentiations of BMSC in both cardiac gene and proteins expressions [10]. In addition, various methods to prepare HA-based hydrogels including oxidized-HA or methacrylated-HA have been investigated for TE of ECM of NP [11–13], respectively.

Various of SF-based membranes, scaffolds or hydrogels have been extensively studied for the applications of TE because of its favorable biological responses, such as weak antigenic effects and inflammatory responses in-vivo [9,14–16]. For example, SF/HA patches laden with hBMSC promoted cardiac repair in a rat myocardial infarction (MI) model [10,16]. Developing suitable mechanical properties for hydrogels is also important to enhance cell proliferations and hBMSC differentiations for using various TE [17,18]. In this regard, the crosslinking tyrosine in SF to produce di-tyrosine bonds by HRP/H_2O_2 enzymatic reactions produced silk elastomers with stiffness that are varied from 0.2 to 10 kPa [17]. However, the stiffness of the aforementioned SF hydrogels was not suitable for the needs of some tissues such as human NP. Although the influences of molecular weights of SF on mechanical properties of SF hydrogels have been reported [17], the influence of total tyrosine contents in SF on the mechanical properties of SF-based hydrogels has not been investigated.

The intervertebral disc (IVD) absorbs shocks by transferring and dissipating loads to the ECM within the superior and inferior discs. IVD degeneration generally causes lower back pain, which is a common health problem. [7,18]. Currently, clinical treatments, such as spinal fusion and partial or total disc replacement cannot fully restore or maintain IVD structures and functions [7,18,19]. Since disc degeneration originates in nucleus pulposus (NP) regions, tissue engineering NP, that may mimic the structure of native NP tissue and possibly fully restore the functionality of healthy IVD discs, is great needed [18,19].

Various biomaterials have been investigated for TE of NP. They are: A. carbohydrate polymers including HA, dextran, chitosan and carboxymethylcellulose (CMC) to produce various hydrogels such as oxidized-HA or methacrylated-HA [11–13], dextran-chitosan-teleostean and methacrylated CMC hydrogels, respectively [7,18,19]; B. proteins including Type II collagen hydrogels, and laminin-based hydrogels, respectively [20,21]; C. hybrid of carbohydrate polymers and proteins, such as crosslinked oxidized HA/gelatin [12] and Type II collagen with a low molecular weight of HA [8,22]. However, the mechanical properties of some of the aforementioned biomaterials were not suitable for TE of NP, which need to be further processed [11,20,21]. For instance, the elastic modulus of the oxidized-HA hydrogels, using adipic acid di-hydrazide (ADH), are much lower than those of the native ECM of the NP [11]. Although hydrogels produced by oxidized HA-gelatin using ADH could improve the elastic modulus of the hydrogels (~11 kPa), the controlling the chemical reactions of HA/gelatin by ADH might not be easily carried out [12]. Recently, dextran-chitosan-teleostean triple-interpenetrating network and methacrylatedC MC hydrogels has been examined in a goat model to support the mechanical functions of degenerative NP [18,19]. However, chitosan and CMC had not yet been approved by the FDA for use in internal organs.

Although the bioactivities of SF or HA/SF patches for cardiac repairs have been shown in-vitro, and in-vivo, respectively [10,16], the mechanical properties for HS-IPN hydrogels, which consisted of various amounts of tyrosine in SF, in terms of vary strains of SF in this study, and the weight ratios of SF to HA have not been investigated. Moreover, the potential of differentiations of induced hBMSC laden-HS-IPN hydrogels to NP cells has not been examined. Although using NHS/EDC to crosslink HA/SF to facilely prepare HA/SF hydrogels was recently reported [23], it was one step of our processes to prepare hydrogels. However, they did not examine the rheological properties of hydrogels. Using HRP/H_2O_2 to crosslink tyramine-substitute HA and 2% of SF to produce hydrogels with varying HA contents has also been reported by Raia et al. [24]. HA and SF-IPN hydrogels produced by them are configured by di-tyramine bonds in HA crosslinked network, di-tyrosine bonds in SF crosslinked network and tyramine-tyrosine bonds in HA/SF-IPN. The tyramine-substitute HA needed to be complexly and chemically synthesized for the research study, which was not a commercially available biomaterial.

To produce HS-IPN hydrogels, HA-crosslinked network hydrogels were first produced by using BDDE solutions to crosslink HA polymers (Scheme 1). The influences of varying strains of SF obtained from *B. mori* cocoons and, at a fixed quantities of HA, varying amounts of SF (expressed in wt. ratios of 5/1~5/7 for HA/SF, respectively) were crosslinked using HRP/H_2O_2 reactions to produce HS-IPN hydrogels. To further increase in-vitro stability of hydrogels, the carboxyl acids of HA in HS-IPN hydrogels were crosslinked with amine groups of added polyethyleneimine (PEI) and HS-IPN using *N*-(3-Dimethyl-amino-propyl)-*N*′-ethylcarbodiimide hydrochloride)/*N*-Hydroxy-succinimide (EDC/NHS) reagents (Scheme 1). The differentiation of hBMSC laden on HS-IPN hydrogels, induced using TGF-β3 to NP cells were examined for TE of NP.

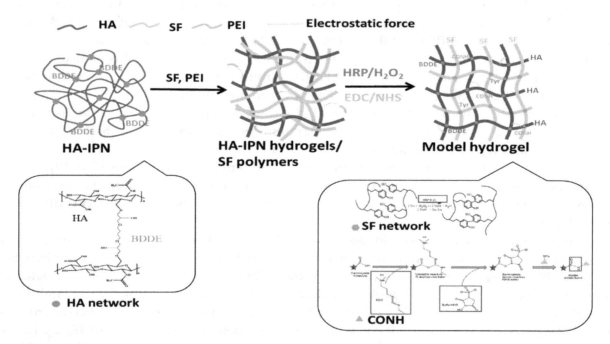

Scheme 1. The schematic procedures to develop HS-IPN hydrogel; HA polymer chains were first crosslinked by BDDE reagent for several hours to produce HA crosslinked network hydrogel (blue dots); SF and a small quantity of PEI (polyethyleneimine) were added into the gel and well blended. SF polymer chains were then crosslinked by HRP/H_2O_2 enzymatic reactions in HA crosslinked network hydrogel to produce large quantities of di-tyrosine bonds, make SF crosslinked hydrogel, (brown dots or Tyr in the model hydrogels) and yield HS-IPN hydrogel. HS-IPN hydrogel and PEI were further crosslinked using EDC/NHS reagents to produce many amide bonds to stabilize the hydrogels.

2. Materials and Methods

2.1. Fabrication of Interpenetrating Network HA-SF Hydrogels

Various strains, strain A, B and C herein, of *B. mori* cocoons were the products of artificially cross breeding specific strains of *B. mori* to produce varying toughness of SF that might vary amino acid configurations of SF, including tyrosine contents. Each strain of *B. mori* was well-bred to control the quality of SF, including low variations in amino acid configurations of SF and its cocoon was gifted from MDARES (Miaoli Agricultural Research and Extension Station, Council of Agriculture, Executive Yuan, Miaoli, Taiwan). In addition, the stain A of SF was used in this study without being further specified.

Solution of SF (MW ~ 185 kDa) of strain A was prepared as described in early reports, such as degumming, de-solving in 9.3 M LiBr and removal of Li$^+$ using dialysis with DI water that were published by the authors' laboratory [10,16]. Briefly, silk cocoons were boiled in 0.02 M Na_2CO for 0.5 h and then rinsed thoroughly in D.I. water to extract the glue-like sericin proteins from silk fibroin, degumming procedures. The extracted SFs were then dissolved in 9.3 M LiBr solution at 60 °C for 4 h to yield a 20% (w/v) solution, which was then dialyzed against D.I. water using a dialysis membrane

(MWCO 6000) (Spectra/Por 1, Repligen Co., Waltham, MA, USA) at room temperature to remove salt for 48 h [10,16].

The MW of SF of strain A, B and C were determined by a SDS-PAGE method (Vertical Electrophoresis System, XCell SureLock^{TM} Mini-Cell, Invitrogen, Waltham, MA, USA) with 150 V for 60 min. The bands of samples and molecular weight ladder (Himark Pre-Stained, Invitrogen, Waltham, MA, USA) were stained by a Coomassie Brilliant Blue R-250 (Sigma Corp., St. Louis, MO, USA). The images of the bands of the samples and standard molecular weight ladder were analyzed by an Image J (Windows version, Java 1.6.0-45, 32 bit mode, NIH, Bethesda, MD, USA). For amino acids analysis, 0.4 mg of dried SF samples (e.g., strand A) were hydrolyzed by 6 N HCL at 115 CC for 24 h. The analysis of amino acids of various samples were conducted by an amino acid analyzer (Hitachi L-8900, Tokyo, Japan).

To prepare HA hydrogel (H gel), HA (MW: 200 kDa, Lifecore Biomedical Inc., Chaska, MN, USA) was dissolved in NaOH at a concentration of 10%. 1,4-Butanediol diglycidyl ether (BDDE, Sigma Corp., St. Louis, MO, USA) was used for cross-linking reactions that was conducted at 37 °C for 6–8 h. The pH value of the solution was then adjusted to about 7.0 by adding HCl solution to terminate the crosslinking reaction [25,26]. After the HA-BDDE crosslinking reaction had been terminated, 0.15 mM PEI (polyethyleneimine, Mw. 25 KDa, Sigma, St. Louis, MO, USA), an appropriate amount of H_2O_2 (about 0.15 mM) and horseradish peroxidase (50 U/mL) (HRP, Sigma, St. Louis, MO, USA) and 5% SF solution were added to the H gels and well blended. SF in the aforementioned solutions were crosslinked and gradually turned to form the second interpenetration network, SF IPN-gels, because large quantities of di-tyrosine bonds of SF were produced by the reactions of HRP/H_2O_2, the enzyme reactions, in H gels for about 1 h at 37 °C [27].

To prepare SF hydrogel (S gel), an appropriate amount of H_2O_2 (about 0.15 mM) and horseradish peroxidase (50 U/mL) (HRP, Sigma, St. Louis, MO, USA) were added to 5% SF solution SF of strain A to induce enzymatic reaction for 1 h at 37 °C [27]. HS-IPN hydrogels were produced by blending the aforementioned H gels and S gels (Scheme 1).

To stabilize the HS-IPN hydrogels, 0.15 mM PEI in the hydrogels was further crosslinked by EDC/NHS (1:1) (N-(3-Dimethyl-amino-propyl)-N'-ethylcarbodiimide hydrochloride, $C_8H_{17}N_3$ HCL, EDC), Sigma Corp., St. Louis, MO, USA)/(N-Hydroxy-succinimide, $C_4H_5NO_3$, NHS), Fluka Chemical Corp., Rochester, NY, USA) for around 30 min. at temperature lower than 20 °C to produce amide bonds of carboxyl groups of HA with amine groups in PEI and in SF, respectively (Scheme 1). In addition, to examine the influences of varying ratios of SF on viscoelastic properties of HS-IPN hydrogels, the weight ratios of HA (4% at final) and SF in the hydrogels were changed to 5:1, 5:3, and 5:7 (e.g., HS1-IPN, HS3-IPN, HS7-IPN (namely, the model hydrogels)), respectively.

2.2. Characterizations of HS-IPN Hydrogels

2.2.1. Attenuated Total Reflectance-Fourier Transform Infrared (ATR-FTIR) Spectra of the Hydrogels

After various HS-IPN hydrogels were freeze-dried at −50 °C, their transmission spectra of the samples were examined using an ATR-FTIR instrument (IRAffinity-1, Shimadzu Co, Kyoto, Japan) with a resolution of 4 cm^{-1} at wave numbers 400–4000 cm^{-1} [10]. Detailed procedures for the measurements can be referred elsewhere. The spectra were analyzed using the built-in standard software package (IRAffinity-1, Shimadzu Co, Kyoto, Japan) [10].

2.2.2. Swelling Ratios of H, S, Crosslinked H/PEI and the Model Hydrogels

To study the swelling ratios of various gels, the samples were placed in phosphate saline (PBS) allowed for unrestricted deformation swelling test. H/PEI crosslinking hydrogels were fabricated after H gels were produced, followed by crosslinked 0.015 mM PEI in the gels by EDC/NHS crosslinking reagent for 30 min. The purpose of producing H/PEI crosslinking hydrogels was to examine swelling property of the gels after amide bonds formation between the carboxyl groups in HA gels and amine groups of PEI. The weights of hydrogels were measured at time of 0, 0.5, 1, 4, 24 and 48 h, W_t, after they were immersed into PBS for the aforementioned time. Their surfaces were gently wiped before they were weighted. The swelling ratio of a hydrogel was calculated by the following equation while W_{net} was the weight of the hydrogel before immersed into PBS:

$$\text{Swelling ratio } (\%) = \frac{W_t - W_{net}}{W_{net}} \times 100\% \tag{1}$$

2.2.3. Morphology of HS-IPN Hydrogels

To observe the morphology and the pore structure of HS-IPN hydrogels, all hydrogels were frozen at −50 °C and then at liquid nitrogen, further dried in vacuum for several days, so-called freeze-dry technique, to produce scaffolds. They were cut by a surgical knife to obtain the cross-section for observing the pore size distributions and characterizing the network in the scaffolds. All samples were coated with Pt and imaged with scanning electron microscopy (SEM, JSM-7600F, JEOL, Tokyo, Japan) at an accelerating voltage of 3–10 kV. The pore sizes of gel were calculated by ImagePro software (IPP7.0, Media Cybernetics, Rockville, MD, USA).

2.3. Rheological Studies of Vary Compositions of HS-IPN Hydrogels

The viscoelastic properties of various hydrogels were determined using a rheometer with an oscillatory mode (Discovery-HR1, TA Instruments, New Castle, DE, USA) at 37 °C. Oscillation frequency sweep tests using a parallel-plate sensor with a gap of 1 mm and loaded by 0.1 mL of HS-IPN hydrogels were carried out by the rheometer. The elastic modulus (G′) and viscous modulus (G″), phase shift angle (δ) and complex modulus (|G*|) of vary compositions for HS-IPN hydrogels were determined at a fixed strain of 0.01 rad with varying angular frequencies of 1–100 rad/s, as reported in other studies [11,12,28].

Compressive Modules of the Model Hydrogels

An MTS Model 858 Bionix Test System (MTS Corp., Eden Prairie, MN, USA) with a 5 kg load cell was used to perform confined compression tests on the porcine NP and l HS7-IPN hydrogels (or the model hydrogels). An acrylic cylinder with a diameter of 10 mm and a height of 30 mm was made and used as the indenter. A hollow, cylindrical acrylic container with an outer diameter of 30 mm, an inner diameter of 10 mm, and a height of 20 mm, was use in the confined compression test. A cavity with a diameter of 25.5 mm and a depth of 2.5 mm was cut into the bottom of the container and a porous plate was placed at the bottom of it. Porcine NP and hydrogel specimens were prepared and used to fill the acrylic container to a height of 5 mm. The confined compression tests were then carried out with five specimens for each material. A preload of 5 N (63.7 kPa) was applied to the indenter for 10 min to obtain a pressure balance on each specimen. In each confined compression test, following the preload phase, the indenter was moved at 0.1 mm/min until a final strain of 5% for the model hydrogels and the porcine NP while that of 15% (large strain compression) performed for the model hydrogels was mainly used for the comparisons with other studies. The load-displacement curve that was thus obtained for each specimen was then converted to a stress–strain (σ–ε) curve. The elastic modulus E was calculated as $E = \sigma/\varepsilon$ for the linear portion of the stress-strain curve.

2.4. Cytotoxicity of the Model Hydrogels

To evaluate the biocompatibility of the model hydrogels, MTT (3-(4,5-dimethylthiazol-2-yl)-2,5-diphenyl-tetrazolium bromide), Sigma–Aldrich, St. Louis, MO, USA) assays were performed to test the cytotoxicity of vary ratios of extraction mediums, taken from supernatants, which was obtained by incubation the hydrogels with L929 fibroblasts for 24 h according to the guidelines of international standard organization (ISO) 10993-5 [29]. L929 fibroblasts were purchased from the Bio-resource Collection and Research Center (BCRC, Hsin-Chu, Taiwan) and cultured in Dulbecco's modified

Eagle's medium that contained 10% horse serum at 37 °C in a 5% CO2 incubator. After L929 cells were cultured with vary volume ratios of extraction mediums to cultural medium (e.g., 1/8, 1/4, 1/2 and 1.0 (or extraction medium only), namely dilution ratios) for 24 h, MTT assay to the cells were carried out to examine the cell viabilities as early report of this lab [29].

2.5. Differentiations of Induced hBMSC to NP in the Model Hydrogels

To evaluate the potential of the model hydrogels for NP regeneration, human bone marrow-derived mesenchymal stem cell (hBMSC) from passage 6~8 were used according to our previous study [10]. The density of hBMSC of 1×10^7 cell/mL was seeded on the surface layer of a 1.0 cm^2 the model hydrogels to produce the cell-laden hydrogels. The viability of hBMSC in the model hydrogels after 3 d of cultivation were stained with Live/Dead or Viability/Cytotoxicity kit (Invitrogen, Waltham, MA, USA) according to the manual instructions. Live/Dead stain of the cells was observed by a laser confocal scanning microscopy (LCSM) (Olympus FV1000, Olympus Corp., Tokyo Japan) [10].

To induce the chondrogenesis of hBMSC in the model hydrogels, after the cell proliferations in the hydrogels for 3 d, the culture medium was changed to chondrogenic differentiation medium containing TGF-β3 which was purchased from Lonza Corp. (Lonza, Gaithersburg, MD, USA). The differentiation medium was changed every two days, and followed by culture for 7 d and 14 d, respectively, for chondrogenesis study of hBMSC.

2.5.1. Immuno-Histochemical (IHC) Analysis of Specific Protein Expressions of the Differentiations of Induced hBMSC to NP

IHC analysis of specific protein expressions the differentiations of hBMSC to NP after induced by TGF-β3 was carried out at 7 d and 14 d. All samples were washed using PBS and fixed by 4% paraformaldehyde. After the fixation, they were moved from the cultural wells and then embedded in paraffin and sectioned into sections with a thickness of 5 μm. The sections were then stained with Alcian blue for examining the depositions of glycosaminoglycan (GAG) and collagen type II, respectively. Cell-free hydrogel was stained as a negative control.

2.5.2. Real-Time PCR for Specific Gene Expressions of the Differentiations of Induced hBMSC to NP

To determine the specific gene expressions of the differentiations of induced hBMSC to NP in the model hydrogels, the ECM-related gene expression, including Aggrecan (AGN), Collagen type I (Col I) and Collagen type II (Col II) were assessed by real-time PCR (StepOnePlus™ Real-Time PCR System, Applied Biosystems, Foster, CA, USA) at 7 d and 14 d, according to other studies [11,12,19]. GAPDH was used as the internal control. The relative gene expression was calculated as $2^{-\Delta\Delta C_t}$.

2.6. Statistics

All calculations are made using SigmaStat statistical software (Jandel Science, San Rafael, CA, USA) [10]. Statistical significance in the Student t-test corresponded to a confidence level of 95%. Data presented are mean ± SD from at least triplicate measurements. Differences were considered statistically significant at $p < 0.05$.

3. Results and Discussion

3.1. Fabricating Fluidity HA Hydrogels by Adjusting Parameters of HA, and Reactions Conditions of BDDE Crosslinking Reactions

HA hydrogels have high water contents because HA contents a large amounts of carboxyl groups and have been used for tissue repairs and in drug delivery systems [11]. However, without chemical modifications such as crosslinking of HA hydrogels, the gels would be easily disassembled in aqueous environment and then lost their mechanical properties which would usually deviate from those of human tissues including NP. Hydrogels for tissue repairs, including IVD repair, crosslinking HA to produce the HA network hydrogel has been widely prepared by oxidized HA or methacrylated HA, in order to improve and sustain the mechanical properties of HA [12–14]. However, it involves complex and delicate chemical reactions. Alternatively, HA network hydrogel was widely fabricated by crosslinking hydroxyl groups of N-acetyl-D-glucosamine (NAG) in HA using the epoxide groups of BDDE at a high pH condition (pH >11). In this study, HA network hydrogels were produced as the aforementioned method with modifying reaction parameters (Scheme 1). For the reactions in which HA was crosslinked by BDDE, several parameters such as the MW and concentrations of HA, the concentrations of BDDE, and the reaction time, would influence various properties of HA hydrogels [2,30]. For example, the uses of various concentrations of BDDE (0.01~20%) under alkaline conditions to crosslink various concentrations of HA with high molecular weight (2.65~10%, MW >10^3 kDa) to produce HA crosslinked hydrogels have been extensively investigated [30]. Here, a low concentration (~2.5%) and a low MW (~200 kDa) of HA was crosslinked using around 2.0% BDDE for 5–7 h to produce HA crosslinked hydrogels (Scheme 1), which had low G' and G" (e.g., 0.24 ± 0.092 kPa and 0.09 ± 0.005 kPa, $n = 3$, respectively) with phase angle, δ, 21.4°, indicating that viscoelastic HA crosslinked hydrogels were high fluidity. Hence, HA crosslinked hydrogels could be mixed well with varying amounts of SF for producing HS-IPN hydrogels.

Using HRP/H_2O_2 Reactions to Crosslink SF and Producing HS-IPN Hydrogels

Although the de-sericin process for SF polymers affects its molecular weight [31], the de-sericin procedures herein were well controlled which ensured the molecular weight of SF was approximately 185 kDa, as determined by SDS-PAGE (data not shown). Since high fluidity of HA crosslinked hydrogels could be well mixed with SF, using HRP/H_2O_2 enzymatic reactions to crosslink Tyr in SF within the hydrogels could homogenously take place to produce SF-IPN hydrogels and, consequently, produce HS-IPN hydrogels (Scheme 1). To further stabilize the HS-IPN hydrogels, 0.15 mM of PEI added into the hydrogels was further crosslinked by EDC/NHS for about 30 min. to produce amide bonds of carboxyl groups of HA with amine groups in PEI and in SF in the hydrogels, respectively. (Scheme 1).

According to the results of rheological study (in Section 3.3), the viscoelastic properties of HS7-IPN hydrogels were similar to matrix for human NP. The HS7-IPN hydrogels were chosen as a model hydrogel for inducing hBMSC to differentiate to NP cells or NP tissue engineering in this study.

3.2. Characterizations of HS-IPN Hydrogels

3.2.1. ATR-FTIR Spectra of HA, SF, and the Model Hydrogels Consisted of Varying Strains of SF

The model hydrogels were characterized by spectra using an ATR-FTIR spectrophotometer (Figure 1). In the Figure 1, the transmission spectra for the peaks of carboxyl groups of H gels such as 1600 and 1402 cm^{-1} were about the same as those of HA polymers. Moreover, the peak of amide II of H gels was 1556 cm^{-1}, which was similar to that of HA polymers. The transmission spectra for the peaks of amide I, II and III groups of S gels were 1640, 1511 and 1230 cm^{-1} that were about the same as SF polymers. The transmission spectra from HS-IPN gels produced from different strains of SF (e.g., strain A, B and C) contained several characteristic peaks as those of H gels and S gels with minor variations. For instance, the peaks of amide II of HS-IPN gels shifted from 1556 cm^{-1} of H gels to 1528, 1522 and 1525 cm^{-1} for strain A, B and C, respectively (Figure 1). The presences of the peak of amide I of HS-IPN hydrogels only minor shifted from 1640 cm^{-1} to 1635 cm^{-1} of S gels, and there was no difference among the model hydrogels consisted of varying strains of SF. Although the quantities of functional groups for varying weight ratios of HA/SF of the HS-IPN hydrogels were different such as HS1-IPN and the model hydrogels, the ATR-FTIR spectra for those hydrogels were similar to those of the model hydrogels and not able to characterize their differences (Figure 1). Hence, other characterizations including rheological properties of the hydrogels needed to be carried out to determine the differences among the model hydrogels consisted of varying strains of SF.

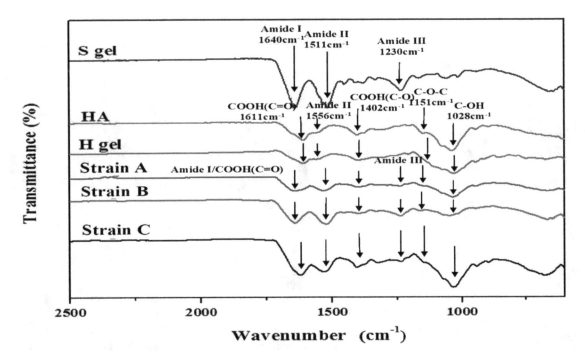

Figure 1. ATR-FTIR transmission spectra of varying functional groups of SF, HA polymers, and H, S and HS7-IPN hydrogels. The characteristic peaks of H hydrogels such as carboxyl groups (COOH), and S hydrogels such as amide I, II and III, respectively, were found in HS-IPN hydrogels without varying by the SF of strains of A, B and C.

3.2.2. The Pore Structures of Scaffolds and Swelling Ratios for H, S and the Model Hydrogels

To examine the pore structures of scaffolds for H, S and the model hydrogels, SEM micrographs of the scaffolds were carried out and presented (Figure 2). The mean pore sizes in H and S scaffolds were approximately 49.36 ± 15.04 μm, and 13.40 ± 1.25 μm, respectively, ($n = 3$). The mean pore size of the scaffolds of S gels was significantly smaller ($p < 0.01$, $n = 3$) than that for the scaffold of H gels (Figure 2). Interestingly, the mean pore size of the scaffolds of the model hydrogels was around 38.96 ± 5.05 μm ($n = 3$) which was between those H and S scaffolds. (Figure 2). In addition, the pore size of the scaffolds of the model hydrogels was suitable for the proliferations of cells, including hBMSC [7].

The swelling ratios of H, S, H/PEI crosslinking and the model hydrogels in PBS were examined and shown (Figure 3). Since H gels were produced by highly hydrophilic polymers, the swelling ratios of the hydrogels fast increased within 12 h till 550% and then slowly increased up to ~700% at 48 h which was significantly higher than other hydrogels ($p < 0.01$, $n = 3$). Interestingly, after the first 6 h of fast swelling stage, the swelling ratios for the model and H/PEI crosslinked hydrogels were in plateau regions (e.g., $185.9 \pm 24.4\%$, $n = 3$) until the end of study, 48 h. However, after swelling in the first 4 h, the swelling ratios for S hydrogels in PBS were about $24.9 \pm 0.9\%$ till 48 h, which were significantly lower than others ($p < 0.01$, $n = 3$) (Figure 3). The swelling ratios of the model hydrogels were very close to those of H/PEI crosslinked hydrogels, and the values were between those for H and S gels [32]. Notably, H/PEI crosslinked hydrogels highly reduced the swelling ratios in PBS compared with those for H gels, indicating that EDC/NHS reactions among HA and PEI in hydrogels would effectively crosslink carboxylic and amine groups to produce the amide bonds within the gels which resulted in significant decreases of the interactions among carboxylic groups of H gels and ambient H_2O (Figure 3). Since EDC/NHS reactions were also carried out for preparing the model hydrogels as those for H/PEI crosslinked hydrogels, the swelling ratios for the model ones were similar to those of those for the later gels although the model hydrogels consisted of SF-IPN. In addition, the swelling ratios for other HS-IPN hydrogels (e.g., HS3-IPN hydrogels) would be similar to those for the model ones.

H hydrogel   **S hydrogel**   **The model hydrogel**

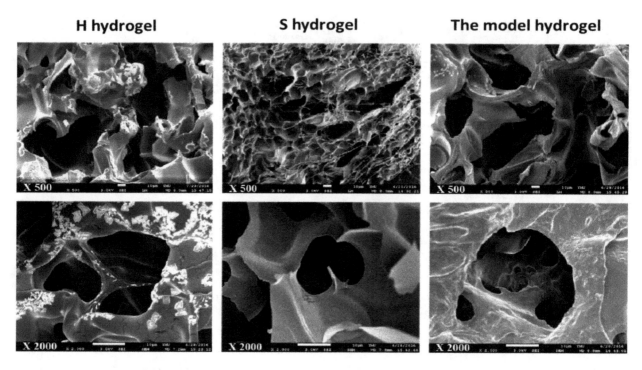

Figure 2. SEM micrographs of H, S, and HS-IPN hydrogels. The pore size of the model hydrogels was 38.96 ± 5.05 μm, ($n = 3$), in between those of H and S hydrogels.

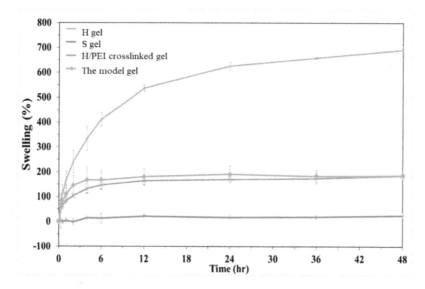

Figure 3. The swelling ratios for H, S, H/PEI crosslinked and the model hydrogels in PBS were shown. The swelling ratios for the H gels were significantly higher than others ($p < 0.01$, $n = 3$), while those for S gels were significantly lower than others ($p < 0.01$, $n = 3$). The swelling ratios of the model hydrogels in PBS were similar to those for crosslinking H/PEI crosslinked hydrogels between those of H and S hydrogels (Data are mean ± SD, $n = 3$).

Notably, the results or data of HS-IPN gels at slow swelling ratio stages were similar to human NP, revealing that they may be suitable for use in TE of NP [7,28]. Moreover, the stability of the model hydrogels in PBS solution was examined by observation and measuring weight loss of the hydrogels after they were immersed into PBS solution for three weeks. The morphology of the model hydrogels was intact without disintegration by observation while approximately 4% of weight loss was found compared to that of the original hydrogels to be immersed in PBS. In addition, the samples became more fluidity towards the end of the four weeks. Notably, the results of HS-IPN gels at slow swelling ratio stages were similar to human NP, revealing that they may be suitable for use in TE of NP [28].

3.3. Varying Strains of SF Influenced the Rheological Properties of HS-IPN Hydrogels

Viscoelastic flow parameters are important mechanical properties to characterize hydrogels for applications of bio-fluids such as lubricant in joints or NP. To determine, G′, G″ and δ values of HS-IPN hydrogels, they are determined using a parallel-plate rheometer operated in an oscillatory mode at a fixed strain (0.01 rad) with various angular frequencies (1–100 rad/s) as used by other studies [11,12,28]. G′ of the model hydrogels consisted of SF of strain A was 4.09 ± 0.32 kPa ($n = 3$, $p < 0.01$) which was significantly higher than those consisted of SF of strains B and C at a fixed angular frequency (Table 1). Since strain A of SF consists of more amount of Tyr than those in strain B and C, it could be assumed that more di-tyrosine bones were formed to crosslink peptides of strain A of SF to produce SF-IPN hydrogels than those in B and C ones. Hence, the rheological properties of the HS-IPN hydrogels were influenced by the strain of SF. The results rheological properties influenced by strains of SF, presented in Table 1, were qualitatively consistent with those UV-excited fluorescent intensities presented in Table 2. According to amino acid analysis for three strains of SF in this study, the total number of amino acids were about 5500, including about 5.1% of Tyr in strain A, while about 4.6% of tyrosine in strains B and C. The Tyr contents for strains of B and C were similar to other reports [27]. Notably, the formations of varying amounts of dityrosine bonds in SF-IPN hydrogels, crosslinked by HRP/H_2O_2 would emit varying intensities of blue fluorescence when the hydrogels were irradiated by UV [29]. The intensity of emitted blue fluorescence of SF-IPN hydrogels for strain A was significantly higher ($n = 4$, $p < 0.01$) than those of strains of B and C (Table 2), which was consistent with that Tyr contents in strain A (e.g., 277 ± 11, $n = 4$) is significantly higher than those of strain B and C (e.g., 255 ± 2, $n = 4$ for B), respectively. Therefore, the influences of strains of SF on the rheological properties of HS-IPN hydrogels possibly resulted from the tyrosine contents in each strain of SF.

Table 1. The viscoelastic parameters of the model hydrogels fabricated from different strains of SF measured at 0.01 rad and 10 rad/s ($n = 3$). The parameters of strain A such as G′ and |G*|, complex shear modulus, were significantly higher than strain B and C.

Materials	Viscoelastic Properties (0.01 Rad, 10 rad/s)			
	G′ (kPa)	G″ (kPa)	|G*| (kPa)	δ (°)
Strain A	4.09 ± 0.32 **	0.59 ± 0.16	4.13 ± 0.34 **	8.14 ± 1.66
Strain B	3.24 ± 0.16	0.50 ± 0.05	3.27 ± 0.16	8.69 ± 0.5
Strain C	3.40 ± 0.19	0.43 ± 0.07	3.43 ± 0.18	7.28 ± 1.19

$** p < 0.01$ or better.

Table 2. The intensity of blue fluorescence of SF hydrogels excited by UV irradiation, in terms of OD values, for different strains of SF at 1.62 mM H_2O_2 [29]. ($n = 4$).

The Intensity of Blue Fluorescence of SF Hydrogels Consisted of Vary Strains of SF with Their Tyrosine Contents		
Sample	OD Value	Tyrosine Content
Strain A	0.80 ± 0.01 **	277 ± 11 **
Strain B	0.67 ± 0.01	255 ± 2
Strain C	0.67 ± 0.02	255 ± 1

$** p < 0.01$.

Influence of the Weight Ratios of SF to HA in HS-IPN Hydrogels on Rheological Properties of Hydrogels

Other than the SF strains, the weight ratios of SF to HA in producing the HS-IPN hydrogels might also influence the viscoelastic properties of the hydrogels. For examining this factor, at a fixed HA content (4%), varying SF concentrations (in wt.%) in producing HS1-IPN to HS7-IPN hydrogels (or the model gels) were carried out at the aforementioned oscillatory conditions (Figure 4A). G' values increased with increasing the concentrations of SF in the HS-IPN hydrogels. Therefore, the model hydrogels had the highest G' and G" values ($p < 0.001$, $n = 3$) among the produced hydrogels herein tested at varying angular frequencies. For example, the G' values of the model hydrogels (e.g., 4.09 kPa at 10 rad/s) were about 2.6 times higher than those of HS1-IPN gels.

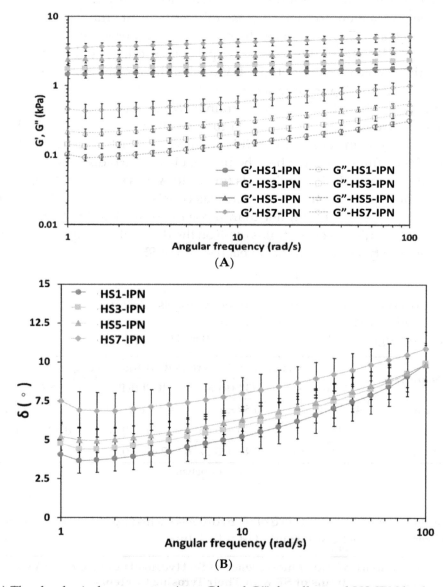

Figure 4. (**A**) The rheological parameters (e.g., G' and G") for all tested HS-IPN hydrogels increased with increasing the weight ratios of SF/HA in producing the hydrogels versus the varying angular frequencies, (**B**) increasing weight ratios of SF/HA in the HS-IPN hydrogels increased the δ values from about 4.2 (e.g., HS1-IPN) hydrogels to 8.0° (e.g., HS7-IPN) hydrogels vs. with varying angular frequencies, respectively (Data are mean ± SD, $n = 3$).

The phase angles, δ values, for all HS-IPN hydrogels were shown (Figure 4B) which increased from around 4.2° to 8.0° at 10 rad/s, respectively, ($n = 3$). The results of δ values indicated that the model hydrogels were less viscoelastic solid than others (e.g., HS1-IPN), respectively. In comparison, the rheological properties for HA crosslinked hydrogels, produced by the same protocols as those for HS-IPN hydrogels, were carried out and had low G' (e.g., 0.24 ± 0.092 kPa, $n = 3$) with phase angle of 21.4°. The results indicated that viscoelastic properties for HA crosslinked hydrogels were high fluidity with a very low elastic modulus. Hence, the results of rheological properties for HA crosslinked hydrogels were not suitable for TE of NP. Notably, the G' and δ values at 10 rad/s for the HS7-IPN hydrogels produced herein were similar to those reported for native NP (5.0~10.3 kPa and 2.5°~35°, respectively) [7,18,28], and therefore, the hydrogels were selected as the model hydrogels for further this investigation.

The G' for IPN hydrogels produced herein were similar to those of laminin-111-PEG hydrogels reported for the matrix of NP [24]. Interestingly, the rheological modulus for G' or |G*|of HS1-IPN hydrogels were fitted the requirements of hydrogels for cardiac repairs. According to the tyrosine contents of varying strains of SF were different (Table 2), and the strains of SF influenced the viscoelastic modules of HS-IPN hydrogels (Table 1). The contents of tyrosine of SF were one of an important factor on determining those modules of HS-IPN hydrogels although the phase angles of the hydrogels might not be the case as those modules (Table 1). Moreover, the model hydrogels contained more concentrations of SF and amounts of di-tyrosine bonds in qualitative in the hydrogels than other HS-IPN hydrogels that resulted in increasing their viscoelastic modulus (Figure 4A and Table 1). However, the amounts of di-tyrosine bonds in each HS-IPN hydrogel were not able to be determined quantitatively. Although, the bonds could be semi-quantitatively evaluated using the intensity of UV-excited blue fluorescence [29].

Although using NHS/EDC to crosslink HA/SF to facilely prepare HA/SF hydrogels was recently reported [23], it was only one step of our processes to prepare hydrogels. However, they did not perform the rheological study for the aforementioned hydrogels [23]. Recently, using sonication and UV photo-polymerization to prepare SF/methacrylated HA or to produce SF-based IPN hydrogels has been reported by Xiao et al. [33], respectively. However, the rheological properties of the hydrogels were not determined. Interestingly, using HRP/H_2O_2 to crosslink tyramine-substitute HA and 2% of SF to produce hydrogels with varying HA contents has been reported by Raia et al. [24]. HA and SF-IPN hydrogels produced in their study are configured by di-tyramine bonds in HA crosslinked network, di-tyrosine bonds in SF crosslinked network and tyramine-tyrosine bonds in HA/SF-IPN, which bonding structures of their hydrogels were distinct from those in HS-IPN hydrogels produced in this study. Hence, the rheological properties of our hydrogels were different from theirs [24]. Moreover, the tyramine-substitute HA needed to be complexly and chemically synthesized for the research which was not a commercially available biomaterial, while the biomaterials were generally commercially available.

3.4. Confined Compressive Modules of the Model Hydrogels

The confined compressive stress of the model hydrogels was conducted on the hydrogels under 5% strain with value of 0.109 ± 0.011 MPa ($n = 3$) which was slightly lower than the for human NP (e.g., around 0.5–1.5 MPa) [7]. Notably, the compressive modulus for the model hydrogels was 2.29 ± 0.05 MPa ($n = 3$) was similar to that of human NP [7]. Although using NHS/EDC to crosslink HA/SF to facilely prepare HA/SF hydrogels was recently reported by Yang et al. [23] that was generally simple to produce the hydrogels than those produced by the procedures for this study, the confined compressive stress of their products was about 12 kPa at 30% of strain, which was about 25 times less than the stress of our model hydrogels (e.g., 314 ± 2.8 kPa, at 15% strain, $n = 3$). Hence, the compressive stress for their hydrogels would not fit the need of human NP.

Recently, Xiao et. al. [34] reported that using sonication and further UV photo-polymerization to prepare SF-IPN and methacrylated HA to produce network SF/HA hydrogels, which had a stiff but brittle SF structure while SF-IPN hydrogels produced herein (Scheme 1) were non-brittle. However, the confined compression modules of their hydrogels were not examined [23,31] as did in this study.

3.5. Cytotoxicity Examinations for the Model Hydrogels

In-vitro cytotoxicity for the model hydrogels were performed according to the requirements of ISO10993-5. Briefly, after L929 cells were cultured with vary concentrations of extraction mediums, for 24 h, and the MTT assay to the cells were carried out to examine the cell viabilities [26]. According to the MTT assay shown in Figure 5, the L929 viability of the group with a dilution ratio of 0.5 exceeded that of the group of extraction medium only which met the requirements of ISO10993-5, revealing that the model hydrogel was biocompatible and suitable for use in TE. The results of the biocompatibility of the hydrogels were similar to other SF-based biomaterials [8,31].

3.6. Differentiations of hBMSC to NP Cells in the Model Hydrogels

To evaluate the differentiation of hBMSCs to NP cells, they were induced by TGF-β3 in the model hydrogels. The morphology of hBMSC, and the accumulations of NP-related ECM deposits in the hydrogels were examined using vimentin stain and immuno-histochemical (IHC) analysis, such as glycosaminoglycan (GAG) and collagen type II stain, respectively. The cell cultivated in cultural wells was also stained as a control group. Figure 6A exhibited that the deposition of glycosaminoglycan (GAG, in blue), one of the main ECMs in the NP, in hBMSC-laden model hydrogels was much more than that in the control [34]. Collagen type II, an important component of ECM in NP tissue, forms a fibrillar network that traps proteoglycan and resists swelling [11,12,20]. The deposition of GAG (in blue) and accumulation of collagen type II (in brown) in the model hydrogels containing differentiated hBMSC increased with increasing the cultural period (Figure 6A,B, respectively). The depositions of the GAG and collagen type II were broad distribution in the model hydrogels, revealing that the hydrogels herein were suitable to the differentiations of hBMSCs to NP cells. Enhancing expressions in GAG and collagen type II of the differentiations of hBMSCs to NP cells in the model hydrogels compared to those expressions in the control group revealed that they promoted the differentiations of hBMSC (Figure 6A,B, respectively). According to those results, the mechanical properties of the model hydrogels compared with hardness matrix (i.e., cultural wells) were suitable to the differentiations of hBMSCs to NP cells.

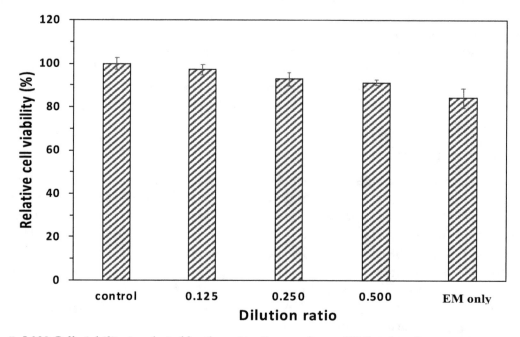

Figure 5. L929 Cell viability incubated by the extraction mediums (EM), taken from supernatants which was incubated with the model hydrogels for 24 h, at different dilute ratios according to the guidelines of ISO 10993-5. The relative L929 viability (%) of the group with a dilution ratio of 0.5 exceeded that of the group of EM only. The results of MTT assay for L929 Cell viability showed that the model hydrogels were biocompatible biomaterials (data are mean ±SD, $n = 3$).

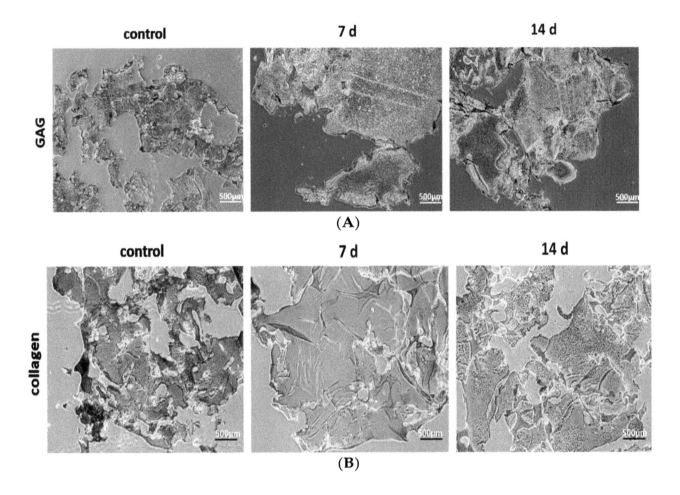

Figure 6. Immuno-histochemical (IHC) analysis of hBMSCs in the model hydrogels after hBMSC were induced by TGF-β3 to differentiate them to NP cells for 7 d and 14 d, respectively. Deposition of (**A**) GAG (in blue) and, (**B**) Collagen type II (in brown) in the model hydrogels were shown which were significantly higher than those of cultivated in cultural wells (e.g., control group).

Aggrecan (AGN), collagen type II (Col II) and collagen type I (Col II) were selected as test genes for examining the differentiations of hBMSCs in the model hydrogels [11,12,21]. The gene expressions of chondrogenesis of hBMSCs, cultured in the model hydrogels, revealed significant up-regulation of both AGN and Col II but significant down-regulation of Col I, compared to those of hBMSCs cultivated in cultural wells (e.g., the control group) for 7 d (Figure 7A–C). The results gene expressions of differentiations of hBMSCs to NP cells presented herein were similar to those of other studies [11–13] although the compositions of their hydrogels were different from this study. AGN and Col II gene expressions increased with the 7 d of differentiations of hBMSCs to NP cells in the model hydrogels that were consistent with the increasing depositions of GAG and collagen type II of IHC stains (Figure 6A,B, respectively). However, the gene expressions of the differentiations of hBMSCs to NP cells in the model hydrogels (Figure 7A–C, respectively) were not further increased in the period of 7 d to 14 d. The possible mechanisms such like reducing the activity of CD44 of HA for the differences of results the differentiations of hBMSCs to NP cells in Figures 6 and 7 at 14 d need to be further studied. Although adjust mechanical properties of HA-based or methacrylated HA-based containing SF hydrogels using varying techniques for other TE studies have been reported by others [11,23,24]. This study reported an examination of the differentiations of hBMSCs to NP cells in HS-IPN hydrogels. Nevertheless, the model hydrogels promoted the differentiations of hBMSCs to NP cells for tissue engineering of NP.

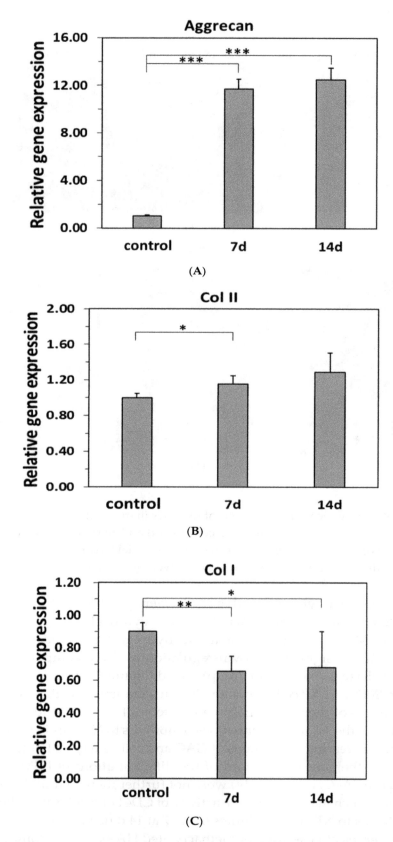

Figure 7. Real-time PCR analysis for the relative quantities of NP-specific gene expressions of inducing differentiation of hBMSC to NP cells on the model hydrogels for 7 d and 14 d. Gene expressions of AGN (**A**), COL II (**B**) and COL I (**C**) for the control group (e.g., ctrl) were the data for 7 d incubated at cultural wells. (* $p < 0.05$, ** $p < 0.01$, ***$p < 0.001$; data presented are mean ±SD, $n = 3$).

4. Conclusions

The roles of SF on characteristics of HS-IPN hydrogels were examined. The schematic procedures for preparing HS-IPN hydrogels were presented in Scheme 1. The pore size of the model hydrogels was 38.96 ± 5.05 μm between those of H and S hydrogels (Figure 2). The swelling ratios for HS-IPN hydrogels (e.g., 185.9 ± 24.4%, $n = 3$) were similar to those for H/PEI hydrogels (Figure 3) which were much less than those for HA hydrogels. The hydrogels composed of SF of strain A showed higher G′ and G″ values than those composed of SF of strain B and C (Tables 1 and 2, respectively). Moreover, the model hydrogels consisted of the highest weight ratios of SF to HA showed significantly higher G′, G″ and |G*|values than other hydrogels consisted of those of low weight ratios. (Figure 4A). The model hydrogel was biocompatible for TE applications (Figure 5). Moreover, the model hydrogels significantly promoted the differentiations of hBMSC to NP cells with increasing the expressions of GAG and collagen type II for 7 d and 14 d (Figure 6). Moreover, the induced hBMSC in the hydrogels increased the gene expressions of AGN and COL II while decreased those of COL I for 7 d of cultivations (Figure 7). Hence, the model hydrogels developed herein were suitable to tissue engineering for NP regeneration.

Author Contributions: Conceptualization, T.-W.C., W.-P.C.; methodology, T.-W.C., P.-W.T., T.-Y.W.; validation, P.-W.T., T.-Y.W.; investigation, T.-Y.W., H.-Y.L.; data curation, P.-W.T., H.-Y.L.; writing—original draft preparation, T.-W.C., W.-P.C.; writing—review and editing, T.-W.C., W.-P.C. All authors have read and agreed to the published version of the manuscript.

Acknowledgments: The authors would like to thank the Miaoli District Agricultural Research and Extension Station, Council of Agriculture, Executive Yuan, Taiwan for providing silk fibroin.

Nomenclature

Nomenclature	Full name
HA	Hyaluronic acid
SF	Silk fibroin
H gel	HA hydrogel
S gel	SF hydrogel
HS1-IPN hydrogel	HA/SF hydrogel with weight ratios of HA to SF was 5:1
HS3-IPN hydrogel	HA/SF hydrogel with weight ratios of HA to SF was 5:3
HS5-IPN hydrogel	HA/SF hydrogel with weight ratios of HA to SF was 5:5
HS7-IPN hydrogel (the model hydrogel)	HA/SF hydrogel with weight ratios of HA to SF was 5:7

References

1. Higuchi, A.; Ling, Q.D.; Chang, Y.; Hsu, S.T.; Umezawa, A. Physical cues of biomaterials guide stem cell differentiation fate. *Chem. Rev.* **2013**, *113*, 3297–3328. [CrossRef] [PubMed]

2. Zhang, J.; Ma, X.; Fan, D.; Zhu, C.; Deng, J.; Hui, J.; Ma, P. Synthesis and characterization of hyaluronic acid/human-like collagen hydrogels. *Mater. Sci. Eng. C* **2014**, *43*, 547–554. [CrossRef] [PubMed]

3. Wu, J.P.J.; Cheng, B.; Roffler, S.R.; Lundy, D.J.; Yen, C.Y.T.; Chen, P.; Lai, J.J.; Pun, S.H.; Stayton, P.S.; Hsieh, P.C.H. Reloadable multidrug capturing delivery system for targeted ischemic disease treatment. *Sci. Transl. Med.* **2016**, *8*, 365ra160. [CrossRef] [PubMed]

4. Lin, C.C.; Metters, A.T. Hydrogels in controlled release formulations: Network design and mathematical modeling. *Adv. Drug Deliv. Rev.* **2006**, *58*, 1379–1408. [CrossRef]

5. Ahrens, M.; Tsantrizos, A.; Donkersloot, P.; Martens, F.; Lauweryns, P.; Le Huec, J.C.; Moszko, S.; Fekete, Z.; Sherman, J.; Yuan, H.A.; et al. Nucleus replacement with the DASCOR disc arthroplasty device: Interim two-year efficacy and safety results from two prospective, non-randomized multicenter European studies. *Spine* **2009**, *34*, 1376–1384. [CrossRef] [PubMed]

6. Joshi, A.; Fussell, G.; Thomas, J.; Hsuan, A.; Lowman, A.; Karduna, A.; Vresilovic, E.; Marcolongo, M. Functional compressive mechanics of a PVA/PVP nucleus pulposus replacement. *Biomaterials* **2006**, *27*, 176–184. [CrossRef]

7. Lanza, R.; Langer, R.; Vancati, J. *Principles of Tissue Engineering*, 3rd ed.; Academic Press: Burlington, VT, USA, 2007.

8.   Tsaryk, R.; Gloria, A.; Russo, T.; Anspach, L.; Santis, R.D.; Ghanaati, S.; Unger, R.E.; Ambrosio, L.; Kirkpatrick, C.J. Collagen-low molecular weight hyaluronic acid semi-interpenetrating network loaded with gelatin microspheres for cell and growth factor delivery for nucleus pulposus regeneration. *Acta Biomater.* **2015**, *20*, 10–21. [CrossRef]

9.   Talebin, S.; Mehrali, M.; Taebnia, N.; Pennisi, C.P.; Kadumudi, F.B.; Foroughi, M.; Hasany, M.; Nikkhah, M.; Akbari, M.; Orive, G.; et al. Self-healing hydrogels: The next paradigm in tissue engineering. *Adv. Sci.* **2019**, *6*, 1801664. [CrossRef]

10. Yang, M.C.; Huang, Y.Y.; Wang, S.S.; Chou, N.K.; Chi, N.H.; Shieh, M.J.; Chang, Y.L.; Chung, T.W. The cardiomyogenic differentiation of rat mesenchymal stem cells on silk fibroin-polysaccharide cardiac patches in vitro. *Biomaterials* **2009**, *30*, 3757–3765. [CrossRef]

11. Su, W.Y.; Chen, Y.C.; Lin, F.H. Injectable oxidized hyaluronic acid/adipic acid di-hydrazide hydrogel for nucleus pulposus regeneration. *Acta Biomater.* **2010**, *6*, 3044–3055. [CrossRef]

12. Chen, Y.C.; Su, W.Y.; Yang, S.H.; Gefen, A.; Lin, F.H. In-situ forming hydrogels composed of oxidized high molecular weight hyaluronic acid and gelatin for nucleus pulposus regeneration. *Acta Biomater.* **2013**, *9*, 5181–5193. [CrossRef] [PubMed]

13. Kim, D.H.; Martin, J.T.; Elliott, D.M.; Smith, L.J.; Mauck, R.L. Phenotypic stability matrix elaboration and functional maturation of nucleus pulposus cells encapsulated in photo-crosslinkable hyaluronic acid hydrogels. *Acta Biomater.* **2015**, *12*, 21–29. [CrossRef] [PubMed]

14. Mehrali, M.; Bagherifard, S.; Akbari, M.; Thakur, A.; Mirani, B.; Mehrali, M.; Hasany, M.; Orive, G.; Das, P.; Dolatshahi-Pirouz, A. Bending electronics with human body: A pathway toward a cybernetic future. *Adv. Sci.* **2018**, *5*, 1700931. [CrossRef] [PubMed]

15. Affas, S.; Schafer, F.-M.; Algarrahi, K.; Cristofaro, V.; Sullivan, P.; Yang, X.; Costa, K.; Sack, B.; Gharaee-Kermani, M.; Macoska, J.A.; et al. Augmentation cystoplasty of diseased porcine bladders with silk fibroin grafts. *Tissue Eng. Part A* **2019**, *25*, 855–866. [CrossRef]

16. Chi, N.H.; Yang, M.C.; Chung, T.W.; Chen, J.Y.; Chou, N.K.; Wang, S.S. Cardiac repair achieved by bone marrow mesenchymal stem cells/silk fibroin/ hyaluronic acid patches in a rat of myocardial infarction model. *Biomaterials* **2012**, *33*, 5541–5550. [CrossRef]

17. Yan, L.P.; Silva-Correia, J.; Ribeiro, V.P.; Miranda-Gonçalves, V.; Correia, C.; da Silva Morais, A.; Sousa, R.A.; Reis, R.M.; Oliveira, A.L.; Oliveira, J.M.; et al. Tumor growth suppression induced by biomimetic silk fibroin hydrogels. *Sci. Rep.* **2016**, *6*, 31037. [CrossRef]

18. Gullbrand, S.E.; Schaer, T.P.; Agarwal, P.; Bendigo, J.R.; Dodge, G.R.; Chen, W.; Elliott, D.M.; Mauck, R.L.; Malhotra, N.R.; Smith, L.J. Translational of an injectable triple-interpenetrating-network hydrogels for intervertebral disc regeneration in a goat model. *Acta Biomater.* **2017**, *60*, 201–209. [CrossRef]

19. Lin, H.A.; Gupta, M.S.; Varma, D.M.; Gilchrist, M.L.; Nicoll, S.B. Lower crosslinking density enhances functional nucleus pulposus-like matrix elaboration by human mesenchymal stem cells in carboxymethylcellulose hydrogels. *J. Biomed. Mater. Res. Part A* **2016**, *104*, 165–177. [CrossRef]

20. Sakai, D.; Mochida, J.; Iwashina, T.; Hiyama, A.; Omi, H.; Imai, M.; Nakal, T.; Ando, K.; Hotta, T. Regenerative effects of transplanting mesenchymal stem cells embedded in aterlo-collagen to the degenerated intervertebral disc. *Biomaterials* **2006**, *27*, 335–345. [CrossRef]

21. Francisco, A.T.; Mancino, R.J.; Bowle, R.D.; Brunger, J.M.; Tainter, D.M.; Chen, Y.T.; Richardson, W.R.; Guilak, F.; Setton, L.A. Injectable laminin-functionalized hydrogel for nucleus pulposus regeneration. *Biomaterials* **2013**, *34*, 7381–7388. [CrossRef]

22. Collin, E.C.; Grad, S.; Zeugolis, D.I.; Vinatier, C.S.; Clouet, J.R.; Guicheux, J.J.; Weiss, P.; Alini, M.; Pandit, A.S. An injectable vehicle for nucleus pulposus cell-based therapy. *Biomaterials* **2011**, *32*, 2862–2870. [CrossRef]

23. Yang, S.Q.; Wang, Q.S.; Tariq, Z.; You, R.C.; Li, X.F.; Li, M.Z.; Zhang, Q. Facile preparation of bioactive silk fibroin/hyaluronic acid hydrogels. *Int. J. Biol. Macromol.* **2018**, *118*, 775–782.

24. Raia, N.R.; Partlow, B.P.; McGill, M.H.; Kimmerling, E.P.; Ghezzi, C.E.; Kaplan, D.L. Enzymatically crosslinked silk-hyaluronic acid hydrogels. *Biomaterials* **2017**, *131*, 58–67. [CrossRef] [PubMed]

25. Kenne, L.; Gohil, S.; Nilsson, E.M.; Karlsson, A.; Ericsson, D.; Helander Kenne, A.; Nord, L.I. Modification and cross-linking parameters in hyaluronic acid hydrogels-definitions and analytical methods. *Carbohydr. Polym.* **2013**, *91*, 410–418. [CrossRef] [PubMed]

26. Tuin, A.; Zandstra, J.; Kluijtmans, S.G.; Bouwstra, J.B.; Harmsen, M.C.; Van Luyn, M.J. Hyaluronic acid-recombinant gelatin gels as a scaffold for soft tissue regeneration. *Eur. Cell Mater.* **2012**, *24*, 320–330. [CrossRef]

27. Su, D.; Yao, M.; Liu, J.; Zhong, Y.; Chen, X.; Shao, Z. Enhancing mechanical properties of silk fibroin hydrogel through restricting the growth of β-sheet domains. *ACS Appl. Mater. Interfaces* **2017**, *9*, 17489–17498. [CrossRef]

28. Iatridis, J.C.; Weidenbaum, M.; Setton, L.A.; Mow, V.C. Is the nucleus pulposus a solid or a fluid? mechanical behaviors of the nucleus pulposus of the human intervertebral disc. *Spine* **1996**, *10*, 1174–1184. [CrossRef]

29. Lee, P.C.; Zan, B.S.; Chen, L.T.; Chung, T.W. Multi-functional PLGA-based Nanoparticles as a Controlled-Release Drug Delivery System for Antioxidant and Therapy. *Int. J. Nanomed.* **2019**, *14*, 1533–1549. [CrossRef]

30. Choi, S.C.; Yoo, M.A.; Lee, S.Y.; Lee, H.J.; Son, D.H.; Jung, J.; Noh, I.; Kim, C.W. Modulation of biomechanical properties of hyaluronic acid hydrogels by crosslinking agents. *J. Biomed. Mater. Res. A* **2015**, *103*, 3072–3080. [CrossRef] [PubMed]

31. Pritchard, E.M.; Hu, X.; Finley, V.; Kuo, C.K.; Kaplan, D.L. Effect of silk protein processing on drug delivery from silk films. *Macromol. Biosci.* **2013**, *13*, 311–320. [CrossRef]

32. Partlow, B.P.; Hanna, C.W.; Rnjas-Kovacina, J.; Moreau, J.E.; Applegate, M.B.; Burke, K.A.; Marelli, B.; Mitropoulos, A.N.; Omenetto, F.G.; Kaplan, D.L. Highly tunable elastomeric silk biomaterials. *Adv. Funct. Mater.* **2014**, *24*, 4615–4624. [CrossRef] [PubMed]

33. Xiao, W.Q.; Qu, X.H.; Li, J.L.; Chen, L.; Tan, Y.F.; Li, J.; Li, B.; Liao, X.L. Synthesis and characterization of cell-laden double-network hydrogels based on silk fibroin and methacrylated hyaluronic acid. *Eur. J. Polym.* **2019**, *118*, 382–392. [CrossRef]

34. Bian, L.; Zhai, D.Y.; Tous, E.; Rai, R.; Mauck, R.L.; Burdick, J.A. Enhanced MSC chondrogenesis following delivery of TGF-"β" 3 from alginate microspheres within hyaluronic acid hydrogels in vitro and in vivo. *Biomaterials* **2011**, *32*, 6425–6434. [CrossRef] [PubMed]

12

PHB is Produced from Glycogen Turn-Over during Nitrogen Starvation in *Synechocystis* sp. PCC 6803

Moritz Koch [1], **Sofía Doello** [1], **Kirstin Gutekunst** [2] **and Karl Forchhammer** [1,*]

[1] Interfaculty Institute of Microbiology and Infection Medicine Tübingen, Eberhard-Karls-Universität Tübingen, 72076 Tübingen, Germany; moritz.koch@uni-tuebingen.de (M.K.); sofia.doello@gmail.com (S.D.)

[2] Department of Biology, Botanical Institute, Christian-Albrechts-University, 24118 Kiel, Germany; kgutekunst@bot.uni-kiel.de

* Correspondence: karl.forchhammer@uni-tuebingen.de

Abstract: Polyhydroxybutyrate (PHB) is a polymer of great interest as a substitute for conventional plastics, which are becoming an enormous environmental problem. PHB can be produced directly from CO_2 in photoautotrophic cyanobacteria. The model cyanobacterium *Synechocystis* sp. PCC 6803 produces PHB under conditions of nitrogen starvation. However, it is so far unclear which metabolic pathways provide the precursor molecules for PHB synthesis during nitrogen starvation. In this study, we investigated if PHB could be derived from the main intracellular carbon pool, glycogen. A mutant of the major glycogen phosphorylase, GlgP2 (*slr1367* product), was almost completely impaired in PHB synthesis. Conversely, in the absence of glycogen synthase GlgA1 (*sll0945* product), cells not only produced less PHB, but were also impaired in acclimation to nitrogen depletion. To analyze the role of the various carbon catabolic pathways (EMP, ED and OPP pathways) for PHB production, mutants of key enzymes of these pathways were analyzed, showing different impact on PHB synthesis. Together, this study clearly indicates that PHB in glycogen-producing *Synechocystis* sp. PCC 6803 cells is produced from this carbon-pool during nitrogen starvation periods. This knowledge can be used for metabolic engineering to get closer to the overall goal of a sustainable, carbon-neutral bioplastic production.

Keywords: cyanobacteria; bioplastic; PHB; sustainable; glycogen; metabolic engineering; Synechocystis

1. Introduction

Cyanobacteria are among the most widespread organisms on our planet. Their ability to perform oxygenic photosynthesis allows them to grow autotrophically with CO_2 as the sole carbon source [1]. Additionally, many cyanobacteria acquired the ability to fix nitrogen, one of the most limiting nutrients [2]. However, many others are not able to fix nitrogen, one of them being the well-studied model organism *Synechocystis* sp. PCC 6803 (hereafter: *Synechocystis*) [3]. Nitrogen starvation starts a well-orchestrated survival process in *Synechocystis*, called chlorosis [4]. During chlorosis, *Synechocystis* degrades not only its photosynthetic machinery, but also accumulates large quantities of biopolymers, namely glycogen and poly-hydroxy-butyrate (PHB) [5]. Glycogen synthesis following the onset of nitrogen starvation serves transiently as a major sink for newly fixed CO_2 [6] before CO_2 fixation is tuned down during prolonged nitrogen starvation. During resuscitation from chlorosis, a specific glycogen catabolic metabolism supports the re-greening of chlorotic cells [7]. By contrast to the pivotal role of glycogen, the function of the polymer PHB remains puzzling, since mutants impaired in PHB synthesis survived and recovered from chlorosis as awild-type [8,9]. Nevertheless, many different cyanobacterial species produce PHB, implying a hitherto unrecognized functional importance [10]. In other microorganisms PHB fulfills various functions during conditions of unbalanced nutrient availability and can also protect cells against low temperatures or redox stress [11–13]. Understanding the intracellular mechanisms that lead to PHB production could help to elucidate the physiological role of this polymer. Regardless of the physiological significance of PHB, this polymer has been recognized as a promising alternative for current plastics, which contaminate terrestrial and aquatic ecosystems [14]. PHB can serve as a basis for completely biodegradable plastics, with properties comparable to petroleum-derived plastics [15,16]. Since *Synechocystis* produces PHB only under nutrient limiting conditions, this phenomenon can be exploited to temporally separate the initial biomass production from PHB production induced by shifting cells to nitrogen limiting conditions [10].

One of the biggest obstacles preventing economic PHB production in cyanobacteria remains the low level of intracellular PHB accumulation [17]. While chemotrophic bacteria are capable of producing more than 80% PHB of their cell dry mass, (e.g., *Cupriavidus necator*), most cyanobacteria naturally produce less than 20% of their cell dry mass [15]. Additionally, their growth rate is too slow to compete with the PHB production in chemotrophic bacteria. There have been many attempts in the past to further improve the intracellular PHB production, often with limited [1] success [18–20]. One of the most successful approaches has been achieved by random mutagenesis, leading to up to 37% PHB of the cell dry mass [21]. However, more directed approaches involving genetic engineering are often limited by a lack of knowledge about how the cells' metabolism works in detail. For example, until today, it was still unknown from which carbon metabolites PHB was derived. There have been several different studies analyzing the intracellular fluxes in cyanobacteria [22]. However, most of them did not analyze the carbon flow during prolonged nitrogen starvation. One of these studies showed that in nitrogen-starved photosynthetically grown cyanobacteria up to 87% of the carbon in PHB is derived from intracellular carbon sources rather than from newly fixed CO_2 [23]. However, until now, it was not clearly resolved which metabolic routes provide the precursors for PHB synthesis. This knowledge would lay the foundation for future metabolic engineering approaches to create overproduction strains. Hence, the goal of this study was to find out where the carbon for the PHB production is coming from and which pathways it is taking until it reaches PHB.

It has been shown that disruption of PHB synthesis results in an increased production of glycogen; however, an overproduction of glycogen did not lead to higher amounts of PHB [24]. Another study that also investigated the accumulation of glycogen in a PHB-free mutant ΔphaC, could not detect any differences in growth or glycogen accumulation [8].

An important aspect in the issue concerning the relation between glycogen and PHB metabolism deals with the contribution of various carbon metabolic pathways for the production of precursors for PHB under conditions of nitrogen limitation. *Synechocystis* is able to catabolize glucose via three parallel operating glycolytic pathways [25] (Figure 1): the Embden-Meyerhof-Parnas (EMP) pathway, the oxidative pentose phosphate (OPP) pathway [26], and the Entner Doudoroff (ED) pathway [25]. When nitrogen-starved cells recover from chlorosis, they require the parallel operating OPP and ED pathways, whereas the EMP pathway seems dispensable [7]. Metabolic analysis of mutants overexpressing the transcriptional regulator *rre37* showed a correlated upregulation of PHB synthesis and EMP pathway genes (*phaAB* and *pfkA*, respectively) [27]. However, so far is has not been investigated, how important these pathways for the production of PHB during nitrogen starvation.

This work started with the initial aim to define whether PHB synthesis depends on the metabolism of glycogen. Since the initial results implied that PHB is strongly affected by glycogen catabolism, we further investigated the importance of the different carbon pathways EMP, ED and OPP for the production of PHB. These findings shall help to further understand the intracellular PHB metabolism in cyanobacteria, which can be used to create more efficient PHB overproduction strains, making the production of PHB as a bioplastic more cost efficient.

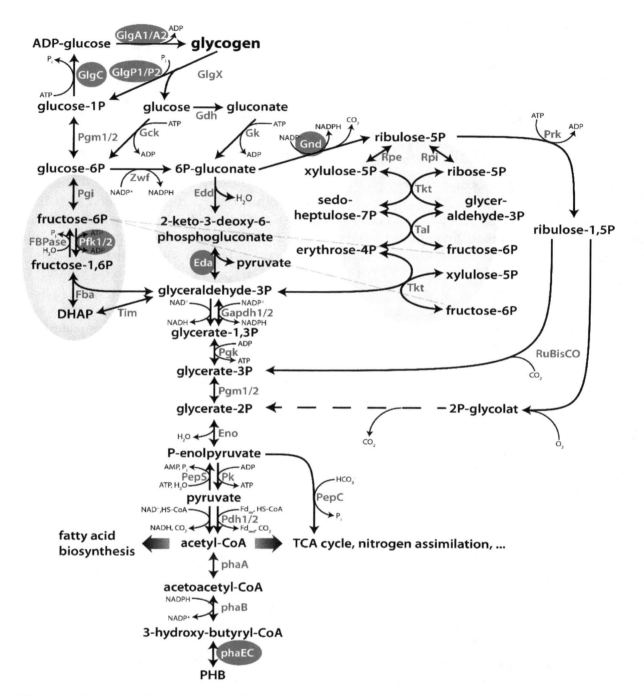

Figure 1. Overview of central metabolism of Synechocystis. Genes which were deleted in this study are highlighted in with a red background. Dotted lines represent several enzymatic reactions. The EMP, ED and OPP (Embden-Meyerhof-Parnas, Entner-Doudoroff, Oxidative Pentose Phosphate) pathways are highlighted in green, blue and yellow, respectively.

2. Results

Following the onset of nitrogen-starvation, large quantities of fixed carbon are stored in *Synechocystis* cells as glycogen granules. Long-term starvation experiments of *Synechocystis* cultures have shown that, while cells are chlorotic, glycogen is slowly degraded, following its initial rapid accumulation but PHB is slowly and steadily accumulating [9]. Considering that chlorotic cells are photosynthetically inactive, these data could indicate a potential correlation between the turn-over of glycogen and the synthesis of PHB. An overview of the metabolic pathways connecting the glycogen pool with PHB is shown in Figure 1. To substantiate the hypothesis that PHB might be derived from glycogen turn-over, we investigated PHB accumulation in various mutant strains, in which key steps in different pathways are interrupted. The respective mutations are shown in Figure 1. All strains used in this work were characterized previously, with their phenotypes, including growth behaviours, described in the respective publications (see Table A1). Furthermore, all mutants used in these studies were fully segregated to ensure clear phenotypes.

2.1. Impact of Glycogen Synthesis on PHB Production

To analyze the role of glycogen synthesis on the production of PHB, we first analyzed the accumulation of these biopolymers during nitrogen starvation in mutants with defects in glycogen synthesis. The double mutant of the two glycogen synthase genes glgA1 (sll0945) and glgA2 (sll1393) is unable to acclimate to nitrogen deprivation and rapidly dies upon shifting cells to nitrogen free BG11^0 medium [8] and, therefore, could not be analyzed. Instead, we used a knockout mutant of the glucose-1-phosphate adenylyltransferase (glgC, slr1176) and two knockout strains of each of the isoforms of the glycogen synthase, glgA1 (sll0945) and glgA2 (sll1393). Yoo et al. [28] reported that the single glgA1 and glgA2 mutants were still able to produce similar amounts of glycogen as the wild-type (WT), since one glycogen synthase is still present, and this seems to be sufficient to reach the wild-type levels of glycogen. However, the structure of the glycogen produced by the two isoforms seemed to slightly differ in chain-length distribution [28]. In that study, no distinguishing phenotype of the two mutant strains had been reported. In the present study, the cultures were shifted to nitrogen free medium BG11^0 and further incubated under constant illumination of 40 μmol photons m^{-2} s^{-1}. Under these experimental conditions, the ΔglgA1 mutant showed an impaired chlorosis reaction, whereas the ΔglgA2 mutant performed chlorosis as the wild-type strain (Figure 2A). To further determine the viability of two weeks nitrogen-starved cells, serial dilutions were dropped on nitrate-supplemented BG11 plates. As shown in Figure 2B, the ΔglgA1 mutant was severely impaired in recovering from nitrogen starvation, whereas ΔglgA2 could recover from chlorosis with the same efficiency as the wild-type (Figure 2B).

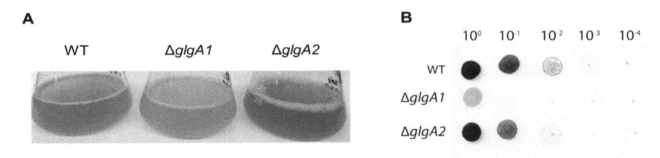

Figure 2. Characterization of the glycogen synthase mutants, ΔglgA1 and ΔglgA2. (**A**) Cultures after five days of nitrogen starvation. (**B**) Recovery assay of chlorotic wild-type (WT) and mutants ΔglgA1 and ΔglgA2, using the drop agar method. Cultures that were nitrogen-starved for 14 days were serially diluted from 1 to 1:10,000 and from each dilution, a drop of 5 μL was plated on BG11 agar and grown for seven days.

During the course of three weeks of nitrogen starvation, the quantities of PHB and glycogen that accumulate in the cells were determined (Figure 3A,B).

In the wild-type, the amount of glycogen already peaked after the first week and slowly decreased in the following two weeks (Figure 3B). As previously reported by Yoo et al. [28], the single ΔglgA1 and ΔglgA2 mutants initially accumulated similar amounts of glycogen to the wild-type, but in contrast to the wild-type, the level of glycogen remained high. The PHB content in the glgA2 mutant was similar to the wild-type for the first seven days of nitrogen starvation, but PHB accumulation slowed down afterwards (Figure 3B). By contrast, the glgA1 mutant was strongly impaired in PHB production. Together, the phenotype of the glgA1 and glgA2 mutants indicates that glycogen synthase GlgA1 plays a much more important role in nitrogen starvation acclimation than GlgA2, although the amount of glycogen produced by these two strains is almost the same. One explanation could be that the subtle differences in the glycogen produced form the two isoenzymes might result in different functions, with GlgA1-produced glycogen being much more relevant for the maintenance metabolism in chlorotic cells and for the resuscitation from chlorosis than glycogen produced by GlgA2. In clear correlation with the redundant role of GlgA2, the *glgA2* mutant was not impaired in PHB synthesis, whereas mutation of the functionally important *glgA1* gene resulted in strongly impaired PHB synthesis.

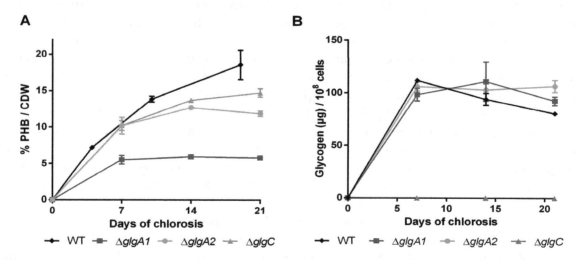

Figure 3. Polyhydroxybutyrate (PHB) content in percentage of cell dry weight (CDW) (**A**) and cellular glycogen content (**B**) of mutants impaired in the glycogen synthesis. Cultures were shifted to nitrogen free medium at day 0 and were subsequently grown for 21 days. Each point represents a mean of three independent biological replicates.

The *glgC* mutant was previously characterized by Grundel et al. [6]. They showed that Δ*glgC* is not able to perform a proper nitrogen-starvation acclimation response: it maintains its pigments while it loses viability, which was also observed in our experiments. Namakoshi et al. [29] showed that this mutant is unable to synthesize glycogen, which is also in line with our results. Under our conditions, unlike previously described by Damrow et al. [8], the *glgC* mutant did not show an increased amount of PHB compared to the WT, but seemed to accumulate less PHB instead (Figure 3A). It has to be noted though that Damrow et al. [8] investigated only one single timepoint, after seven days of nitrogen starvation. In addition, these results should be treated with care, since PHB content is normalized to cell dry weight, which shrinks in the *glgC* mutant due to progressive cell lysis. Consequently, the cell density was severely diminished at the end of the experiment (OD$_{750}$ of 0.51, compared to ~1.15 of other mutants and the wild-type). The differences between our study and that of Damrow et al. [8] thus may result from differences in cell lysis rather than from differences in PHB synthesis. When the relatively low cell density of the *glgC* mutant is considered, it produces much less PHB per volume compared to the wild-type.

2.2. Impact of Glycogen Degradation on PHB Production

If glycogen turn-over would result in PHB accumulation during chlorosis, synthesis of PHB should be abrogated when glycogen degradation is impaired. To test this assumption, mutants in catabolic glycogen phosphorylase genes (*glgP*) were investigated with the same methods as described above. Glycogen can be degraded by the two glycogen phosphorylase isoenzymes, encoded by *glgP1* (slr1356) and *glgP2* (slr1367) [7]. A detailed study by Doello et al. [7] showed that GlgP2 is the main enzyme responsible for glycogen degradation during resuscitation from nitrogen chlorosis. Knocking out GlgP1 (Δ*glgP1*) does not affect the efficiency of recovery, whereas knocking out GlgP2 (Δ*glgP2*) or both phosphorylases (Δ*glgP1/2*) completely impairs the ability to degrade glycogen [7]. Here, we investigated glycogen and PHB accumulation during three weeks of nitrogen starvation in these glycogen phosphorylases mutants.

Although the initial amount of glycogen was higher in the *glgP1* mutant compared to the WT (Figure 4B), the amount decreased during the course of the experiment. By contrast, no glycogen degradation occurred in the Δ*glgP2* and Δ*glgP1/2* double mutant. This correlates with the specific requirement of chlorotic cells for GlgP2 for resuscitation from nitrogen starvation, as it has been previously described [7].

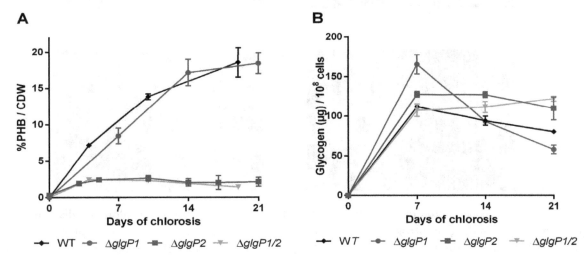

Figure 4. PHB content in percentage of cell dry weight (CDW) (**A**) and glycogen content (**B**) of mutants impaired in the glycogen degradation. Cultures were shifted to nitrogen free medium at day 0 and were subsequently grown for 21 days. Each point represents a mean of three independent biological replicates.

Intriguingly, the different mutants showed a drastic difference in the amounts of PHB being produced (Figure 4A): While the Δ*glgP1* strain produced similar amounts of PHB as the wild-type, a strong decrease was observed for the Δ*glgP2* strain. The same phenotype was observed for the double knockout mutant, indicating that origin of the effect is based on the absence of *glgP2*. The PHB synthesis phenotypes were further confirmed by fluorescence microscopy after staining PHB with Nile red (Figure 5). PHB granules appear as bright red fluorescing intracellular granular structures. In agreement with the results from PHB quantification by HPLC analysis, the Δ*glgP1* strain showed similar amounts and distribution of PHB granules than the wild-type. By contrast, only very small granules, if at all visible, could be detected in the strains Δ*glgP2* and Δ*glgP1/2*.

Altogether, the inability of the mutants Δ*glgP2* and Δ*glgP1/2* to accumulate PHB demonstrates unequivocally that glycogen catabolism through GlgP2 is required for the ongoing PHB synthesis during prolonged nitrogen starvation.

2.3. Impact of Mutations in Carbon Catabolic Pathway on PHB Production

The experiments outlined above revealed that specific glycogen synthesizing or degrading enzymes have a strong effect on the amounts of PHB being produced and that glycogen turn-over via GlgP2 provides the carbon skeletons for PHB synthesis. To investigate how the released glucose phosphate molecules are metabolized downstream of glycogen, knockouts of the three most important glycolytic routes [25] were checked for their PHB and glycogen production during chlorosis. While the strain Δ*eda* (*sll0107*) lacks the ability to metabolize molecules via the ED pathway, Δ*gnd* is not able to use the OPP pathway. Additionally, the strain Δ*pfk1/2* lacks both phosphofructokinases, which causes an interruption of the EMP pathway. Also, the individual knockouts of both isoforms, Δ*pfk1* and Δ*pfk2*, were investigated.

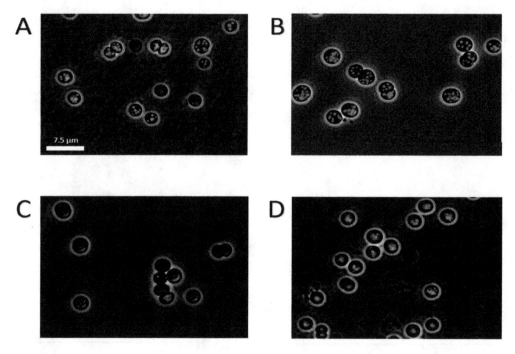

Figure 5. Fluorescence microscopic picture of Nile-red stained PHB granules in chlorotic cells. Cultures where grown for 14 days in nitrogen depleted medium BG11^0. Shown is an overlay of phase contrast with a CY3 channel of the WT (**A**), Δ*glgP1* (**B**), Δ*glgP2* (**C**) and Δ*glgP1/2* (**D**). Scale bar corresponds to 7.5 μm.

Again, the different mutant strains and a WT control were grown for three weeks under nitrogen deprived conditions and PHB and glycogen content was quantified (Figure 6).

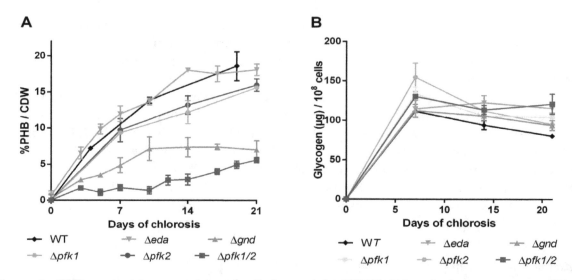

Figure 6. PHB content in percentage of cell dry weight (CDW) (**A**) and glycogen content (**B**) of mutants with disrupted carbon pathway. Cultures were shifted to nitrogen free medium at day 0 and were subsequently grown for 21 days. Each point represents a mean of three independent biological replicates.

Distortion of the ED pathway (Δ*eda*) did not result in any PHB phenotype different from the WT and the glycogen content remained high during the course of the experiment. In the Δ*gnd* mutant, a slower increase of PHB than in the WT was observed within the first ten days (Figure 6A) and PHB accumulation subsequently ceased. When the EMP pathway was blocked (Δ*pfk1/2*), only very little PHB was produced in the first two weeks of the chlorosis. Thereafter though, PHB production slightly increased and finally reached similar levels as in the Δgnd mutant. The total amount of glycogen over the time of chlorosis did not decrease in this mutant. The single Δpfk mutants showed PHB contents similar to the WT, indicating that the two isoenzymes are able to replace each other's function. Taken together, it appears that EMP and OPP pathways contribute to PHB production, whereas the ED pathway does not play a role.

2.4. Impact of PHB Formation on Glycogen Synthesis

In order to check how the PHB production affects the accumulation of glycogen, a PHB-free mutant, namely $\Delta phaEC$, was checked for its production of carbon polymers (Figure 7).

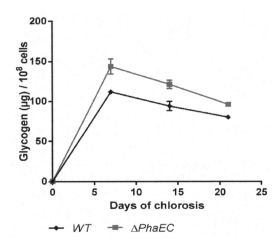

Figure 7. Glycogen content of wild-type and mutant lacking the PHB synthase genes (*PhaEC*). Cultures were shifted to nitrogen free medium at day 0 and were subsequently grown for 21 days. Each point represents a mean of three independent biological replicates.

As expected, the mutant was unable to synthesize PHB (data not shown [8]). Compared to the WT, the mutant produced moderately higher amounts of glycogen and degrades it slightly faster, so that at the end of the experiments, the glycogen levels were quite similar.

3. Discussion

As recently shown by isotope labeling experiments [23], the majority of the carbon from PHB is coming from intracellular metabolites, which contribute around 74% to the carbon within PHB. Additionally, a random mutagenized strain, which is an overproducer of PHB, shows also a strongly accelerated decay of glycogen [30]. Here, we provide clear evidence that the intracellular glycogen pool and its products provide the carbon metabolites for PHB synthesis during nitrogen starvation. In the absence of glycogen degradation, as it is the case in the $\Delta glgP2$ and the $\Delta glgP1/2$ double mutant, PHB synthesis is almost completely abrogated. In the $\Delta glgP2$ mutant, the remaining GlgP1 enzyme is apparently not efficiently catabolizing glycogen, which agrees with its lack of function for the resuscitation from chlorosis. By contrast, GlgP2 is required for glycogen catabolism and resuscitation from chlorosis [7]. From these data, it is reasonable to hypothesize that during chlorosis, GlgP2 slowly degrades glycogen, and the degradation products end up in the PHB pool.

Like the two glycogen phosphorylase isoenzymes, the two glycogen synthase isoenzymes GlgA1 and GlgA2 appear to have specialized functions. Even though both $\Delta glgA1$ and $\Delta glgA2$ synthesized similar amounts of glycogen, deletion of *glgA1* resulted in a mutant with reduced bleaching, viability and PHB content whereas $\Delta glgA2$ was less affected. This indicates that glycogen produced by GlgA1 is important for resuscitation and growth. On the other hand, GlgA2 produced glycogen appeared less important or its function is yet unknown. Previous publications did not see such a difference, which may be explained by a much shorter time of nitrogen starvation used in their study [6]. Taken together, those mutants (*glgP2* and $\Delta glgA1$) that were less viable under nitrogen starvation, did also synthesize less PHB. It remains to be demonstrated, if the different impact that GlgA1 and GlgA2 exert on viability and PHB synthesis originates from the slightly different branching patterns [28] of the glycogen, that they synthesize.

In the ΔglgC mutant, PHB is formed, although no glycogen is produced (Figure 3), which seems to contradict the hypothesis of glycogen-derived PHB [8,28]. Taking into account the above results, it appears likely that the carbon metabolites used for PHB can under certain conditions bypass the glycogen pool. When glgC is knocked out, glucose-1P (and its precursor, glucose-6P) cannot be further converted into ADP-glucose and may accumulate. The glucose-phosphates could then be downstream metabolized to glyceraldehyde-3P and further converted into PHB. By contrast, when glgA1 is mutated, the newly fixed carbon can be converted by GlgC to ADP-glucose and subsequently enters the GlgA2-synthesized inactive glycogen pool, where it cannot be further metabolized into PHB. A similar connection between PHB and glycogen has already been described in other organisms (Sinorhizobium meliloti), where PHB levels were lower in a mutant lacking glgA1 [31]. Since the GlgA1/GlgA2 double mutant rapidly dies upon nitrogen starvation [6], the impact of the complete absence of glycogen due to glycogen synthase deficiency on PHB synthesis cannot be experimentally tested.

Furthermore, we observed that a slower degradation of glycogen often correlates with a low-PHB-phenotype, as seen in the case of ΔglgA1, ΔglgP2, ΔglgP1/2 and Δpfk1/2 (Figures 3, 4 and 6, respectively). This further supports the hypothesis, that only when glycogen gets degraded during the process of chlorosis, PHB is formed. In two cases, ΔglgP1/2 and Δpfk1/2, the amount of glycogen was even increasing during the later course of nitrogen starvation. This hints towards an ongoing glycogen formation during nitrogen chlorosis, which gets only visible once glycogen degradation is disturbed. Apparently, glycogen metabolism is much more dynamic than presumed from the relative static pool size observed in the wild-type. A steady glycogen synthesis may be counterbalanced by ongoing degradation, together resulting in only a slow net change of its pool size.

The residual metabolism in nitrogen starved chlorotic cells [9,32] is probably required to ensure long-term survival through repair of essential biomolecules such as proteins, DNA and RNA, osmoregulation, regulated shifts in metabolic pathways and the preparation for a quick response as soon as nutrients are available again [33,34]. According to these needs, non-growing starved cells still require a constant supply of ATP, reduction equivalents and the ability to produce cellular building blocks for survival. In line with this, we observed that PHB production was mainly achieved via the EMP pathway, which has the highest ATP yield among the three main carbon catalytic pathways (EMP, OPP, ED). In addition, we found that the OPP pathway is involved in PHB synthesis as well. This pathway provides metabolites for biosynthetic purposes as the repair of biomolecules for maintenance. By contrast, deletion of the ED pathway, which has a lower ATP yield in comparison to the EMP pathway and is physiologically probably most important in connection with photosynthesis and the Calvin–Benson cycle [7,25], did not impair PHB production under nitrogen starvation. Nevertheless, the glycogen levels did not decrease in the Δgnd mutant, implying that mutation of the ED pathway affects the dynamics of glycogen turn-over discussed above.

Finally, we observed that the various carbon catabolic pathways have different functional importance for PHB production, in a time-dependent manner: While the mutant Δgnd (blocking the OPP pathway) produced PHB in the first phase of the experiment but later stopped its synthesis, the Δpfk1/2 mutant (impaired in the EMP pathway) was initially blocked in PHB accumulation but later started to produce it (Figure 4). Interestingly, the time point, at which the Δgnd mutant stopped PHB production matched its start in the Δpfk1/2 mutant. This could indicate a consecutive role of EMP and OPP pathway during nitrogen chlorosis. Although the exact function of the EMP pathway remains unknown, we show here for the first time a phenotype of a cyanobacterial mutant lacking this pathway. This suggested to also investigate the deletion of the individual knockouts. Deletion of only one of the Pfk isoenzymes (pfk1: sll1196 and pfk2: sll0745) resulted in mutants that produced about 80% of the PHB of WT cells, whereas the PHB production of the double mutant Δpfk1/2 was severely reduced. Pfk1 and Pfk2 can thus obviously compensate for the loss of the respective other, even though PHB production is highest if both enzymes are present. This observation is well in line with transcriptomic studies, which detected an increase in expression of both Pfk isoenzymes during nitrogen starvation [5]. The observation that both EMP and OPP pathway are of importance during arrested growth under nitrogen starvation is in agreement with earlier investigations that reported the upregulation of the sugar catabolic genes *pfk1*, *pfk2*, *zwf*, *gnd* and *gap1* concomitantly with glycogen accumulation [5,35]. EMP and OPP pathway thus support PHB production in non-growing, nitrogen-starved cells, whereas ED and OPP pathway are most important during resuscitation form nitrogen chlorosis after feeding the cells with nitrate [7].

The mutant Δ*phaEC* did not produce any PHB and degraded glycogen similarly to the WT (Figure 7). This indicates that there is no direct feedback between these two polymer pools. In the absence of PHB synthesis, metabolites form glycogen degradation could be leaked by overflow reactions. Under unbalanced metabolic situations, it has been shown that cyanobacteria can excrete metabolites into the medium to control their intracellular energy status [6,36,37]. In any case, this result demonstrates that PHB and glycogen do not compete for CO_2 fixation products, but glycogen is epistatic over PHB synthesis.

Previous studies showed that, under nitrogen starvation, certain genes are upregulated, which are under the regulation of SigE, a group 2 σ factor [38]. Among these genes are glycogen degrading enzymes like *glgP1* and *glgP2* and *glgX*, but also the key enzymes for the pathways further downstream, namely *pfk* and *gnd*. The fact that all these genes are expressed simultaneously with the genes of the PHB synthesis [9], demonstrate that all relevant transcripts of the key enzymes required for the conversion from glycogen to PHB are present during nitrogen starvation. Our finding that PHB is mainly synthesized from glycogen degradation during nitrogen chlorosis is supported by a recent study, where a *Synechocystis* sp. PCC 6714 strain with enhanced PHB accumulation was created by random mutagenesis. Transcriptome analysis revealed that this strain exhibits an increased expression of glycogen phosphorylase [21]. This indicates that manipulation of glycogen metabolism may be a key for improved PHB synthesis.

Gaining further insights into the intracellular carbon fluxes could provide more information on how PHB production is regulated. Once the regulation is understood, this knowledge could be used to redirect the large quantities of glycogen towards PHB. This knowledge could be used in metabolic engineering approaches to either completely reroute the carbon from glycogen (making up more than 60% of the CDW) to PHB, for example by overexpression of glycogen degrading enzymes, or even from inorganic carbon to PHB directly. Therefore, the new insights from this work can be exploited for biotechnological applications to further increase the amounts of PHB being produced in cyanobacteria.

4. Materials and Methods

4.1. Cyanobacterial Cultivation Conditions

For standard cultivation, *Synechocystis* sp. PCC 6803 cells were grown in 200 mL BG_{11} medium, supplemented with 5 mM $NaHCO_3$ [39]. A list of the used strains of this study is provided in Table A1. Two different wild-type strains, a Glc sensitive and a Glc tolerant one, were used. Both strains showed the same behavior during normal growth as well as during chlorosis. Appropriate antibiotics were added to the different mutants to ensure the continuity of the mutation. The cells were cultivated at 28 °C, shaking at 120 rpm and constant illumination of 40–50 µmol photons m^{-2} s^{-1}. Nitrogen starvation was induced as described previously [40]. In short, exponentially growing cells (OD 0.4–0.8) were centrifuged for 10 min at 4000× g. The cells were washed in 100 mL of BG_0 (BG_{11} medium without $NaNO_3$) before they were centrifuged again. The resulting pellet was resuspended in BG_0 until it reached an OD of 0.4.

4.2. Microscopy and Staining Procedures

To observe PHB granules within the cells, 100 µL of cyanobacterial cells were centrifuged (1 min at 10,000× g) and 80 µL of the supernatant discarded. Nile Red (10 µL) was added and used to resuspend the pellet in the remaining 20 µL of the supernatant. From these mixtures, 10 µL were taken and applied on an agarose coated microscope slide to immobilize the cells. The Leica DM5500B microscope (Leica, Wetzlar, Germany) was used with a 100×/1.3 oil objective for fluorescence microscopy. To detect Nile red stained PHB granules, an excitation filter BP 535/50 was used, together with a suppression filter BP 610/75. A Leica DFC360FX (Leica, Wetzlar, Germany)) was used for image acquisition.

4.3. PHB Quantification

PHB content within the cells was determined as described previously [41]. Roughly 15 mL of cells were harvested and centrifuged at 4000× g for 10 min at 25 °C. The resulting pellet was dried for 3 h at 60 °C in a speed-vac (Christ, Osterode, Germany), before 1 mL of concentrated H_2SO_4 was added and boiled for 1 h at 100 °C to break up the cells and to convert PHB to crotonic acid. From this, 100 µL were taken and diluted in 900 µL 0.014 M H_2SO_4. To remove cell debris, the samples were centrifuged for 10 min at 10,000× g, before 500 µL of the supernatant were transferred to 500 µL 0.014 M H_2SO_4. After an additional centrifugation step with the same conditions as above, the supernatant was used for HPLC analysis on a Nucleosil 100 C 18 column (Agilent, Santa Clara, CA, USA) (125 by 3 mm). As a liquid phase, 20 mM phosphate buffer (pH 2.5) was used. Commercially available crotonic acids was used as a standard with a conversion ratio of 0.893. The amount of crotonic acid was detected at 250 nm.

4.4. Glycogen Quantification

Intracellular glycogen content was measured by harvesting 2 mL of cyanobacterial culture. The cells were washed twice with 1 mL of ddH_2O. Afterwards the pellet was resuspended in 400 µL KOH (30% w/v) and incubated for 2 h at 95 °C. For the subsequent glycogen precipitation, 1200 µL ice cold ethanol (final concentration of 70%) were added. The mixture was incubated at –20 °C for 2–24 h. Next, the solution was centrifuged at 4 °C for 10 min at 10,000× g. The pellet was washed twice with 70% and 98% ethanol and dried in a speed-vac for 20 min at 60 °C. Next, the pellet was resuspended in 1 mL of 100 mM sodium acetate (pH 4.5) and 8 µL of an amyloglucosidase solution (4.4 U/µL) were added. For the enzymatic digest, the cells were incubated at 60 °C for 2 h. For the spectrometical glycogen determination, 200 µL of the digested mixture was used and added to 1 mL of O-toluidine-reagent (6% O-toluidine in 100% acetic acid). The tubes were incubated for 10 min at 100 °C. The samples were cooled down on ice for 3 min, before the OD_{635} was measured. The final result was normalized to the cell density at OD_{750}, where $OD_{750} = 1$ represents 10^8 cells. A glucose standard curve was used to calculate the glucose contents in the sample from their OD540.

4.5. Drop Agar Method

Serial dilutions of chlorotic cultures were prepared (10^0, 10^{-1}, 10^{-2}, 10^{-3}, 10^{-4} and 10^{-5}) starting with an OD_{750} of 1. Five microliters of these dilutions were dropped on solid BG_{11} agar plates and cultivated at 50 µmol photons m^{-2} s^{-1} and 27 °C for 7 days.

Author Contributions: Conceptualization, M.K. and K.F.; Methodology, M.K., S.D. and K.F.; Investigation, M.K. and S.D.; Writing-Original Draft Preparation, M.K., S.D. and K.F.; Writing-Review & Editing, M.K., S.D., K.G. and K.F.; Supervision, K.F.; Project Administration, M.K. and K.F.

Acknowledgments: We would like to thank Yvonne Zilliges for providing the mutants ΔglgA1, ΔglgA2 and ΔglgC. Furthermore, we thank Eva Nussbaum for maintaining the strain collection and Andreas Kulick for assistance with HPLC analysis.

Appendix A

Table A1. List of used strains.

Strain	Background	Relevant Marker of Genotype	Reference
Synechocystis sp. PCC 6803 GS	GS	-	Pasteur culture collection
Synechocystis sp. PCC 6803 GT	GT	-	Chen et al. 2016
Δ*glgA1*	GT	*sll0945::kmR*	Gründel et al. 2012
Δ*glgA2*	GT	*sll1393::cmR*	Gründel et al. 2012
Δ*glgC*	GT	*slr1176::cmR*	Damrow et al. 2012
Δ*glgP1*	GS	*sll1356::kmR*	Doello et al. 2018
Δ*glgP2*	GS	*slr1367::spR*	Doello et al. 2018
Δ*glgP1/2*	GS	*sll1356::kmR, slr1367::spR*	Doello et al. 2018
Δ*eda*	GT	*sll0107::gmR*	Chen et al. 2016
Δ*gnd*	GT	*sll0329::gmR*	Chen et al. 2016
Δ*pfkB1/2*	GT	*sll1196::kmR, sll0745::spR*	Chen et al. 2016
Δ*phaEC*	GS	*slr1829, slr1830::kmR*	Klotz et al. 2016

References

1. Soo, R.M.; Hemp, J.; Parks, D.H.; Fischer, W.W.; Hugenholtz, P. On the origins of oxygenic photosynthesis and aerobic respiration in Cyanobacteria. *Science* **2017**, *355*, 1436–1440. [CrossRef] [PubMed]
2. Vitousek, P.M.; Howarth, R.W. Nitrogen Limitation on Land and in the Sea—How Can. It Occur. *Biogeochemistry* **1991**, *13*, 87–115. [CrossRef]
3. Kaneko, T.; Tanaka, A.; Sato, S.; Kotani, H.; Sazuka, T.; Miyajima, N.; Sugiura, M.; Tabata, S. Sequence analysis of the genome of the unicellular cyanobacterium *Synechocystis* sp. strain PCC6803. I. Sequence features in the 1 Mb region from map positions 64% to 92% of the genome. *DNA Res.* **1995**, *2*, 153–166. [CrossRef]
4. Forchhammer, K.; Schwarz, R. Nitrogen chlorosis in unicellular cyanobacteria—a developmental program for surviving nitrogen deprivation. *Environ. Microbiol.* **2018**. [CrossRef] [PubMed]
5. Osanai, T.; Oikawa, A.; Shirai, T.; Kuwahara, A.; Iijima, H.; Tanaka, K.; Ikeuchi, M.; Kondo, A.; Saito, K.; Hirai, M.Y. Capillary electrophoresis-mass spectrometry reveals the distribution of carbon metabolites during nitrogen starvation in *Synechocystis* sp. PCC 6803. *Environ. Microbiol.* **2014**, *16*, 512–524. [CrossRef] [PubMed]
6. Grundel, M.; Scheunemann, R.; Lockau, W.; Zilliges, Y. Impaired glycogen synthesis causes metabolic overflow reactions and affects stress responses in the cyanobacterium *Synechocystis* sp. PCC 6803. *Microbiology* **2012**, *158*, 3032–3043. [CrossRef] [PubMed]
7. Doello, S.; Klotz, A.; Makowka, A.; Gutekunst, K.; Forchhammer, K. A Specific Glycogen Mobilization Strategy Enables Rapid Awakening of Dormant Cyanobacteria from Chlorosis. *Plant Physiol.* **2018**, *177*, 594–603. [CrossRef]
8. Damrow, R.; Maldener, I.; Zilliges, Y. The Multiple Functions of Common Microbial Carbon Polymers, Glycogen and PHB, during Stress Responses in the Non-Diazotrophic Cyanobacterium *Synechocystis* sp. PCC 6803. *Front. Microbiol.* **2016**, *7*, 966. [CrossRef]
9. Klotz, A.; Georg, J.; Bučinská, L.; Watanabe, S.; Reimann, V.; Januszewski, W.; Sobotka, R.; Jendrossek, D.; Hess, W.R.; Forchhammer, K. Awakening of a Dormant Cyanobacterium from Nitrogen Chlorosis Reveals a Genetically Determined Program. *Curr. Biol.* **2016**, *26*, 2862–2872. [CrossRef]
10. Ansari, S.; Fatma, T. Cyanobacterial Polyhydroxybutyrate (PHB): Screening, Optimization and Characterization. *PLoS ONE* **2016**, *11*, e0158168. [CrossRef]
11. Nowroth, V.; Marquart, L.; Jendrossek, D. Low temperature-induced viable but not culturable state of Ralstonia eutropha and its relationship to accumulated polyhydroxybutyrate. *FEMS Microbiol. Lett.* **2016**, *363*, fnw249. [CrossRef]

12. Batista, M.B.; Teixeira, C.S.; Sfeir, M.Z.T.; Alves, L.P.S.; Valdameri, G.; Pedrosa, F.O.; Sassaki, G.L.; Steffens, M.B.R.; de Souza, E.M.; Dixon, R.; et al. PHB Biosynthesis Counteracts Redox Stress in Herbaspirillum seropedicae. *Front. Microbiol.* **2018**, *9*, 472. [CrossRef] [PubMed]

13. Urtuvia, V.; Villegas, P.; González, M.; Seeger, M. Bacterial production of the biodegradable plastics polyhydroxyalkanoates. *Int. J. Biol. Macromol.* **2014**, *70*, 208–213. [CrossRef] [PubMed]

14. Li, W.C.; Tse, H.F.; Fok, L. Plastic waste in the marine environment: A review of sources, occurrence and effects. *Sci. Total Environ.* **2016**, *566*, 333–349. [CrossRef] [PubMed]

15. Drosg, B.; Gattermayr, F.; Silvestrini, L. Photo-autotrophic Production of Poly(hydroxyalkanoates) in Cyanobacteria. *Chem. Biochem. Eng. Q.* **2015**, *29*, 145–156. [CrossRef]

16. Martin, K.; Lukas, M. Cyanobacterial Polyhydroxyalkanoate Production: Status Quo and Quo Vadis? *Curr. Biotechnol.* **2015**, *4*, 464–480.

17. Singh, A.; Sharma, L.; Mallick, N.; Mala, J. Progress and challenges in producing polyhydroxyalkanoate biopolymers from cyanobacteria. *J. Appl. Phycol.* **2017**, *29*, 1213–1232.

18. Lau, N.S.; Foong, C.P.; Kurihara, Y.; Sudesh, K.; Matsui, M. RNA-Seq Analysis Provides Insights for Understanding Photoautotrophic Polyhydroxyalkanoate Production in Recombinant *Synechocystis* sp. *PLoS ONE* **2014**, *9*, e86368. [CrossRef]

19. Khetkorn, W.; Incharoensakdi, A.; Lindblad, P.; Jantaro, S. Enhancement of poly-3-hydroxybutyrate production in *Synechocystis* sp. PCC 6803 by overexpression of its native biosynthetic genes. *Bioresour. Technol.* **2016**, *214*, 761–768. [CrossRef]

20. Carpine, R. Genetic engineering of *Synechocystis* sp. PCC6803 for poly-β-hydroxybutyrate overproduction. *Algal Res. Biomass Biofuels Bioprod.* **2017**, *25*, 117–127. [CrossRef]

21. Kamravamanesh, D.; Kovacs, T.; Pflügl, S.; Druzhinina, I.; Kroll, P.; Lackner, M.; Herwig, C. Increased poly-beta-hydroxybutyrate production from carbon dioxide in randomly mutated cells of cyanobacterial strain *Synechocystis* sp. PCC 6714: Mutant generation and characterization. *Bioresour. Technol.* **2018**, *266*, 34–44. [CrossRef]

22. Steuer, R.; Knoop, H.; Machné, R. Modelling cyanobacteria: From metabolism to integrative models of phototrophic growth. *J. Exp. Bot.* **2012**, *63*, 2259–2274. [CrossRef]

23. Dutt, V.; Srivastava, S. Novel quantitative insights into carbon sources for synthesis of poly hydroxybutyrate in Synechocystis PCC 6803. *Photosynth. Res.* **2018**, *136*, 303–314. [CrossRef]

24. Rajendran, V.; Incharoensakdi, A. Disruption of polyhydroxybutyrate synthesis redirects carbon flow towards glycogen synthesis in *Synechocystis* sp. PCC 6803 overexpressing glgC/glgA. *Plant Cell Physiol.* **2018**, *59*, 2020–2029.

25. Chen, X.; Schreiber, S.; Appel, J.; Makowka, A.; Fähnrich, B.; Roettger, M.; Hajirezaei, M.R.; Sönnichsen, F.D. The Entner-Doudoroff pathway is an overlooked glycolytic route in cyanobacteria and plants. *Proc. Natl. Acad. Sci. USA* **2016**, *113*, 5441–5446. [CrossRef]

26. Yu, J.; Liberton, M.; Cliften, P.F.; Head, R.D.; Jacobs, J.M.; Smith, R.D.; Koppenaal, D.W.; Brand, J.J.; Pakrasi, H.B. Synechococcus elongatus UTEX 2973, a fast growing cyanobacterial chassis for biosynthesis using light and CO2. *Sci. Rep.* **2015**, *5*, 8132. [CrossRef] [PubMed]

27. Osanai, T.; Oikawa, A.; Numata, K.; Kuwahara, A.; Iijima, H.; Doi, Y.; Saito, K.; Hirai, M.Y. Pathway-level acceleration of glycogen catabolism by a response regulator in the cyanobacterium Synechocystis species PCC 6803. *Plant Physiol.* **2014**, *164*, 1831–1841. [CrossRef] [PubMed]

28. Yoo, S.H.; Lee, B.H.; Moon, Y.; Spalding, M.H.; Jane, J.L. Glycogen Synthase Isoforms in *Synechocystis* sp. PCC6803: Identification of Different Roles to Produce Glycogen by Targeted Mutagenesis. *PLoS ONE* **2014**, *9*, e91524. [CrossRef]

29. Namakoshi, K.; Nakajima, T.; Yoshikawa, K.; Toya, Y.; Shimizu, H. Combinatorial deletions of glgC and phaCE enhance ethanol production in *Synechocystis* sp. PCC 6803. *J. Biotechnol.* **2016**, *239*, 13–19. [CrossRef]

30. Kamravamanesh, D.; Slouka, C.; Limbeck, A.; Lackner, M.; Herwig, C. Increased carbohydrate production from carbon dioxide in randomly mutated cells of cyanobacterial strain *Synechocystis* sp. PCC 6714: Bioprocess understanding and evaluation of productivities. *Bioresour. Technol.* **2019**, *273*, 277–287. [CrossRef] [PubMed]

31. Wang, C.X.; Saldanha, M.; Sheng, X.; Shelswell, K.J.; Walsh, K.T.; Sobral, B.W.; Charles, T.C. Roles of poly-3-hydroxybutyrate (PHB) and glycogen in symbiosis of Sinorhizobium meliloti with *Medicago* sp. *Microbiology* **2007**, *153*, 388–398. [CrossRef] [PubMed]

32. Sauer, J.; Schreiber, U.; Schmid, R.; Völker, U.; Forchhammer, K. Nitrogen starvation-induced chlorosis in Synechococcus PCC 7942. Low-level photosynthesis as a mechanism of long-term survival. *Plant Physiol.* **2001**, *126*, 233–243. [CrossRef] [PubMed]

33. Lever, M.A.; Rogers, K.L.; Lloyd, K.G.; Overmann, J.; Schink, B.; Thauer, R.K.; Hoehler, T.M.; Jørgensen, B.B. Life under extreme energy limitation: A synthesis of laboratory- and field-based investigations. *FEMS Microbiol. Rev.* **2015**, *39*, 688–728. [CrossRef]

34. Kempes, C.P.; van Bodegom, P.M.; Wolpert, D.; Libby, E.; Amend, J.; Hoehler, T. Drivers of Bacterial Maintenance and Minimal Energy Requirements. *Front. Microbiol.* **2017**, *8*, 31. [CrossRef]

35. Osanai, T.; Azuma, M.; Tanaka, K. Sugar catabolism regulated by light- and nitrogen-status in the cyanobacterium *Synechocystis* sp. PCC 6803. *Photochem. Photobiol. Sci.* **2007**, *6*, 508–514. [CrossRef] [PubMed]

36. Cano, M.; Holland, S.C.; Artier, J.; Burnap, R.L.; Ghirardi, M.; Morgan, J.A.; Yu, J. Glycogen Synthesis and Metabolite Overflow Contribute to Energy Balancing in Cyanobacteria. *Cell Rep.* **2018**, *23*, 667–672. [CrossRef]

37. Benson, P.J.; Purcell-Meyerink, D.; Hocart, C.H.; Truong, T.T.; James, G.O.; Rourke, L.; Djordjevic, M.A.; Blackburn, S.I.; Price, G.D. Factors Altering Pyruvate Excretion in a Glycogen Storage Mutant of the Cyanobacterium, Synechococcus PCC7942. *Front. Microbiol.* **2016**, *7*, 475. [CrossRef]

38. Osanai, T.; Kanesaki, Y.; Nakano, T.; Takahashi, H.; Asayama, M.; Shirai, M.; Kanehisa, M.; Suzuki, I.; Murata, N.; Tanaka, K. Positive regulation of sugar catabolic pathways in the cyanobacterium *Synechocystis* sp. PCC 6803 by the group 2 sigma factor sigE. *J. Biol. Chem.* **2005**, *280*, 30653–30659. [CrossRef]

39. Rippka, R.; Deruelles, D.; Waterbury, J.B.; Herdman, M.; Stanier, R.Y. Generic Assignments, Strain Histories and Properties of Pure Cultures of Cyanobacteria. *J. Gen. Microbiol.* **1979**, *111*, 1–61. [CrossRef]

40. Schlebusch, M.; Forchhammer, K. Requirement of the Nitrogen Starvation-Induced Protein Sll0783 for Polyhydroxybutyrate Accumulation in *Synechocystis* sp. Strain PCC 6803. *Appl. Environ. Microbiol.* **2010**, *76*, 6101–6107. [CrossRef]

41. Taroncher-Oldenburg, G.; Nishina, K.; Stephanopoulos, G. Identification and analysis of the polyhydroxyalkanoate-specific beta-ketothiolase and acetoacetyl coenzyme A reductase genes in the cyanobacterium *Synechocystis* sp. strain PCC6803. *Appl. Environ. Microbiol.* **2000**, *66*, 4440–4448. [CrossRef] [PubMed]

13

Amphiphilic Polymeric Micelles Based on Deoxycholic Acid and Folic Acid Modified Chitosan for the Delivery of Paclitaxel

Liang Li [1], Na Liang [2], Danfeng Wang [1], Pengfei Yan [1], Yoshiaki Kawashima [3], Fude Cui [4] and Shaoping Sun [1,*]

[1] Key Laboratory of Chemical Engineering Process & Technology for High-efficiency Conversion, College of Heilongjiang Province, School of Chemistry and Material Science, Heilongjiang University, Harbin 150080, China; lliang1991001@163.com (L.L.); dfwang626@163.com (D.W.); yanpf@vip.sina.com (P.Y.)
[2] Key Laboratory of Photochemical Biomaterials and Energy Storage Materials, Heilongjiang Province, College of Chemistry & Chemical Engineering, Harbin Normal University, Harbin 150025, China; liangna528@163.com
[3] Department of Pharmaceutical Engineering, School of Pharmacy, Aichi Gakuin University, Nagoya 464-8650, Japan; sykawa123@163.com
[4] School of Pharmacy, Shenyang Pharmaceutical University, Shenyang 110016, China; syphucuifude@163.com
* Correspondence: sunshaoping@hlju.edu.cn

Abstract: The present investigation aimed to develop a tumor-targeting drug delivery system for paclitaxel (PTX). The hydrophobic deoxycholic acid (DA) and active targeting ligand folic acid (FA) were used to modify water-soluble chitosan (CS). As an amphiphilic polymer, the conjugate FA-CS-DA was synthesized and characterized by Proton nuclear magnetic resonance (^{1}H-NMR) and Fourier-transform infrared spectroscopy (FTIR) analysis. The degree of substitutions of DA and FA were calculated as 15.8% and 8.0%, respectively. In aqueous medium, the conjugate could self-assemble into micelles with the critical micelle concentration of 6.6×10^{-3} mg/mL. Under a transmission electron microscope (TEM), the PTX-loaded micelles exhibited a spherical shape. The particle size determined by dynamic light scattering was 126 nm, and the zeta potential was +19.3 mV. The drug loading efficiency and entrapment efficiency were 9.1% and 81.2%, respectively. X-Ray Diffraction (XRD) analysis showed that the PTX was encapsulated in the micelles in a molecular or amorphous state. In vitro and in vivo antitumor evaluations demonstrated the excellent antitumor activity of PTX-loaded micelles. It was suggested that FA-CS-DA was a safe and effective carrier for the intravenous delivery of paclitaxel.

Keywords: chitosan; deoxycholic acid; folic acid; amphiphilic polymer; micelles; paclitaxel

1. Introduction

Paclitaxel (PTX) is an important clinical chemotherapeutic drug that exhibits strong antitumour activity against a variety of cancer types. However, the low solubility of PTX due to its bulky polycyclic structure hampers its clinical application [1]. Many attempts have been made to find less toxic and better-tolerated carriers to increase the solubility of PTX for intravenous delivery, such as nanoparticles, dendrimers, liposomes and nanosuspensions [2–5]. In recent years, polymeric micelles have attracted growing interest due to their attractive characteristics, such as their excellent solubilization ability, small size, high stability, prolonged circulation time, low toxicity, ability to evade scavenging by the mononuclear phagocyte system (MPS), high biocompatibility and efficient accumulation in tumor tissues via an enhanced permeability and retention (EPR) effect [6,7]. The micelles have a unique core–shell structure with hydrophobic segments as the internal core and hydrophilic segments as the outer shell. The internal core provides a storeroom for poorly water-soluble drugs, and the outer shell allows the retention of the stability of micelles in aqueous medium and provides the opportunity to target the delivery of antitumor drugs to the tumor by further modification [8,9].

To date, numerous amphiphilic block or graft copolymers have been synthesized and applied as micellar drug delivery systems [10]. Among them, chitosan (CS) has been extensively studied for its biocompatibility, non-toxicity and biodegradability [11,12]. Moreover, the abundant active amine and hydroxyl groups in CS could offer many opportunities for chemical modification. In recent years, several chitosan-based PTX delivery systems have been developed, such as N-mercapto acetyl-N'-octyl-O, N''-glycol chitosan micelles [13], 3,6-O,O'-dimyristoyl chitosan micelles [14], folic acid–cholesterol–chitosan micelles [15], PTX conjugated trimethyl chitosan nanoparticles [16], palmitoyl chitosan nanoparticles [17], N-succinyl-chitosan nanoparticles [18] and PTX-loaded chitosan nanoparticles prepared by the nano-emulsion method [19]. However, chitosan with high molecular weight has poor solubility in aqueous medium at neutral pH, which limits its medical and pharmaceutical applications. In contrast, water-soluble chitosan with a low molecular weight and high degree of deacetylation is a superior candidate for amphiphilic copolymer synthesis [20]. In the present study, the water-soluble chitosan was used as the hydrophilic part of the copolymer to form a micellar system for PTX delivery.

Deoxycholic acid (DA) is a typical bile acid that is secreted from the gallbladder to emulsify fats and other hydrophobic compounds [21]. As an endogenous compound with a lipophilic nature, the introduction of DA to CS could adjust the hydrophilicity/hydrophobicity balance of the conjugate and would not lead to any serious toxicity [22]. DA has been approved as an excellent pharmaceutical additive for injection [23].

Molecular ligands were often grafted onto drug carriers to develop tumor-targeted drug delivery systems. It has been reported that folate receptors are over-expressed in many types of cancers, while almost undetectable in healthy tissues [24]. The folic acid-modified nanocarriers could improve therapeutic efficacy via folate receptor-mediated active targeting. The antitumor efficiency could be significantly enhanced by synergetic active and passive tumor targeting [25].

Based on the above, in present study, a biocompatible nanocarrier based on deoxycholic acid and folic acid-modified chitosan (FA-CS-DA) was designed for targeting the delivery of PTX. The synthesis, characterization and self-assembly of FA-CS-DA and the characterization and in vitro/in vivo antitumor activity of PTX-loaded micelles were studied in detail.

2. Results and Discussion

2.1. Preparation of FA-CS-DA

The synthesis of FA-CS-DA was performed via the amide bond formation between the amino groups of CS and the carboxyl groups of DA and FA. As shown in Figure 1, the FA-CS-DA was synthesized by a two-step reaction. First, the intermediate CS-DA was prepared by the conjugation of carboxylic groups of DA with the primary amino groups of CS. Then, an FA molecule was introduced by attaching the carboxyl groups of FA to the remaining terminal amino of CS-DA. EDC and NHS were used in both reactions to active the carboxyl groups [26]. The unreacted DA and FA, as well as any by-product, were removed by dialysis. The degree of substitutions (DS) of DA and FA were calculated as 15.8% and 8.0%, respectively.

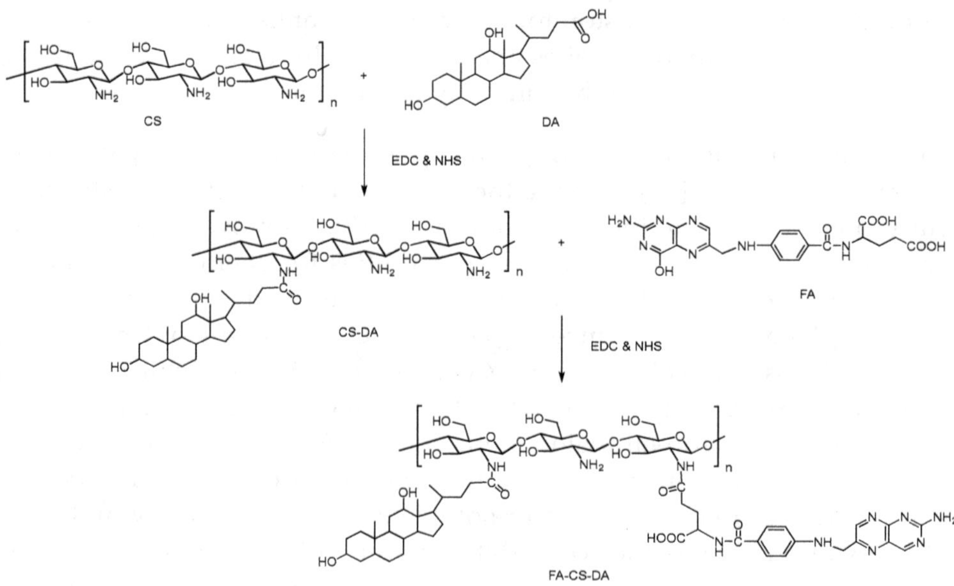

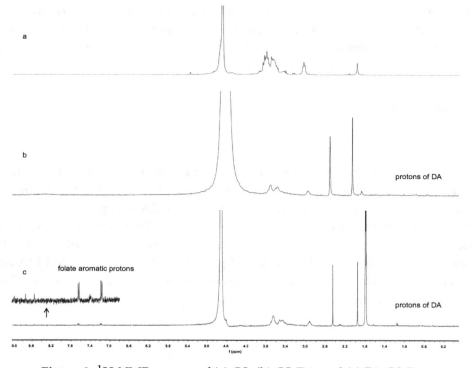

Figure 1. Scheme of the synthesis of FA-CS-DA (folic acid–chitosan–deoxycholic acid).

2.2. Characterization of FA-CS-DA

2.2.1. Proton Nuclear Magnetic Resonance (^{1}H-NMR) Characterization

To confirm the conjugate formation, the ^{1}H-NMR spectra of CS, CS-DA and FA-CS-DA are shown in Figure 2. Compared with CS, new peaks appeared in the range of 0.5–2.5 ppm in the spectrum of CS-DA and were assigned to the –CH$_3$ and –CH$_2$– protons of DA, which indicated the successful introduction of DA. Furthermore, FA-CS-DA showed characteristic signals attributed to the protons of FA at 7.20, 7.66 and 8.55 ppm. More specifically, the signal at 8.55 ppm was assigned to the proton of the pterin ring of FA, and signals at 7.66 and 7.20 ppm corresponded to the aromatic protons of FA. The aforementioned results revealed that both DA and FA were successfully grafted onto the backbone of CS.

Figure 2. ^{1}H-NMR spectra of (**a**) CS, (**b**) CS-DA and (**c**) FA-CS-DA.

2.2.2. Fourier-Transform Infrared (FTIR) Characterization

FTIR analysis was used to further confirm the successful synthesis of FA-CS-DA. As presented in Figure 3, for CS, the signal at 1637 cm^{-1} was attributed to the C–O stretching vibration of C=O group of the amide I band, and the peak at 1517 cm^{-1} was assigned to the N–H bending vibration of the amide II band. In the spectrum of CS-DA, new peaks at 2925 and 2864 cm^{-1} were due to the C–H stretching vibration of methylene of DA. The enhancement of peak intensity at 3476 cm^{-1} suggested the increase of hydroxyl groups after the grafting of DA. For FA-CS-DA, the new signal at 1698 cm^{-1} was assigned to the unreacted carboxyl groups in FA, which implied the introduction of FA. All these differences indicated the formation of FA-CS-DA.

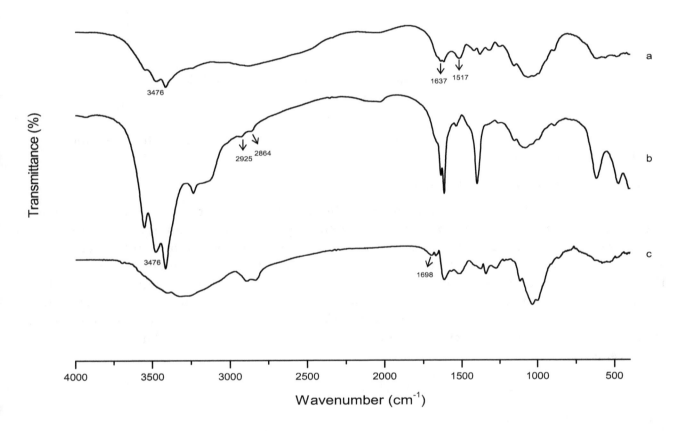

Figure 3. FTIR spectra of (**a**) CS, (**b**) CS-DA and (**c**) FA-CS-DA.

2.2.3. Critical Micelle Concentration (CMC) of FA-CS-DA

CMC is the lowest concentration for the amphiphilic polymer to form micelles in aqueous medium, and it is an important parameter that indicates the stability of micelles. In this study, pyrene was used as a fluorescence probe to measure the CMC of FA-CS-DA. At low concentrations, the polymer molecules existed in a single-stranded form, and the fluorescence intensity remained constant. Once the concentration was higher than CMC, the polymer molecules formed micelles, and pyrene was solubilized in the hydrophobic core of micelles. As a result, the fluorescence intensity increased significantly. Moreover, the intensity of the third energy peak (383 nm, I_3) increased more dramatically than the first peak (373 nm, I_1) [27]. The intensity ratio of I_1/I_3 was used as an indicator of the polarity of the environment, and the variation of I_1/I_3 against the logarithm of polymer concentration is shown in Figure 4. The CMC of FA-CS-DA could be determined from the point of inflection, and the value was calculated to be 6.6×10^{-3} mg/mL. The low value suggested the high stability of FA-CS-DA micelles.

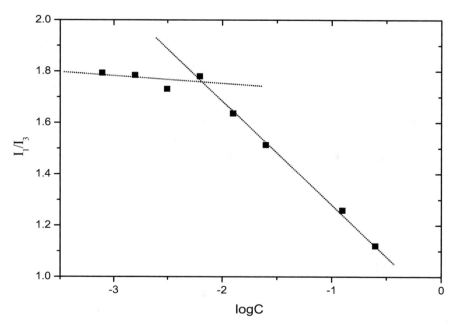

Figure 4.   Variation of the fluorescence intensity ratio of I_1/I_3 against the logarithm of FA-CS-DA concentration.

2.3. Preparation of PTX-Loaded FA-CS-DA Micelles

As the polymer FA-CS-DA had an amphiphilic structure, including hydrophobic segments of DA, hydrophilic segments of CS and tumor targeting ligands of FA, in an aqueous environment, it could self-assemble into micelles. In this study, FA-CS-DA was dispersed in distilled water, and the micelles were prepared under ultrasonication without the addition of any emulsifier or stabilizer. The drug loading into the hydrophobic domains occurred simultaneously with the formation of micelles via either hydrophobic–hydrophobic interactions or Van der Waals interactions between PTX molecules and the hydrophobic groups of the polymer [28]. For the optimized drug-loaded micelles, the drug encapsulation efficiency and the drug loading capacity were calculated to be 81.2% and 9.1%, respectively.

2.4. Characterization of PTX-Loaded FA-CS-DA Micelles

2.4.1. Particle Size and Zeta Potential

It was well known that the particle size and size distribution of nanoparticles could dramatically affect the fate of the particles. Nanoparticles in the range of 10–200 nm could reduce the reticuloendothelial system (RES) uptake and enhance the endocytic uptake in tumors via the EPR effect [29]. The mean particle size of PTX-loaded FA-CS-DA micelles determined by the DLS method was 126 nm, with a polydispersity index (PDI) of 0.256, and this was larger than the bare ones (78 nm, PDI of 0.232), which indicated the encapsulation of PTX into the micelles.

Zeta potential is often used to indicate the stability of particle systems. With high zeta potential, the particles could repel each other and prevent aggregation, therefore enhancing the stability of the solution. The resultant zeta potential values of bare and PTX-loaded FA-CS-DA micelles were +29.1 mV and +19.3 mV, respectively. The relatively high positive potential was attributed to the ionized amino groups of CS. It was reported that the positively charged particles could enhance the endocytosis by cells [30].

2.4.2. Transmission Electron Microscopy (TEM) Observation

TEM was used to directly visualize the size and morphology of the micelles. The TEM micrograph of the PTX-loaded micelles presented in Figure 5 showed that the FA-CS-DA was capable of forming polymeric micelles, and the micelles had a near-spherical shape with narrow distribution. Furthermore, the size obtained by TEM was smaller than that measured by DLS, which was due to the different states of the particles in the measurements, i.e., the dried state and the hydrated state, respectively. More exactly, the outer shell of the micelles could be collapsed during the process in TEM experiment [31].

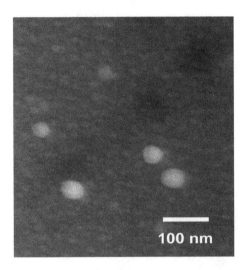

Figure 5. Transmission electron microscopy (TEM) image of paclitaxel (PTX)-loaded FA-CS-DA micelles.

2.4.3. X-Ray Diffraction (XRD) Analysis

XRD analysis was conducted to confirm the existence state of PTX in the polymeric micelles. As shown in Figure 6, the XRD diagram of PTX presented several peaks at 2θ of 5.53°, 8.87°, 10.04°, 11.14° and 12.53°, and there were a large number of small peaks in the range of 15° to 30°. For blank micelles, there were no typical crystal peaks in the pattern. The physical mixture of PTX and bare micelles still showed the typical crystal peaks of PTX with weaker intensity. However, the PTX-loaded micelles had a similar spectrum to the blank micelles and there were no PTX peaks. It was implied that PTX was entrapped in the FA-CS-DA micelles in an amorphous or molecular state, which might lead to better absorption of the drug.

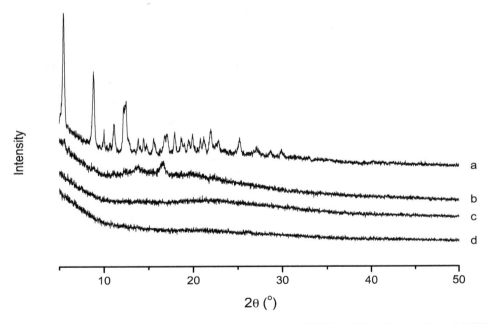

Figure 6. XRD spectra of (**a**) PTX, (**b**) a physical mixture of PTX and blank micelles, (**c**) PTX-loaded micelles and (**d**) blank micelles.

2.5. In Vitro Cytotoxicity Study

The in vitro cytotoxicity of the PTX-loaded micelles was evaluated by a standard MTT assay against MCF-7 cells. As illustrated in Figure 7, more than 99% of the cells were alive after the treatment of blank micelles even with high concentrations, which suggested the nontoxicity of the vehicle. For PTX formulations, the cytotoxicity was concentration-dependent. When the drug concentration increased, the cell viability decreased, which implied that a sufficient exposure level was important for the drug to kill the cells effectively. Moreover, it was exciting to see that the PTX-loaded micelles exhibited higher cytoxicity than the free PTX in dimethyl sulfoxide (DMSO). This might be explained by the effect of the FA-CS-DA micellar vehicle. For MCF-7 cells with over-expressed folate receptors on the surface [32], more FA-CS-DA micelles could be internalized into the cells via the receptor-mediated endocytosis. It could be speculated that FA-CS-DA might be a potential drug carrier for PTX.

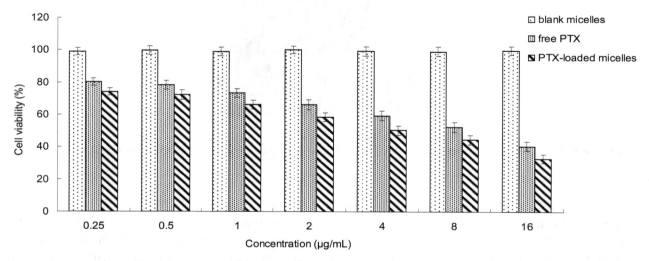

Figure 7. In vitro cytotoxicity of PTX-loaded FA-CS-DA micelles.

2.6. In Vivo Tumor Growth Inhibition Study

The tumor growth inhibition study was performed to further evaluate the in vivo antitumor activity of PTX-loaded FA-CS-DA micelles. As shown in Figure 8, the tumors excised from the mice in the PTX-loaded micelles group were significantly smaller than those from the normal saline group, and the TIR was as high as 78.1%. The outstanding antitumor efficacy could be explained by the following facts: first, there were a number of FA receptors on the surface of the tumor, and the PTX-loaded FA-CS-DA micelles could accumulate in the tumor tissues and then internalize into the tumor cells via the folate receptor-mediated active targeting. Second, via the electrostatic interaction, the positively charged micelles could efficiently bind to the tumor surface, which exhibited a high negative charge [33]. Moreover, the particles with positive charge were more likely to penetrate into the tumor cells [34]. Third, the particle size of <200 nm could facilitate the tumor accumulation of micelles by the EPR effect. In summary, all of these could increase the PTX concentration in the tumor, therefore obtaining excellent antitumor efficacy.

Figure 8. Tumors excised from the mice after intravenous injection treatment.

3. Materials and Methods

3.1. Materials

Chitosan (CS, Mw of 30 kDa, deacetylation degree > 90%) was obtained from Kittolife Co., Ltd., Seoul, Korea. Paclitaxel (PTX) was purchased from Natural Field Biological Technology Co., Ltd., Xi'an, China. Deoxycholic acid (DA), folic acid (FA), 1-Ethyl-3-(3-dimethylaminopropyl) carbodiimide hydrochloride (EDC) and N-hydroxysuccinimide (NHS) were supplied by Aladdin Industrial Co., Shanghai, China. Pyrene, 2,4,6-trinitrobenzene sulfonic acid (TNBS) and 3-(4,5-dimethylthiazol-2-yl)-2,5-diphenyl tetrazolium bromide (MTT) were supplied by Sigma Chemical Co., St. Louis, MO, USA. Dulbecco's modified Eagle's medium (DMEM), penicillin–streptomycin mixture and fetal bovine serum (FBS) were obtained from Gibco BRL, Carlsbad, CA, USA. All other chemicals and solvents were of analytical or chromatographic grade and used without further purification. Distilled water or Milli-Q water was used in all experiments.

3.2. Animals and Cell Lines

Specific pathogen-free mice weighing 20 ± 2 g were supplied by the Laboratory Animal Center of Harbin Medical University, Harbin, China. MCF-7 cells (human breast cancer cells) and H22 cells (mouse hepatocellular carcinoma cells) were kindly donated by the Department of Pharmacology, Harbin Medical University. All animal procedures were performed in compliance with the animal care protocols approved by the Animal Ethics Committee of Harbin Medical University (16 March 2018, No. 201803160028).

3.3. Synthesis of FA-CS-DA

FA-CS-DA was synthesized via the reaction of carboxyl groups of DA and FA with amino groups of chitosan under the catalyzation of EDC and NHS. The CS-DA was prepared as follows. Firstly, 1.0024 g of chitosan was dissolved in 50 mL of distilled water as solution A. The solution of DA (0.6012 g) with EDC (0.3426 g) and NHS (0.4113 g) in 50 mL of methanol was stirred for 1.5 h as solution B. Then solution B was added drop-wisely into solution A and stirred for 24 h at room temperature. After that, the reaction mixture was added to 50 mL of methanol. The resultant was filtered, and the filtrate was transferred into the mixture of methanol, water and ethanol at the ratio of 1:1:1 (v/v), then centrifuged at 6000 rpm for 20 min to remove unreacted DA and other impurities. The product of CS-DA was dried under vacuum at 40 °C.

For the synthesis of FA-CS-DA, a certain amount of CS-DA (0.14 g) was dissolved in 50 mL of acetate–acetate buffer solution (pH 4.7) as solution C. The solution of FA (0.04 g) with EDC (0.035 g) and NHS (0.042 g) in 20 mL of DMSO was stirred for 1 h as solution D. The solution D was added drop-wisely into solution C and stirred at an ambient temperature in dark condition. Twenty-four hours later, the resultant was dialyzed against an excess amount of distilled water (MWCO of 30 kDa) for 48 h for purification and then lyophilized to get the product of FA-CS-DA.

3.4. Characterization of FA-CS-DA

3.4.1. ^{1}H-NMR Characterization

^{1}H-NMR spectra were recorded on a Bruker Avance NMR spectrometer (AV-400, Bruker, Switzerland) operated at 400 MHz to analyze the structure of FA-CS-DA. The mixture of deuterated water (D_2O) and DMSO-d6 was used to prepared native and modified chitosan solution, respectively.

3.4.2. FTIR Characterization

The formation of FA-CS-DA was further verified by FTIR spectra that recorded with KBr pellets using FTIR spectrometer (Tensor II, Bruker, Switzerland) in the range from 4000 to 400 cm^{-1}, with a resolution of 2 cm^{-1}.

3.4.3. Measurement of the Degree of Substitution (DS)

The DS is defined as the number of DA (or FA) groups per 100 sugar units of CS. The DS of DA was determined by measuring the free amino groups of native chitosan and CS-DA via the TNBS method [35]. The DS of FA was measured by an ultraviolet-visible (UV-vis) spectrophotometer. Specifically, FA-CS-DA was dissolved in the mixture of water and DMSO at the volume ratio of 1:1, and the absorbance was determined by an UV-vis spectrophotometry (UV mini-1240, Shimadzu, Kyoto, Japan) at 363 nm. To get the standard curve for FA, a series of FA solutions with increasing amounts of FA were prepared.

3.4.4. Determination of Critical Micelle Concentration (CMC)

The CMC of FA-CS-DA was determined by measuring the fluorescence intensity of pyrene using a fluorescence spectrometer (F-2500 FL Spectrophotometer, Hitachi Ltd., Tokyo, Japan). In brief, 10 µL of pyrene in acetone solution was added into a series of 10 mL volumetric flasks, and the acetone was evaporated in dark. Then, the FA-CS-DA solutions with different concentrations were added to each flask, and the final concentration of pyrene was 6×10^{-7} mol/L. The samples were sonicated for 1 h, and then remained undisturbed and kept from light to reach equilibrium. The emission spectra of pyrene were recorded from 350 to 500 nm at the excitation wavelength of 337 nm. The slit widths for excitation and emission were set at 5 and 2.5 nm, respectively. The fluorescence intensity ratio of the first peak (I_1, 373 nm) to the third peak (I_3, 384 nm) was analyzed for determination of CMC.

3.5. Preparation of PTX-Loaded FA-CS-DA Micelles

The PTX-loaded FA-CS-DA micelles were prepared by ultrasonication using a probe-type sonicator (JY92-II, Ningbo Scientz Biotechnology Co., Ltd., Ningbo, China) [20]. Firstly, FA-CS-DA was dispersed in distilled water. Then the PTX in acetone solution was added quickly, and the mixture was sonicated at 224 W for 10 min with 2 s active/3 s duration. The mixture was dialyzed against distilled water for 5 h by using a dialysis bag (MWCO of 30 kDa) to remove unloaded PTX. The resultant was subsequently freeze-dried (Freeze Dryer Model FD-1A-50, Bioyikang Experimental Instrument Co., Ltd., Beijing, China) to get the PTX-loaded FA-CS-DA micelles powder.

3.6. Characterization of PTX-Loaded FA-CS-DA Micelles

3.6.1. Measurement of Particle Size and Zeta Potential

A dynamic light scattering method was applied to measure the particle size, zeta potential and size distribution of the samples using the Zetasizer® 3000 (Malvern Instruments, Southborough, MA, USA).

3.6.2. TEM Observation

TEM observation of the PTX-loaded micelles was performed using transmission electron microscopy (H-7650, Hitachi Ltd., Tokyo, Japan) operated at 60 kV. Before observation, an aqueous droplet of the sample was deposited on a copper grid coated with carbon. After 4 min, the grid was tapped with filter paper to remove surface water, followed by air drying and negatively stained with 2% phosphotangstic acid.

3.6.3. XRD Analysis

XRD analysis was used to study the existence state of PTX after loaded into micelles. Samples were measured by an X-ray diffractometer (Geigerflex, Rigaku Co., Tokyo, Japan) using Cu Kα radiation source at 30 kV and 30 mA. The relative intensity was recorded in the range of 5–50° (2θ) at scanning speed of 4°/min and step size of 0.02°.

3.7. Determination of Drug Loading and Drug Encapsulation Efficiency

The amount of PTX encapsulated into the micelles was determined as follows: a certain amount of lyophilized drug-loaded micelles was weighted, and then acetonitrile was added, followed by ultrasonication at 200 W for 10 min to extract PTX from the core of the micelles. After filtered through a 0.22 μm microporous filter, the PTX concentration was measured by HPLC method using a mobile phase delivery pump (LC-10ATVP, Shimadzu, Japan) and a Diamonsil^{TM} C_{18} reverse-phase column (200 mm × 4.6 mm, 5 μm, Dikma Technologies Inc., Beijing, China). The mobile phase was the mixture of acetonitrile and water at a volume ratio of 70:30. The column temperature was set at 30 °C. The detection wavelength was 227 nm, and the flow rate was 1 mL/min. The drug loading (DL%) and drug encapsulation efficiency (EE%) were calculated with the following formulas.

$$DL\% = W_{encapsulated} / W_{micelles} \times 100\% \tag{1}$$

$$EE\% = W_{encapsulated} / W_{fed} \times 100\% \tag{2}$$

where $W_{encapsulated}$, W_{fed} and $W_{micelles}$ represented the weight of PTX encapsulated in the micelles, the PTX fed initially, and the weight of PTX-loaded micelles, respectively.

3.8. In Vitro Cytotoxicity Study

The cytotoxicity of PTX-loaded micelles was investigated by MTT method. H22 cells in the logarithmic growth phase were seeded in a 96-well microtitre plate at the density of 1×10^5 cells/well and cultured in DMEM with 10% (v/v) fetal bovine serum and 1% penicillin–streptomycin at 37 °C in a humidified incubator with 5% CO_2. When the cells became 75% confluent, free PTX, blank FA-CS-DA micelles and PTX-loaded FA-CS-DA micelles with a PTX concentration ranging from 0.25 to 16 μg/mL was added, respectively. After incubation for 24 h, the cell viability was determined. At a predetermined time, the supernatant of each well was aspirated off and replaced with fresh medium, and then 10 μL of MTT solution was added. With the action of active mitochondrial dehydrogenase in live cells, the dissolved MTT could be converted to water-insoluble purple formazan crystals. After incubation for another 4 h, the unreacted MTT was discarded, followed by the addition of 100 μL of DMSO to dissolve the formazan crystals. The absorbance was analyzed by a BioRad microplate reader (Bio-Rad 680, Bio-Rad Laboratories, Hercules, CA, USA) at 490 nm. Untreated control cells were set as 100% viable. Cell viability was calculated by the following equation:

$$Cell\ viability\ (\%) = A_{sample} / A_{control} \times 100\% \tag{3}$$

where A_{sample} and $A_{control}$ were the absorbance of cells exposed to the sample and the absorbance of untreated cells, respectively.

3.9. In Vivo Antitumor Activity Study

The in vivo antitumor activity of PTX-loaded FA-CS-DA micelles was evaluated in H22 tumor-bearing mice. To establish the tumor model, mice were inoculated subcutaneously in the right armpit with 5×10^6 cells (0.2 mL) [36]. When the tumor xenografts became palpable, the mice were randomly divided into 2 groups (n = 6) and treated with physiological saline and PTX-loaded FA-CS-DA micelles (15 mg/kg), respectively. Samples were administered via the tail vein once every 3 days for 4 times. After 12 days of treatment, the mice were sacrificed and the tumors were harvested and weighted. The antitumor activity of the PTX-loaded micelles was expressed by the tumor inhibition rate (TIR), which could be calculated using Equation (4).

$$TIR = (1 - W_1 / W_2) \times 100\% \tag{4}$$

where W_1 and W_2 represented the average tumor weight of PTX-loaded micelles group and normal saline group, respectively.

3.10. Statistical Analysis

Each experiment was performed in triplicate. Values were expressed as mean $\pm$ standard deviation. Statistical data analysis was performed using the Student's *t*-test with the significance level set at $p < 0.05$.

4. Conclusions

In the present study, an amphiphilic chitosan derivative of FA-CS-DA was synthesized via amidation reaction. The polymer could self-assemble into micelles in an aqueous milieu and showed good solubilization ability for PTX. The developed micellar system had excellent features, such as high stability upon dilution, high drug loading and encapsulation efficiency, small particle size and excellent cytotoxicity to tumor cells in vitro and in vivo. It could be concluded that the FA-CS-DA was a promising micellar carrier for PTX delivery, and further, more detailed studies will be performed.

Author Contributions: S.S. conceived and designed the experiments. L.L. performed the experiments, analyzed the data and wrote the paper. D.W. performed the in vitro cytotoxicity study. N.L. performed the in vivo tumor growth inhibition study and revised the paper. P.Y., Y.K. and F.C. gave us much useful advice and some pieces of guidance.

References

1. Sofias, A.M.; Dunne, M.; Storm, G.; Allen, C. The battle of "nano" paclitaxel. *Adv. Drug Deliv. Rev.* **2017**, *122*, 20–30. [CrossRef] [PubMed]
2. Eloy, J.O.; Petrilli, R.; Topan, J.F.; Antonio, H.M.R.; Barcellos, J.P.A.; Chesca, D.L.; Serafini, L.N.; Tiezzi, D.G.; Lee, R.J.; Marchetti, J.M. Co-loaded paclitaxel/rapamycin liposomes: Development, characterization and in vitro and in vivo evaluation for breast cancer therapy. *Colloids Surf. B Biointerfaces* **2016**, *141*, 74–82. [CrossRef] [PubMed]
3. Tatiparti, K.; Sau, S.; Gawde, K.A.; Iyer, A.K. Copper-free 'click' chemistry-based synthesis and characterization of carbonic anhydrase-IX anchored albumin-paclitaxel nanoparticles for targeting tumor hypoxia. *Int. J. Mol. Sci.* **2018**, *19*, 838. [CrossRef] [PubMed]
4. Li, Y.; Zhao, X.; Zu, Y.; Zhang, Y. Preparation and characterization of paclitaxel nanosuspension using novel emulsification method by combining high speed homogenizer and high pressure homogenization. *Int. J. Pharm.* **2015**, *490*, 324–333. [CrossRef] [PubMed]
5. Yang, H. Targeted nanosystems: Advances in targeted dendrimers for cancer therapy. *Nanomedicine* **2016**, *12*, 309–316. [CrossRef] [PubMed]
6. Cagel, M.; Tesan, F.C.; Bernabeu, E.; Salgueiro, M.J.; Zubillaga, M.B.; Moretton, M.A.; Chiappetta, D.A. Polymeric mixed micelles as nanomedicines: Achievements and perspectives. *Eur. J. Pharm. Biopharm.* **2017**, *113*, 211–228. [CrossRef] [PubMed]
7. Liang, N.; Sun, S.; Gong, X.; Li, Q.; Yan, P.; Cui, F. Polymeric micelles based on modified glycol chitosan for paclitaxel delivery: Preparation, characterization and evaluation. *Int. J. Mol. Sci.* **2018**, *19*, 1550. [CrossRef] [PubMed]
8. Cong, Z.; Shi, Y.; Wang, Y.; Wang, Y.; Niu, J.E.; Chen, N.; Xue, H. A novel controlled drug delivery system based on alginate hydrogel/chitosan micelle composites. *Int. J. Biol. Macromol.* **2018**, *107*, 855–864. [CrossRef] [PubMed]
9. Shi, C.; Zhang, Z.; Wang, F.; Luan, Y. Active-targeting docetaxel-loaded mixed micelles for enhancing antitumor efficacy. *J. Mol. Liq.* **2018**, *264*, 172–178. [CrossRef]
10. Biswas, S.; Kumari, P.; Lakhani, P.M.; Ghosh, B. Recent advances in polymeric micelles for anti-cancer drug delivery. *Eur. J. Pharm. Sci.* **2016**, *83*, 184–202. [CrossRef] [PubMed]
11. Yang, Y.; Wang, S.; Wang, Y.; Wang, X.; Wang, Q.; Chen, M. Advances in self-assembled chitosan nanomaterials for drug delivery. *Biotechnol. Adv.* **2014**, *32*, 1301–1316. [CrossRef] [PubMed]
12. Ahsan, S.M.; Thomas, M.; Reddy, K.K.; Sooraparaju, S.G.; Asthana, A.; Bhatnagar, I. Chitosan as biomaterial in drug delivery and tissue engineering. *Int. J. Biol. Macromol.* **2018**, *110*, 97–109. [CrossRef] [PubMed]

13. Huo, M.; Fu, Y.; Liu, Y.; Chen, Q.; Mu, Y.; Zhou, J.; Li, L.; Xu, W.; Yin, T. N-mercapto acetyl-N′-octyl-O, N″-glycol chitosan as an efficiency oral delivery system of paclitaxel. *Carbohydr. Polym.* **2018**, *181*, 477–488. [CrossRef] [PubMed]

14. Silva, D.S.; Almeida, A.; Prezotti, F.; Cury, B.; Campana-Filho, S.P.; Sarmento, B. Synthesis and characterization of 3,6-O,O′- dimyristoyl chitosan micelles for oral delivery of paclitaxel. *Colloids Surf. B Biointerfaces* **2017**, *152*, 220–228. [CrossRef] [PubMed]

15. Cheng, L.C.; Jiang, Y.; Xie, Y.; Qiu, L.L.; Yang, Q.; Lu, H.Y. Novel amphiphilic folic acid-cholesterol-chitosan micelles for paclitaxel delivery. *Oncotarget* **2017**, *8*, 3315–3326. [CrossRef] [PubMed]

16. He, R.; Yin, C. Trimethyl chitosan based conjugates for oral and intravenous delivery of paclitaxel. *Acta Biomater.* **2017**, *53*, 355–366. [CrossRef] [PubMed]

17. Mansouri, M.; Nazarpak, M.H.; Solouk, A.; Akbari, S.; Hasani-Sadrabadi, M.M. Magnetic responsive of paclitaxel delivery system based on SPION and palmitoyl chitosan. *J. Magn. Magn. Mater.* **2017**, *421*, 316–325. [CrossRef]

18. Skorik, Y.A.; Golyshev, A.A.; Kritchenkov, A.S.; Gasilova, E.R.; Poshina, D.N.; Sivaram, A.J.; Jayakumar, R. Development of drug delivery systems for taxanes using ionic gelation of carboxyacyl derivatives of chitosan. *Carbohydr. Polym.* **2017**, *162*, 49–55. [CrossRef] [PubMed]

19. Gupta, U.; Sharma, S.; Khan, I.; Gothwal, A.; Sharma, A.K.; Singh, Y.; Chourasia, M.K.; Kumar, V. Enhanced apoptotic and anticancer potential of paclitaxel loaded biodegradable nanoparticles based on chitosan. *Int. J. Biol. Macromol.* **2017**, *98*, 810–819. [CrossRef] [PubMed]

20. Liang, N.; Sun, S.; Li, X.; Piao, H.; Piao, H.; Cui, F.; Fang, L. α-Tocopherol succinate-modified chitosan as a micellar delivery system for paclitaxel: Preparation, characterization and in vitro/in vivo evaluations. *Int. J. Pharm.* **2012**, *423*, 480–488. [CrossRef] [PubMed]

21. Heřmánková, E.; Žák, A.; Poláková, L.; Hobzová, R.; Hromádka, R.; Širc, J. Polymeric bile acid sequestrants: Review of design, in vitro binding activities, and hypocholesterolemic effects. *Eur. J. Med. Chem.* **2018**, *144*, 300–317. [CrossRef] [PubMed]

22. Hofmann, A.F.; Hagey, L.R. Bile acids: Chemistry, pathochemistry, biology, pathobiology, and therapeutics. *Cell. Mol. Life Sci.* **2008**, *65*, 2461–2483. [CrossRef] [PubMed]

23. Liu, M.; Du, H.; Zhai, G. Self-assembled nanoparticles based on chondroitin sulfate-deoxycholic acid conjugates for docetaxel delivery: Effect of degree of substitution of deoxycholic acid. *Colloids Surf. B Biointerfaces* **2016**, *146*, 235–244. [CrossRef] [PubMed]

24. Dhas, N.L.; Ige, P.P.; Kudarha, R.R. Design, optimization and in-vitro study of folic acid conjugated-chitosan functionalized PLGA nanoparticle for delivery of bicalutamide in prostate cancer. *Powder Technol.* **2015**, *283*, 234–245. [CrossRef]

25. Scomparin, A.; Salmaso, S.; Eldar-Boock, A.; Ben-Shushan, D.; Ferber, S.; Tiram, G.; Shmeeda, H.; Landa-Rouben, N.; Leor, J.; Caliceti, P.; et al. A comparative study of folate receptor-targeted doxorubicin delivery systems: Dosing regimens and therapeutic index. *J. Control. Release* **2015**, *208*, 106–120. [CrossRef] [PubMed]

26. Park, C.; Vo, C.L.N.; Kang, T.; Oh, E.; Lee, B.J. New method and characterization of self-assembled gelatin–oleic nanoparticles using a desolvation method via carbodiimide/N-hydroxysuccinimide (EDC/NHS) reaction. *Eur. J. Pharm. Biopharm.* **2015**, *89*, 365–373. [CrossRef] [PubMed]

27. Băran, A.; Stîngă, G.; Anghel, D.-F.; Iovescu, A.; Tudose, M. Comparing the spectral properties of pyrene as free molecule, label and derivative in some colloidal systems. *Sens. Actuators B Chem.* **2014**, *197*, 193–199. [CrossRef]

28. Zhang, C.; Qu, G.; Sun, Y.; Wu, X.; Yao, Z.; Guo, Q.; Ding, Q.; Yuan, S.; Shen, Z.; Ping, Q.; et al. Pharmacokinetics, biodistribution, efficacy and safety of N-octyl-O-sulfate chitosan micelles loaded with paclitaxel. *Biomaterials* **2008**, *29*, 1233–1241. [CrossRef] [PubMed]

29. Acharya, S.; Sahoo, S.K. PLGA nanoparticles containing various anticancer agents and tumour delivery by EPR effect. *Adv. Drug Deliv. Rev.* **2011**, *63*, 170–183. [CrossRef] [PubMed]

30. Wang, H.; Zuo, Z.; Du, J.; Wang, Y.; Sun, R.; Cao, Z.; Ye, X.; Wang, J.; Leong, K.W.; Wang, J. Surface charge critically affects tumor penetration and therapeutic efficacy of cancer nanomedicines. *Nano Today* **2016**, *11*, 133–144. [CrossRef]

31. Kim, C.; Lee, S.C.; Kang, S.W.; Kwon, I.C.; Kim, Y.H.; Jeong, S.Y. Synthesis and the micellar characteristics of poly(ethylene oxide)–deoxycholic acid conjugates. *Langmuir* **2000**, *16*, 4792–4797. [CrossRef]

32. Wang, F.; Chen, Y.; Zhang, D.; Zhang, Q.; Zheng, D.; Hao, L.; Liu, Y.; Duan, C.; Jia, L.; Liu, G. Folate-mediated targeted and intracellular delivery of paclitaxel using a novel deoxycholic acid-O-carboxymethylated chitosan–folic acid micelles. *Int. J. Nanomed.* **2012**, *7*, 325–337.

33. Yen, H.; Young, Y.; Tsai, T.; Cheng, K.; Chen, X.; Chen, Y.; Chen, C.; Young, J.; Hong, P. Positively charged gold nanoparticles capped with folate quaternary chitosan: Synthesis, cytotoxicity, and uptake by cancer cells. *Carbohydr. Polym.* **2018**, *183*, 140–150. [CrossRef] [PubMed]

34. Fröhlich, E. The role of surface charge in cellular uptake and cytotoxicity of medical nanoparticles. *Int. J. Nanomed.* **2012**, *7*, 5577–5591. [CrossRef] [PubMed]

35. Bernkop-Schnürch, A.; Krajicek, M.E. Mucoadhesive polymers as platforms for peroral peptide delivery and absorption: Synthesis and evaluation of different chitosan-EDTA conjugates. *J. Control. Release* **1998**, *50*, 215–223. [CrossRef]

36. Liang, N.; Sun, S.; Hong, J.; Tian, J.; Fang, L.; Cui, F. In vivo pharmacokinetics, biodistribution and antitumor effect of paclitaxel-loaded micelles based on α-tocopherol succinate-modified chitosan. *Drug Deliv.* **2016**, *23*, 2651–2660. [PubMed]

14

Processability and Degradability of PHA-Based Composites in Terrestrial Environments

Patrizia Cinelli *, Maurizia Seggiani *, Norma Mallegni, Vito Gigante and Andrea Lazzeri

Department of Civil and Industrial Engineering, University of Pisa, Largo Lucio Lazzarino 2, 56122 Pisa, Italy; norma.mallegni@gmail.com (N.M.); vito.gigante@dici.unipi.it (V.G.); andrea.lazzeri@unipi.it (A.L.)
* Correspondence: patrizia.cinelli@unipi.it (P.C.); maurizia.seggiani@unipi.it (M.S.)

Abstract: In this work, composites based on poly(3-hydroxybutyrate-3-hydroxyvalerate) (PHB-HV) and waste wood sawdust (SD) fibers, a byproduct of the wood industry, were produced by melt extrusion and characterized in terms of processability, thermal stability, morphology, and mechanical properties in order to discriminate the formulations suitable for injection molding. Given their application in agriculture and/or plant nursery, the biodegradability of the optimized composites was investigated under controlled composting conditions in accordance with standard methods (ASTM D5338-98 and ISO 20200-2004). The optimized PHB-HV/SD composites were used for the production of pots by injection molding and their performance was qualitatively monitored in a plant nursery and underground for 14 months. This study presents a sustainable option of valuation of wood factory residues and lowering the production cost of PHB-HV-based compounds without affecting their mechanical properties, improving their impact resistance and biodegradability rates in terrestrial environments.

Keywords: biocomposites; natural fibers; poly(3-hydroxybutyrate-3-hydroxyvalerate); biodegradation; impact properties

1. Introduction

Petroleum-based plastics are light, strong, durable, and demonstrate good resistance to degradation [1]. They offer a wide range of applications in the domestic, medical, and industrial fields in the form of single-use gears, packaging, furniture, machine chassis, and accessories to improve life quality [2]. For these reasons, approximately 150 million tons of plastic are consumed worldwide each year and this consumption is expected to continue growing until 2020 [3].

Large-scale dependence on petroleum-derived plastics leads to serious pollution problems; the methodologies used for plastic waste disposal are challenging. In landfills, the degradation rates are tremendously slow [4,5]. Incineration generates harmful by-products if advanced combustion designs, optimized operating practices, and effective emission-control technologies are not adopted. Chemical and mechanical recycling is not always possible, requires an advanced collection system, and causes a deterioration of the properties of plastic materials compromising their reuse [6,7].

Given this scenario, with the aim of decreasing plastic environmental impact, the use of bio-based and/or biodegradable polymers, such as polylactide acid (PLA), aliphatic polyesters, and polyhydroxyalkanoates (PHAs), having similar physicochemical properties as conventional plastics represent a valuable solution to the current plastic pollution problem [8–11]. However, the replacement of nondegradable with biodegradable plastics requires the complete knowledge of the biodegradability of these plastics and their blends in controlled and uncontrolled environments to have a real positive environmental impact. It is necessary to know their biodegradability and biodegradation rates both in managed (industrial and home composting, anaerobic digestion) and unmanaged (soil, oceans and rivers) environments for postconsumer management. Only a few biopolymers (PHAs and thermoplastic starch, TPS) are degradable in a wide range of managed and unmanaged environmental conditions; whereas most of them (e.g., PLA, polycaprolactone (PCL), poly(butylene succinate) (PBS), and polyhydroxyoctanoate (PHO)) degrade only in a narrower set of environmental conditions [12]. Narancic et al. [12] identified synergies but also antagonisms between polymers in blends that affect their biodegradability and biodegradation rates in different environments. Consequently, biodegradable plastics such as PHAs that degrade in a wide variety of controlled and uncontrolled environments are the most promising candidates for terrestrial and marine applications where their release into the environment does not cause plastic pollution. PHAs are a family of polyesters of several R-hydroxyalkanoic acids, synthetized by several microorganisms in the presence of excess carbon and when essential nutrients such as oxygen, nitrogen, or phosphorus are limiting or after pH shift [13–18]. PHAs have thermoplastic properties similar to those of polypropylene, good mechanical properties, and excellent biodegradability in various ecosystems [19–25]. The most common PHAs are the poly([R]-3-hydroxybutyrate) (PHB) and its copolyester with [R]-3-hydroxyvalerate (PHB-HV), which is well suited for food packaging [21]. Despite their good properties and excellent biodegradability, their relatively high cost (€7–12/kg) [14], compared to other biopolymers such as poly-lactic acid (PLA) (€2.5–3/kg) [26], has limited their use in commodities such as packaging and service items, restricting their use to high-value applications in the medical and pharmaceutical sectors. Efforts have been made to incorporate low-value materials, such as starch, into PHAs in order to reduce the cost of the final products [27–29]. Waste lignocellulose fibers, highly-available and at low-cost, sourced from agricultural and industrial crops, have been investigated as fillers in PHA-based composites [30–34].

In some previous cases, the developed composites displayed promising mechanical and physical properties. Natural fiber-reinforced composites show higher degradation when subjected to outdoor applications compared with composites with synthetic fibers [35]. Biodegradation of a composite occurs with the degradation of its individual components as well as with the loss of interfacial strength between them [36]. The weak interfacial bonding between highly polar natural fiber and non-polar matrix can lead to a reduction in final properties of the composite, ultimately hindering their industrial usage. Different methods have been applied to improve the compatibility and interfacial bond strength, including the use of various surface modification techniques [37,38]. It is harder to obtain high degrees of alignment with natural fibers than for synthetic fibers, since during the extrusion process, the long natural fibers tend to twist randomly [39]. This behavior compromises the mechanical properties of the composite as most of the natural fibers are not aligned parallel to the direction of the applied load.

Composites containing sawdust fibers and petroleum-based plastics have been well-studied. For example, they were processed by Sombatsompop et al. [40,41] with polypropylene (PP) and polyvinylchloride (PVC), revealing a reduction in the overall strength and toughness of the composites with increasing the wood fiber content. In a previous work [42], composites based on PHB-HV and fibers of *Posidonia oceanica*, a dominant Mediterranean seagrass, were successfully produced by melt extrusion and their degradability was investigated in sea water. The results showed an increase in the impact resistance of the composites with increasing fiber content. The presence of fibers favored the physical disintegration of the composite, increasing its biodegradation rate under simulated and real marine conditions.

In the present work, composites based on PHB-HV and waste sawdust fibers (SD), derived from the wood industry, were produced by extrusion and characterized in terms of processability, rheological, and mechanical properties. Given their use in terrestrial applications, biodegradation tests were completed on the developed composites under simulated composting conditions in accordance with standard methods and we preliminarily evaluated the degradability in soil of PHB-HV-SD-based molded specimens (pots with thickness of 1 mm).

2. Results and Discussion

2.1. Composite Processing

The torque measurements of polymeric melts characterize their flow behavior and reflect the trend of viscosity, as demonstrated by Melik et al. [43]. The objective of these measurements was to quantify the processing behavior of the investigated composites and, in particular, to evaluate the effect of the fibers on the melt fluidity. The average torque-time curves obtained at 170 °C and a rotor speed of 100 rpm with a HAAKE MiniLab are reported in Figure 1 for the PCA and PHB-HV/SD composites.

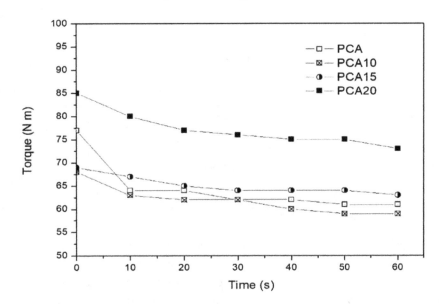

Figure 1. Torque-time curves obtained at 170 °C.

As shown, the incorporation of the SD fibers in the PHB-HV-based compound increased the torque and, consequently, the energy required for the melt mixing. With up to 15 wt % of fibers, there was a moderate increase in torque with respect to the pure matrix, whereas for PCA20, the torque values almost doubled, significantly affecting the processability of the biocomposite.

2.2. Composite Characterization

The thermogravimetric (TG) and their derivative (DTG) curves of SD fibers, PHB-HV, ATBC, and the developed composites are shown in Figure 2.

SD fibers showed an initial weight loss at around 100 °C, attributable to loss in the residual humidity. Then, the sharp decrease observed at 250–350 °C can be related to the degradation of hemicellulose, cellulose, and lignin. Hemicellulose decomposes easily with respect to the other components. Typically, the pyrolysis of hemicellulose occurs between 200 and 280 °C, resulting in formation of CO, CO_2, condensable vapors, and organic acids [44].

Therefore, the thermal degradation of the SD fibers occurs at temperatures higher than 200 °C, confirming their suitability of being processed with thermoplastic polymer matrices, such as PHB-HV, without thermal degradation occurring.

PHB-HV showed the onset of degradation at about 260 °C with maximum weight loss rate at 305 °C and no residue was recorded above 350 °C. For all of the produced composites, the thermal degradation started at temperatures above 200 °C, with the main degradation peaks occurring above 250 °C.

The composite mechanical properties are shown in Table 1.

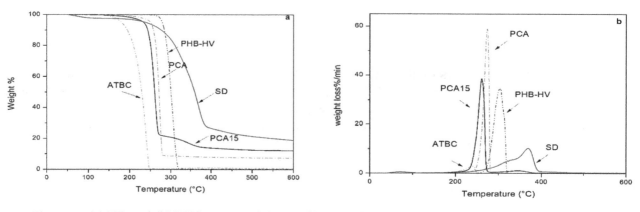

Figure 2. (**a**) TG and (**b**) DTG curves of the SD fibres, PHB-HV, ATBC, and developed composites.

Table 1. Mechanical properties of the composites with different SD fiber contents.

Sample	Young's Modulus (GPa)	Stress at Break (MPa)	Elongation (%)	Charpy Impact Energy (kJ/m^2)
PCA	2.64 ± 0.28	25.62 ± 2.11	2.14 ± 0.50	3.57 ± 0.36
PCA10	2.35 ± 0.24	21.02 ± 0.94	2.05 ± 0.28	6.17 ± 0.24
PCA15	2.52 ± 0.15	18.52 ± 0.84	1.35 ± 0.14	12.24 ± 0.50
PCA20	2.94 ± 0.35	20.93 ± 1.57	1.35 ± 0.13	5.91 ± 0.40

The values are the mean ± SD of at least five determinations.

The results of tensile tests showed that the elastic modulus was almost constant, except in the case of PCA20 where a higher fiber content led to a significant increase in stiffness, causing a decrease in break stress and elongation. This behavior is typical of composites with poor or no compatibility between the components, preventing the stress transfer phenomena from occurring and the filler particle becomes a stress concentrator, leading to brittle fracture [45].

To better understand these results, SEM analysis was performed on the cross-section of the specimens before the tensile test and on the fractured sections obtained after testing (Figure 3).

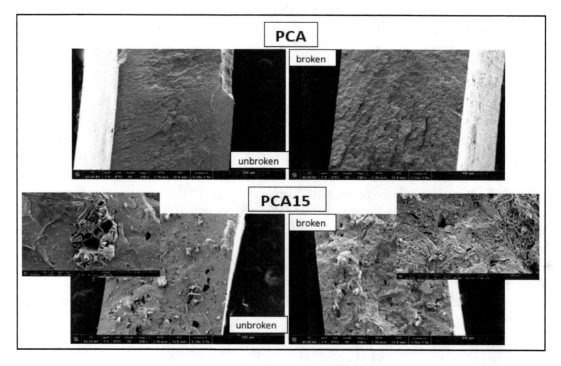

Figure 3. SEM images of the cross-sections of the PCA and PCA15 specimens before (unbroken samples) and after tensile tests (broken samples).

The unbroken PCA15 specimen showed a good dispersion of SD fibers that are homogeneously distributed within the thermoplastic matrix. In the broken specimen section, it was possible to observe a significant fiber pullout. This means that the interfacial interactions between the fibers and the matrix were not sufficiently strong to maintain their cohesion during the tensile test.

Natural fibers are rich in cellulose, hemicellulose, lignin, and pectin; consequently, they tend to be active polar hydrophilic materials, whereas polymeric materials are nonpolar and show considerable hydrophobicity. The hydrophilic nature of natural fibers reduces the adhesion to a hydrophobic matrix and, as a result, a loss in strength may be induced. To prevent this, the fiber surface can be modified by several methods proposed in the literature, such as graft copolymerization of monomers onto the fiber surface, the use of maleic anhydride copolymers, alkyl succinic anhydride, stearic acid, etc. [38,46,47]. Therefore, a compatibilization method may be necessary to produce composites with tailored tensile properties.

Even without compatibilizers, interesting results were obtained from the impact test, in which the PCA15 compound showed a higher impact resistance compared with the matrix without fibers. Factors such as fiber/matrix de-bonding, fiber and/or matrix fracture, and fiber pull out improve the impact performance. Fiber fracture dissipates less energy compared to fiber pull-out, which is common in composites with strong interfacial bonds. A high impact energy is a sign of weak fiber/matrix bond [48,49]. In this case, the absorbed energy value increased up to 15 wt. % SD fibers, and then decreased at higher loadings, which caused an increase in material brittleness [50,51].

2.3. Biodegradability in Lab-Scale Terrestrial Environments

2.3.1. Mineralization

Figure 4 shows the aerobic biodegradation curves obtained on the lab-scale. As shown, after six months, the PCA15 composite reached a mineralization percentage of about 78%, higher than those of the control sample (filter paper) and the composite without fibers (58%). This behavior can be explained by the presence of the fibers favoring the disintegration of the sample, increasing its susceptibility to microbial attack.

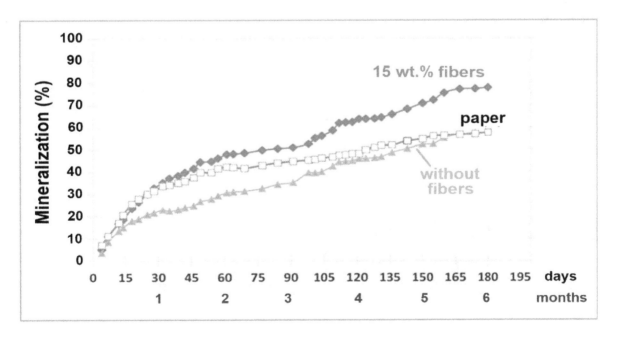

Figure 4. Mineralization curves under simulated terrestrial environmental conditions.

2.3.2. Disintegration

The average results of the disintegration test for each sample are reported in Table 2.

Table 2. Average percentage of disintegration after 90 days under simulated composting conditions.

Sample	Sample No.	Disintegration (%)	Average Disintegration (%)
PCA	1	87.2	92.6
	2	94.3	
	3	96.4	
PCA10	1	83.4	93.2
	2	100.0	
	3	96.2	
PCA15	1	96.3	94.2
	2	88.6	
	3	97.7	

After 90 days, all the samples examined showed greater than 90% disintegration. The composite containing the higher amounts of fibers presented the highest degrees of disintegration, showing that the presence of fibers inside the matrix facilitated the degradation of the material according to the results of the mineralization test.

Plasticizers, including ATBC, can accelerate the disintegration of polymeric matrices under composting conditions due to the greater mobility of the polymer chains, as observed by Arrieta et al. [52] for PHB-based blends. Narancic et al. [12] evidenced, in some cases, synergies between biodegradable polymers in blends that improve their biodegradation rates. A possible explanation of the synergy observed in some environments (e.g., home composting or anaerobic digestion) for PLA-PCL (80/20), PHB–PCL (60/40), and PCL–TPS (70/30) blends was that the addition of amorphous PCL, by lowering the crystallinity of the blends, improves their degradation rate.

Consequently, and also in this case, ATBC may have contributed, lowering the crystallinity of the PHB-HV together with the fibers to accelerate the disintegration of the specimens, thus increasing the biodegradation rate of the PHB-HV.

Degradation of Pots in Soil

The performance and degradation of 180 pots (produced by injection molding with thickness of 1 mm and weight of about 115 g) was evaluated in three different environments: 60 were placed in a greenhouse (15 pots for each composite: PCA, PCA10, and PCA15, and 15 pots of PP, as reference); 60 pots were located outside on a cloth and not in contact with the soil (15 pots for each material); and 60 pots were buried into the soil up to the upper profile, leaving only the plants outside (Figure 5).

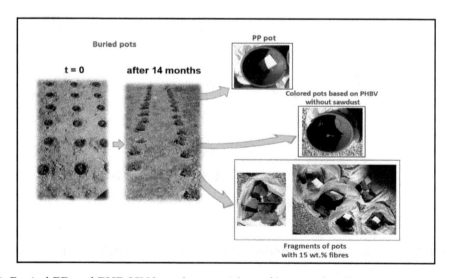

Figure 5. Buried PP and PHB-HV based pots without fibers and with 15 wt % sawdust fibers.

After 14 months of experimentation, the PP pots showed no evident signs of degradation, as expected, whereas the pots based on PHB-HV showed initial signs of degradation, in particular those containing 15 wt % sawdust fibers. The plant growth performance was the same in all pots (PHB-HV, PHB-HV/SD, PP).

Figure 5 shows the results obtained with the buried pots: the PP pots remained intact as expected. The pots based on PHB-HV without fibers were slightly damaged on the bottom, while the pots containing 15 wt % SD fibers were totally fragmented and markedly degraded, confirming the results of the lab-scale degradation tests.

3. Material and Methods

3.1. Materials

PHB-HV (PHI002) with 5 wt % valerate content was supplied in pellets by Naturplast® (Caen, France); it is characterized by a density of 1.25 g/cm^3 and melt flow index (190 °C, 2.16 kg) of 15–20 g/10 min. According to the supplier data sheet, PHB-HV is a semi crystalline polymer with a glass transition temperature of around 5 °C and a melting temperature of around 155 °C.

Sawdust fibers (SD) were obtained from a soft wood-processing company. SD was dried at 80° C in a vacuum oven for at least 24 h, then milled using a lab-scale grinder and screened with a 500-µm sieve.

To improve the flexibility and processability of the composites with high content of fiber, a bio-based plasticizer, acetylbutylcitrate (ATBC), was used. ATBC was purchased from Tecnosintesi® (Bergamo, Italy); it is a colorless liquid, soluble in organic solvents, produced by acetylation of tri-n-butylcitrate, having a density of 1.05 g/cm^3, and a molecular weight of 402.5 g/mol. ATBC is widely used as food packaging as it is non-toxic.

Micro-sized calcium carbonate particles, produced from a high purity white marble, with mean particle size (D$_{50}$) of 2.6 µm and top cut diameter (D$_{98}$) of 15 µm, cubic shape (aspect ratio = 1), and density of 2.7 g/cm^3, provided by OMYA® (Avenza Carrara, Italy) with trade name OMYACARB 2-AZ, were used as rigid inorganic filler. CaCO$_3$ is one of the most commonly used fillers in thermoplastics to reduce both their production cost and to improve their properties, such as stiffness, impact strength, processability, and thermal and dimensional stability [53].

3.2. Composite Preparation

The composites (whose compositions by weight are reported in Table 3) were prepared by mixing the various components at 170 °C for one minute in a Thermo Scientific Haake microcompounder (HAAKE™, Karlsruhe, Germany) with a twin-conic screw system at 100 rpm speed. The labels PCA, PCA10, PCA15 and PCA20 indicate the composite without SD fibers (PHB-HV/CaCO$_3$/ATBC 80/10/10 w/w/w%) and with 10, 15, and 20 wt % SD fibers, respectively.

Table 3. Composite formulations.

| Composite | Weight Percentage | | | |
	PHB-HV	CaCO$_3$	ATBC	Sawdust Fibers
PCA	80	10.0	10.0	0
PCA10	72	9.0	9.0	10
PCA15	68	8.5	8.5	15
PCA20	64	8.0	8.0	20

The miniLab compounder was equipped with a backflow channel (Figure 6) in which the recirculation of the melt occurs. During the melt mixing, the torque was recorded to evaluate the fiber effect on the melt fluidity. The extruded filaments were cut to obtain pellets used for the successive thermal characterizations, lab-scale biodegradation, and disintegration tests.

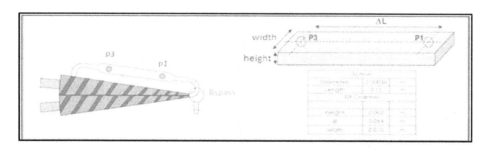

Figure 6. MiniLab Backflow channel and twin-screw conic system.

For each formulation, tensile test specimens (Haake type 3 dog-bone bar) and Charpy Impact test samples (80 × 10 × 4 mm parallelepiped) were produced by feeding the molten material from the minilab into a mini-injection press (Thermo Scientific Haake MiniJet II, HAAKE™, Karlsruhe, Germany) at 210 bar with a mold heated at 60 °C.

Industrial compounding of the selected formulations was carried out using a COMAC EBC 50HT extruder (COMAC, Milan, Italy) with a temperature profile from 135 °C in the feeding zone to 170 °C at the screw terminal zone. The resultant pellets were then used to produce items such as pots using an injection molding machine (Negri Bossi VS250, Negri Bossi, Milan, Italy) at Femto Engineering (Florence, Italy).

3.3. Composite Characterization

The thermal properties of the raw materials and composites in pellet form were evaluated by thermogravimetric analysis (TGA). These measurements were carried out, in duplicate, on 20 mg of sample using a Q500 TGA (TA Instruments; New Castle, DE, USA), under nitrogen flow (100 mL/min), at a heating speed of 10 °C/min from room temperature to 600 °C.

Tensile tests were performed on the injection molded Haake Type 3 (557-2290) dog-bone tensile bars in accordance with ASTM D 638. Stress-strain tests were carried out at room temperature, at a crosshead speed of 10 mm/min, by an Instron 5500R universal testing machine (Canton, MA, USA) equipped with a 10-kN load cell, and interfaced with a computer running MERLIN Instron software, version 4.42 S/N–014733H (Instron®, Canton, MA, USA).

The Charpy impact tests were completed at room temperature on 2-mm V-notched specimens using a 15 J Charpy pendulum (CEAST 9050, Instron®, Canton, MA, USA) following the standard method ISO 179:2000. For each mechanical test, at least five replicates were carried out.

The morphology of the composite tensile specimens, before and after the tensile test, was investigated by scanning electron microscopy (SEM) using a FEI Quanta 450 FEG SEM (Thermo Fisher Scientific, Waltham, MA, USA). The SEM analysis was performed under high-vacuum conditions because air can inhibit the electron beam. The undamaged samples were frozen under liquid nitrogen and then fractured. The surfaces were metalized with gold using a Sputter Coater Edward S150B to avoid charge build up.

3.4. Biodegradation in Terrestrial Environment

3.4.1. Mineralization Test in Compost

Given the application of composites in terrestrial environments, a mineralization test in compost was carried out following a laboratory method [10] based on ASTM D5338 [54]. The setup of the system is shown in Figure 7. The test was carried out in glass vessels (1 L capacity) containing a layer of a mixture of 3 g compost, 4 g perlite, and 0.5 g of sample in form of pellets, wetted with 15 mL distilled water, in order to operate with a weight ratio of sample/compost of 1:6 in accordance with ASTM D5338. Finally, the mixtures were sandwiched between two layers consisting of 5 g perlite wetted with 20 mL of water. Perlite (tradename Agrilit 2), a hygroscopic aluminum silicate, was supplied in granules of 1–2 mm by Perlite Italiana Srl (Milan, Italy). Perlite is able to hold 3 to 4 times its weight in water, largely used in horticultural applications where it provides aeration and optimal moisture conditions for plant growth. In the mineralization test, the perlite mixed with the compost maintained the hydration of the medium and increased the contact surface between the sample and the inoculum of the compost.

The used compost was derived from the organic fraction of solid municipal waste, kindly supplied by Cermec SpA, Consorzio Ecologia e Risorse (Massa Carrara, Italy). Compost had 23.8 wt % carbon content, 2.2 wt % nitrogen content, with a weight C/N ratio of 10.7.

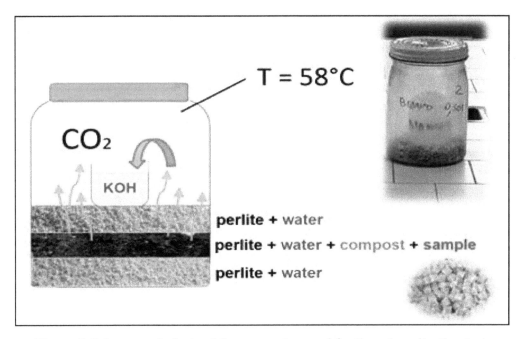

Figure 7. Scheme and photo of the apparatus used for the mineralization test.

A 100 mL beaker was inserted in each test vessel containing 50 mL of 0.1 M KOH. The basic solution trapped the carbon dioxide evolved from the sample (represented schematically by the green arrows in Figure 7) on the basis of the following reaction (Equation (1)):

$$2KOH + CO_2 \rightarrow K_2CO_3 + H_2O \tag{1}$$

Every 2–5 days, the absorbing solution was titrated with 0.1 M HCl after addition of 0.5 N $BaCl_2$ to avoid the presence of soluble carbonates.

The amount of evolved carbon dioxide was evaluated using the following equation:

$$\text{mg } CO_2 = (V_{KOH} \cdot C_{KOH} - V_{HCl} \cdot C_{HCl}) \cdot \frac{44}{2} \tag{2}$$

where $mgCO_2$ is the CO_2 evolved in mg from sample in a single test session; V_{KOH} is the volume in mL of KOH (50 mL) in flask at the beginning of test session; V_{HCl} is the volume in mL of HCl needed for titration of KOH at the end of test session; C_{KOH} and C_{HCl} are the concentrations of KOH and HCl in mol/L, respectively; 44 is molecular weight of CO_2; and 1/2 is the stoichiometric coefficient (mol CO_2/mol KOH) of the CO_2 absorption reaction, as shown in Equation (1).

The mineralization extent of each sample was calculated as neat percentage (corrected for the inoculum endogenous emission from the blank vessels) of the overall theoretical CO_2 ($ThCO_2$) amount, calculated on the basis of the initial organic carbon content of the testing sample:

$$\text{Mineralization\%} = \frac{\Sigma mgCO_2 - \Sigma mgCO_{2,blank}}{ThCO_2} \cdot 100 \tag{3}$$

The mineralization test was carried out in triplicate and the average values are reported.

3.4.2. Disintegration Test

According to ISO 20200 [55], the determination of the disintegration degree of polymeric material under simulated composting conditions [56] was carried out on PCA, PCA10, and PCA15 using a synthetic compost prepared by mixing the different components listed in Table 4. Then, tap water was added to the mixture to adjust its final water content to 50 wt % in total.

Samples in the form of hot-pressed films (about 2.5 g) with the dimensions 2.5 × 2.5 cm and thickness of 1 mm (Figure 8) were used. Before mixing with the synthetic compost, the samples were dried in an oven at 40 ± 2 °C under vacuum for the time needed to reach constant mass.

The test was carried out in flasks made of polypropylene, with dimensions of 30 × 20 × 10 cm (l, w, h), with two 5 mm holes (6.5 cm from the bottom) to provide gas exchange between the inner atmosphere and the outside environment. In each flask, 500 g of synthetic compost, mixed with three pieces of sample, were placed on the bottom, forming a homogeneous layer.

Table 4. Compost composition.

Material	Dry Mass %
Sawdust	40
Rabbit-feed	30
Ripe compost	10
Corn starch	10
Saccharose	5
Cornseed oil	4
Urea	1
Total	100

Figure 8. Sample used for the disintegration test.

The flasks were covered with a lid ensuring a tight seal to avoid excessive evaporation and placed in an air-circulation oven at a constant temperature of 58 ± 2 °C for 90 days. Following the procedure reported in ISO 20200, the gross mass of the flasks filled with the mixture was determined at the beginning of the composting process and they were weighed at determined times and, if needed, the initial mass was restored totally or in part by adding distilled water. At the end of the test, the lid of each flask was removed and the flasks were placed in the air-circulation oven at 58 ± 2 °C for 48 h to dry the content. Then, the compost of each reactor was sieved using three standard sieves (ISO 3310 [57]): 10 mm, 5 mm, and finally a 2 mm sieve. Any residual pieces of sample that did not pass through the sieves were collected, cleaned to remove the compost that covered them, and dried in an oven at 40 ± 2 °C under vacuum to constant mass. The total material recovered from the sieving procedure was considered to be non-disintegrated material. So, the degree of disintegration was calculated, in percent, using the Equation (4):

$$\text{Disintegration (\%)} = \frac{m_{i,sample} - m_{f,sample}}{m_{i,sample}} \cdot 100 \qquad (4)$$

where $m_{i,sample}$ and $m_{f,sample}$ represent the initial sample mass and the final dry mass of the sample recovered after sieving, respectively.

The disintegration test was carried out in triplicate on each sample.

4. Conclusions

In this work, we explored the use of waste wood fibers in combination with PHB-HV to manufacture biodegradable wood plastic composites in accordance with the circular economy principles.

We produced composites based on PHB-HV and different amounts (10, 15, and 20 wt %) of soft wood sawdust, a byproduct of the wood industry, by extrusion in the presence of appropriate amounts of ATBC as a plasticizer and calcium carbonate as an inorganic filler.

Using appropriate amounts of ATBC, smooth processing was achieved for up to 20 wt % f fibers, despite the reduction in the melt fluidity observed with increasing fiber loading. The tensile modulus remained almost constant up to a 15 wt % fiber content, whereas the tensile strength and the elongation slightly decreased by increasing the fiber content up to 20 wt %. The impact resistance of the composites increased markedly with increasing SD amounts: the Charpy's impact energy increased from 3.6 (without fiber) to 12.2 kJ/m^2 for the composite with 15 wt % fiber. The results of the mineralization and disintegration tests in compost showed that the developed composites are compostable in accordance with EN 13427:2000, and the presence of fibers favored the physical disintegration of the composite, increasing the biodegradation rate of the polymeric matrix. After six months in compost, the composite with 15 wt % fiber showed a biodegradability of 78% compared with 58% for the composites without fibers and the paper (control sample). After 14 months, the pots based on PHB-HV/SD retained adequate mechanical performance and physical integrity inside the plant nursery and the plant growth performance was the same as that in traditional polypropylene pots, but the buried pots containing 15 wt % fibers were completely fragmented compared with those without fibers, confirming the accelerating effect of the fibers on the degradation of the polymeric matrix in soil.

In conclusion, the developed composites based on PHB-HV and waste wood sawdust are particularly suitable for production by the extrusion of relatively low-cost items such as pots, which are compostable and biodegradable in soil and usable in agriculture or plant nursery.

Author Contributions: P.C. and M.S. conceived and designed the experiments; N.M. performed the experiments; N.M. and V.G. analyzed the data; P.C., M.S. and V.G. wrote the paper; A.L. revised the paper.

Acknowledgments: The authors would like to thank the Tuscany Region for the financial support of the project PHA through the fund POR FESR 2014–2020 (Bando 2—RSI 2014) (Grant number: 3389.30072014.068000241). We would also like to thank Zefiro and Femto Engineering for the scale-up of the process, Rosa Galante for support in carrying out the biodegradation and disintegration tests, in terrestrial environments and at the lab-scale, respectively, and Randa Ishak for the SEM analysis.

References

1. Song, J.H.; Murphy, R.J.; Narayan, R.; Davies, G.B.H. Biodegradable and compostable alternatives to conventional plastics. *Philos. Trans. R. Soc. B Biol. Sci.* **2009**, *364*, 2127–2139. [CrossRef] [PubMed]
2. Muthuraj, R.; Misra, M.; Mohanty, A.K. Studies on mechanical, thermal, and morphological characteristics of biocomposites from biodegradable polymer blends and natural fibers. In *Biocomposites*; Elsevier: Amsterdam, The Netherlands, 2015; pp. 93–140, ISBN 9781782423737.
3. Koronis, G.; Silva, A.; Fontul, M. Green composites: A review of adequate materials for automotive applications. *Compos. Part B Eng.* **2013**, *44*, 120–127. [CrossRef]
4. Khanna, S.; Srivastava, A.K. On-line Characterization of Physiological State in Poly(β-Hydroxybutyrate) Production by Wautersia eutropha. *Appl. Biochem. Biotechnol.* **2009**, *157*, 237–243. [CrossRef] [PubMed]
5. Castilho, L.R.; Mitchell, D.A.; Freire, D.M.G. Production of polyhydroxyalkanoates (PHAs) from waste materials and by-products by submerged and solid-state fermentation. *Bioresour. Technol.* **2009**, *100*, 5996–6009. [CrossRef] [PubMed]

6.   Hopewell, J.; Dvorak, R.; Kosior, E. Plastics recycling: Challenges and opportunities. *Philos. Trans. R. Soc. B Biol. Sci.* **2009**, *364*, 2115–2126. [CrossRef] [PubMed]

7.   Jose, J.; George, S.M.; Thomas, S. Recycling of polymer blends. In *Recent Developments in Polymer Recycling*; Kluwer Academic Publishers: Dordrecht, The Netherlands, 2011; Volume 37, pp. 187–214, ISBN 978-81-7895-524-7.

8.   Murariu, M.; Dubois, P. PLA composites: From production to properties. *Adv. Drug Deliv. Rev.* **2016**, *107*, 17–46. [CrossRef] [PubMed]

9.   Kourmentza, C.; Plácido, J.; Venetsaneas, N.; Burniol-Figols, A.; Varrone, C.; Gavala, H.N.; Reis, M.A.M. Recent Advances and Challenges towards Sustainable Polyhydroxyalkanoate (PHA) Production. *Bioengineering* **2017**, *4*, 55. [CrossRef]

10. Chiellini, E.; Cinelli, P.; Chiellini, F.; Imam, S.H. Environmentally Degradable Bio-Based Polymeric Blends and Composites. *Macromol. Biosci.* **2004**, *4*, 218–231. [CrossRef]

11. Yu, L.; Dean, K.; Li, L. Polymer blends and composites from renewable resources. *Prog. Polym. Sci.* **2006**, *31*, 576–602. [CrossRef]

12. Narancic, T.; Verstichel, S.; Reddy Chaganti, S.; Morales-Gamez, L.; Kenny, S.T.; De Wilde, B.; Padamati, R.B.; O'Connor, K.E. Biodegradable Plastic Blends Create New Possibilities for End-of-Life Management of Plastics but They Are Not a Panacea for Plastic Pollution. *Environ. Sci. Technol.* **2018**, *52*, 10441–10452. [CrossRef]

13. Law, K.H.; Cheng, Y.C.; Leung, Y.C.; Lo, W.H.; Chua, H.; Yu, H.F. Construction of recombinant Bacillus subtilis strains for polyhydroxyalkanoates synthesis. *Biochem. Eng. J.* **2003**, *16*, 203–208. [CrossRef]

14. Bugnicourt, E.; Cinelli, P.; Lazzeri, A.; Alvarez, V. Polyhydroxyalkanoate (PHA): Review of synthesis, characteristics, processing and potential applications in packaging. *Express Polym. Lett.* **2014**, *8*, 791–808. [CrossRef]

15. Steinbüchel, A.; Lütke-Eversloh, T. Metabolic engineering and pathway construction for biotechnological production of relevant polyhydroxyalkanoates in microorganisms. *Biochem. Eng. J.* **2003**, *16*, 81–96. [CrossRef]

16. Koller, M.; Marsalek, L.; de Sousa Dias, M.M.; Braunegg, G. Producing microbial polyhydroxyalkanoate (PHA) biopolyesters in a sustainable manner. *New Biotechnol.* **2017**, *37*, 24–38. [CrossRef] [PubMed]

17. Salehizadeh, H.; Van Loosdrecht, M.C.M. Production of polyhydroxyalkanoates by mixed culture: Recent trends and biotechnological importance. *Biotechnol. Adv.* **2004**, *22*, 261–279. [CrossRef] [PubMed]

18. Anjum, A.; Zuber, M.; Zia, K.M.; Noreen, A.; Anjum, M.N.; Tabasum, S. Microbial production of polyhydroxyalkanoates (PHAs) and its copolymers: A review of recent advancements. *Int. J. Biol. Macromol.* **2016**, *89*, 161–174. [CrossRef] [PubMed]

19. Numata, K.; Hideki, A.; Tadahisa, I. Biodegradability of Poly(hydroxyalkanoate) Materials. *Materials* **2009**, *2*, 1104–1126. [CrossRef]

20. Breulmann, M.; Künkel, A.; Philipp, S.; Reimer, V.; Siegenthaler, K.O.; Skupin, G.; Yamamoto, M. Polymers, Biodegradable. In *Ullmann's Encyclopedia of Industrial Chemistry*; VCH Publishers: New York, NY, USA, 2009.

21. Philip, S.; Keshavarz, T.; Roy, I. Polyhydroxyalkanoates: Biodegradable polymers with a range of applications. *J. Chem. Technol. Biotechnol.* **2007**, *82*, 233–247. [CrossRef]

22. Padovani, G.; Carlozzi, P.; Seggiani, M.; Cinelli, P. PHB-Rich Biomass and BioH 2 Production by Means of Photosynthetic Microorganisms. *Chem. Eng. Trans.* **2016**, *49*, 55–60. [CrossRef]

23. Deroiné, M.; Le Duigou, A.; Corre, Y.-M.; Le Gac, P.-Y.; Davies, P.; César, G.; Bruzaud, S. Seawater accelerated ageing of poly(3-hydroxybutyrate-co-3-hydroxyvalerate). *Polym. Degrad. Stab.* **2014**, *105*, 237–247. [CrossRef]

24. Volova, T.G.; Boyandin, A.N.; Vasiliev, A.D.; Karpov, V.A.; Prudnikova, S.V.; Mishukova, O.V.; Boyarskikh, U.A.; Filipenko, M.L.; Rudnev, V.P.; Bá Xuân, B.; et al. Biodegradation of polyhydroxyalkanoates (PHAs) in tropical coastal waters and identification of PHA-degrading bacteria. *Polym. Degrad. Stab.* **2010**, *95*, 2350–2359. [CrossRef]

25. Musioł, M.; Sikorska, W.; Janeczek, H.; Wałach, W.; Hercog, A.; Johnston, B.; Rydz, J. (Bio)degradable polymeric materials for a sustainable future—Part 1. Organic recycling of PLA/PBAT blends in the form of prototype packages with long shelf-life. *Waste Manag.* **2018**, 1–8. [CrossRef]

26. Aliotta, L.; Cinelli, P.; Coltelli, M.B.; Righetti, M.C.; Gazzano, M.; Lazzeri, A. Effect of nucleating agents on crystallinity and properties of poly (lactic acid) (PLA). *Eur. Polym. J.* **2017**, *93*, 822–832. [CrossRef]

27. Reis, K.C.; Pereira, J.; Smith, A.C.; Carvalho, C.W.P.; Wellner, N.; Yakimets, I. Characterization of polyhydroxybutyrate-hydroxyvalerate (PHB-HV)/maize starch blend films. *J. Food Eng.* **2008**, *89*, 361–369. [CrossRef]

28. Godbole, S.; Gote, S.; Latkar, M.; Chakrabarti, T. Preparation and characterization of biodegradable poly-3-hydroxybutyrate–starch blend films. *Bioresour. Technol.* **2003**, *86*, 33–37. [CrossRef]

29. Zhang, M.; Thomas, N.L. Preparation and properties of polyhydroxybutyrate blended with different types of starch. *J. Appl. Polym. Sci.* **2009**, *116*, 688–694. [CrossRef]

30. Seggiani, M.; Cinelli, P.; Mallegni, N.; Balestri, E.; Puccini, M.; Vitolo, S.; Lardicci, C.; Lazzeri, A. New Bio-Composites Based on Polyhydroxyalkanoates and Posidonia oceanica Fibres for Applications in a Marine Environment. *Materials* **2017**, *10*, 326. [CrossRef]

31. Seggiani, M.; Cinelli, P.; Verstichel, S.; Puccini, M.; Anguillesi, I.; Lazzeri, A. Development of Fibres-Reinforced Biodegradable Composites. *Chem. Eng. Trans.* **2015**, *43*, 1813–1818. [CrossRef]

32. Imam, S.H.; Cinelli, P.; Gordon, S.H.; Chiellini, E. Characterization of Biodegradable Composite Films Prepared from Blends of Poly(Vinyl Alcohol), Cornstarch, and Lignocellulosic Fiber. *J. Polym. Environ.* **2005**, *13*, 47–55. [CrossRef]

33. Seggiani, M.; Cinelli, P.; Geicu, M.; Elen, P.M.; Puccini, M.; Lazzeri, A. Microbiological valorisation of bio-composites based on polylactic acid and wood fibres. *Chem. Eng. Trans.* **2016**, *49*, 127–132. [CrossRef]

34. Chiellini, E.; Cinelli, P.; Imam, S.H.; Mao, L. Composite Films Based on Biorelated Agro-Industrial Waste and Poly(vinyl alcohol). Preparation and Mechanical Properties Characterization. *Biomacromolecules* **2001**, *2*, 1029–1037. [CrossRef] [PubMed]

35. Pandey, J.K.; Nagarajan, V.; Mohanty, A.K.; Misra, M. 1—Commercial potential and competitiveness of natural fiber composites. In *Woodhead Publishing Series in Composites Science and Engineering*; Misra, M., Pandey, J.K., Mohanty, A.K.B.T.-B., Eds.; Woodhead Publishing: Cambridge, UK, 2015; pp. 1–15, ISBN 978-1-78242-373-7.

36. Azwa, Z.N.; Yousif, B.F.; Manalo, A.C.; Karunasena, W. A review on the degradability of polymeric composites based on natural fibres. *Mater. Des.* **2013**, *47*, 424–442. [CrossRef]

37. Herrera-Franco, P.J.; Valadez-Gonzalez, A. A study of the mechanical properties of short natural-fiber reinforced composites. *Compos. Part B Eng.* **2005**, *36*, 597–608. [CrossRef]

38. Satyanarayana, K.G.; Arizaga, G.G.C.; Wypych, F. Biodegradable composites based on lignocellulosic fibers—An overview. *Prog. Polym. Sci.* **2009**, *34*, 982–1021. [CrossRef]

39. Gigante, V.; Aliotta, L.; Phuong, V.T.; Coltelli, M.B.; Cinelli, P.; Lazzeri, A. Effects of waviness on fiber-length distribution and interfacial shear strength of natural fibers reinforced composites. *Compos. Sci. Technol.* **2017**, *152*, 129–138. [CrossRef]

40. Sombatsompop, N.; Yotinwattanakumtorn, C.; Thongpin, C. Influence of type and concentration of maleic anhydride grafted polypropylene and impact modifiers on mechanical properties of PP/wood sawdust composites. *J. Appl. Polym. Sci.* **2005**, *97*, 475–484. [CrossRef]

41. Sombatsompop, N.; Chaochanchaikul, K. Average mixing torque, tensile and impact properties, and thermal stability of poly(vinyl chloride)/sawdust composites with different silane coupling agents. *J. Appl. Polym. Sci.* **2005**, *96*, 213–221. [CrossRef]

42. Seggiani, M.; Cinelli, P.; Balestri, E.; Mallegni, N.; Stefanelli, E.; Rossi, A.; Lardicci, C.; Lazzeri, A. Novel Sustainable Composites Based on Poly(hydroxybutyrate-co-hydroxyvalerate) and Seagrass Beach-CAST Fibers: Performance and Degradability in Marine Environments. *Materials* **2018**, *11*, 772. [CrossRef]

43. Melik, H.D.; Schechtman, A.L. Biopolyester melt behavior by torque rheometry. *Polym. Eng. Sci.* **1995**, *35*, 1795–1806. [CrossRef]

44. Baysal, E.; Deveci, I.; Turkoglu, T.; Toker, H. Thermal analysis of oriental beech sawdust treated with some commercial wood preservatives. *Maderas. Cienc. Tecnol.* **2017**, *19*, 329–338. [CrossRef]

45. Bledzki, A.K.; Reihmane, S.; Gassan, J. Properties and modification methods for vegetable fibers for natural fiber composites. *J. Appl. Polym. Sci.* **1996**, *59*, 1329–1336. [CrossRef]

46. Renner, K.; Kenyó, C.; Móczó, J.; Pukánszky, B. Micromechanical deformation processes in PP/wood composites: Particle characteristics, adhesion, mechanisms. *Compos. Part A Appl. Sci. Manuf.* **2010**, *41*, 1653–1661. [CrossRef]

47. Gurunathan, T.; Mohanty, S.; Nayak, S.K. A review of the recent developments in biocomposites based on natural fibres and their application perspectives. *Compos. Part A Appl. Sci. Manuf.* **2015**, *77*, 1–25. [CrossRef]

48. Graupner, N.; Müssig, J. A comparison of the mechanical characteristics of kenaf and lyocell fibre reinforced poly(lactic acid) (PLA) and poly(3-hydroxybutyrate) (PHB) composites. *Compos. Part A Appl. Sci. Manuf.* **2011**, *42*, 2010–2019. [CrossRef]

49. Ganster, J.; Fink, H.-P. Novel cellulose fibre reinforced thermoplastic materials. *Cellulose* **2006**, *13*, 271–280. [CrossRef]

50. Ku, H.; Wang, H.; Pattarachaiyakoop, N.; Trada, M. A review on the tensile properties of natural fiber reinforced polymer composites. *Compos. Part B Eng.* **2011**, *42*, 856–873. [CrossRef]

51. Imre, B.; Pukánszky, B. Recent advances in bio-based polymers and composites: Preface to the BiPoCo 2012 Special Section. *Eur. Polym. J.* **2013**, *49*, 1146–1150. [CrossRef]

52. Arrieta, M.P.; Samper, M.D.; López, J.; Jiménez, A. Combined Effect of Poly(hydroxybutyrate) and Plasticizers on Polylactic acid Properties for Film Intended for Food Packaging. *J. Polym. Environ.* **2014**, *22*, 460–470. [CrossRef]

53. Lin, Y.; Chan, C.-M. *3—Calcium Carbonate Nanocomposites. Advances in Polymer Nanocomposites, Types and Applications*; Series in Composites Science and Engineering; Woodhead Publishing: Cambridge, UK, 2012; pp. 55–90, ISBN 978-1-84569-940-6.

54. American Society for Testing and Materials ASTM D5338. *Standard Test Method for Determining Aerobic Biodegradation of Plastic Materials under Controlled Composting Conditions*; American Society for Testing and Materials: West Conshohocken, PA, USA, 1998.

55. International Organization for Standardization ISO 20200. *Plastics Determination of the Degree of Disintegration of Plastic Materials under Simulated Composting Conditions in a Laboratory-Scale Test*; International Organization for Standardization: Geneva, Switzerland, 2015.

56. Vaverková, M.; Toman, F.; Adamcová, D.; Kotovicová, J. Study of the Biodegrability of Degradable/ Biodegradable Plastic Material in a Controlled Composting Environment. *Ecol. Chem. Eng. S* **2012**, *19*, 347–358. [CrossRef]

57. International Organization for Standardization ISO 3310. *Test Sieves—Technical Requirements and Testing—Part 1: Test Sieves of Metal Wire Cloth*; International Organization for Standardization: Geneva, Switzerland, 2016; pp. 1–15.

15

Noble Metal Composite Porous Silk Fibroin Aerogel Fibers

Alexander N. Mitropoulos [1,2,*], **F. John Burpo** [1,*], **Chi K. Nguyen** [1], **Enoch A. Nagelli** [1], **Madeline Y. Ryu** [1], **Jenny Wang** [1], **R. Kenneth Sims** [3], **Kamil Woronowicz** [1] and **J. Kenneth Wickiser** [1]

[1]   Department of Chemistry and Life Science, United States Military Academy, West Point, NY 10996, USA;
chi.nguyen@westpoint.edu (C.K.N.); enoch.nagelli@westpoint.edu (E.A.N.);
madeline.ryu@westpoint.edu (M.Y.R.); jenny.wang@westpoint.edu (J.W.);
kamil.woronowicz@westpoint.edu (K.W.); ken.wickiser@westpoint.edu (J.K.W.)

[2]   Department of Mathematical Sciences, United States Military Academy, West Point, NY 10996, USA

[3]   Department of Civil and Mechanical Engineering, United States Military Academy, West Point, NY 10996, USA; robert.sims@westpoint.edu

*   Correspondence: alexander.mitropoulos@gmail.com (A.N.M.); john.burpo@westpoint.edu (F.J.B.)

Abstract: Nobel metal composite aerogel fibers made from flexible and porous biopolymers offer a wide range of applications, such as in catalysis and sensing, by functionalizing the nanostructure. However, producing these composite aerogels in a defined shape is challenging for many protein-based biopolymers, especially ones that are not fibrous proteins. Here, we present the synthesis of silk fibroin composite aerogel fibers up to 2 cm in length and a diameter of ~300 μm decorated with noble metal nanoparticles. Lyophilized silk fibroin dissolved in hexafluoro-2-propanol (HFIP) was cast in silicon tubes and physically crosslinked with ethanol to produce porous silk gels. Composite silk aerogel fibers with noble metals were created by equilibrating the gels in noble metal salt solutions reduced with sodium borohydride, followed by supercritical drying. These porous aerogel fibers provide a platform for incorporating noble metals into silk fibroin materials, while also providing a new method to produce porous silk fibers. Noble metal silk aerogel fibers can be used for biological sensing and energy storage applications.

Keywords: biopolymer; silk fibroin; aerogel; fiber; nanomaterials; nanoparticles; noble metals; gold; platinum; palladium

1. Introduction

Biopolymers provide unique applications in advanced technology where degradation combined with natural materials are required. In nature, biopolymers take several forms, such as films, fibers, gels, and sponges, which are optimized for their required applications [1,2]. However, producing the equivalent forms with the desired qualities in regenerated biopolymers has been challenging, especially making fibers with controlled diameters and porosity [2]. Nevertheless, working with regenerated biopolymer solutions can enhance properties such as the tensile strength and porosity [3]. Furthermore, regenerated biopolymer solutions can be used as a structural network that can be combined with other materials to synthesize composites not found in nature [4,5]. Starting with biopolymers dissolved in polar solvents, such as water or alcohols, can be useful in providing varied biopolymer conformational folding to enhance the desired properties.

Silk fibroin, purified from the *Bombyx mori* silk caterpillar, is a well-established protein that is processable into fibers, [6,7] films, [8–11] foams, particles, [12] hydrogels, [13,14], and, recently, aerogels after supercritical drying with CO_2 [15–17]. The production of silk fibroin-based materials requires a detailed understanding of the solvent-mediated dielectric environment to induce the hierarchical self-assembly responsible for the mechanical properties resulting from protein primary structure and molecular assembly [18]. Particularly, in silk fibroin, there are three major folding secondary structure motifs, including random coils, alpha helices, or beta-sheets, which control the silk's strength [19]. Specifically, silk fibroin molecular folding can be controlled by the solvent, which forces its primary structure to arrange into the above mentioned secondary structures [18,20].

Alcohols as solvents are useful to stabilize specific secondary structures of silk fibroin in aqueous environments while denaturing the native tertiary conformation [21]. Because the properties of silk fibroin depend strongly on the preparation conditions (fluid environment), the choice of solvent affects the overall quality of the bulk material and the formed nanostructure [3]. 1,1,1,3,3,3-Hexafluoro-propan-2-ol (HFIP) is one of the most applicable solvents for the stabilization of silk's secondary structures, particularly the alpha-helical conformation [21–26]. The high polarity of HFIP as an alcohol allows it to stabilize the silk helical state by decreasing the polarity of the protein chains. This results in local hydrogen bonds that stabilize amphiphilic helical conformations, producing a silk alcogel [21]. The protein concentration is a significant factor in unfolded conformations as the silk secondary structure switches between random coil and alpha-helix depending on the different molecular interactions that can occur [21].

Silk fibroin exposed to HFIP or other alcohols has been used to make gel materials, which can be used for fracture fixation systems or artificial fibers after convective drying [18,26,27]. Dissolving silk in HFIP to cast different forms, particularly gel materials, before convective drying confers the mechanical properties of the folded protein structure and maintains the assembled bulk solid structure in the alcogel formation [27]. Additionally, controlling the nanostructure of the silk fibroin, nanofibrils has been achieved with supercritical CO_2 drying (SCCO$_2$) [17]. Supercritical drying ensures the porosity of the material and maintains the high surface area and low density [28–30]. Using this drying method results in molecular conformational changes of the HFIP–silk fiber, which produce stronger, high surface area materials that have potential for novel biomedical applications.

Furthermore, enhancing the properties of biopolymer gels, such as metallization for catalysis and sensing, can be achieved by equilibrating with gold, palladium, and platinum noble metal complexes, which, after electrochemical reduction, result in nanoparticle growth on the biopolymer nanofibrils [4,5]. Using silk fibroin as the material of choice to add conductive noble metal nanoparticles would enhance the versatility of this mechanically robust and biocompatible material.

Here, we demonstrate the preparation of a composite material consisting of noble metal nanoparticles attached to HFIP-treated porous silk aerogel fibers forming a composite material. Maintaining a constant concentration but changing the type of noble metal ion species determines the extent of the nanoparticle growth on the silk nanofibril surface along with the resulting percentage of metal content in the final aerogel fiber. This allows for material variation based on nobility compared with concentration. The controllable bulk geometries improve this platform to allow it to be used for biofibers. Lastly, utilizing different noble metals can extend the applicability of these biopolymer nanofibrils for other applications, such as catalysis, energy storage, and sensing.

2. Materials and Methods

2.1. Silk Fibroin Fiber Aerogel Synthesis

Silk fibroin solution was prepared as previously described [31]. *B. mori* silkworm cocoons were boiled for 30 minutes in a solution of 0.02 M Na_2CO_3 to remove the sericin glycoprotein. The extracted fibroin was rinsed in deionized water and dried at ambient conditions for 12 h. The dried fibroin was dissolved in 9.3 M LiBr solution at 60 °C for 3 h. The solution was dialyzed against deionized water using a dialysis cassette (Slide-a-Lyzer, Pierce, molecular weight cut-off (MWCO) 3.5 kDa) at room temperature for 2 days until the solution reached a concentration of approximately 60 mg/mL. The obtained solution was purified by centrifugation (20 min at $11,000 \times g$) to remove large aggregates.

For the HFIP aerogels, reconstituted silk fibroin was frozen and lyophilized, resulting in a dried material. The dried silk was stored in ambient conditions to prevent any rehydration of the lyophilized silk fibroin. The lyophilized silk was resolubilized in 1,1,1,3,3,3-Hexafluoro-propan-2-ol (HFIP) (Matrix Scientific, Columbia, SC, USA) to generate a 40 mg/mL solution. The concentration is critical as solutions with lower concentrations produce aggregated gels. The solubilized HFIP silk solution was stored at ambient temperatures until used in the gel-forming process.

The silk aerogels were prepared using silicon tubing with an inner diameter of 1.5 mm (McMaster-Carr, Robbinsville, NJ, USA) and filled with HFIP/silk solution. After the HFIP/silk solution was put into the silicon tubing, the HFIP-silk was submerged in 200 proof ethanol (Fisher Scientific, Waltham, MA, USA). Ethanol diffusion into the HFIP-silk proceeded for 24 h to induce physical crosslinking and form a free-standing fiber gel. Additional ethanol rinses were performed to displace the HFIP after the silk gel fiber was removed from the silicon tubing.

After physical crosslinking, the silk fiber gels were rinsed in deionized water and equilibrated in 100 mM of sodium tetrachloropalladate (II) (Na_2PdCl_4), potassium tetrachloroplatinate (II) (K_2PtCl_4), or gold chloride trihydrate ($HAuCl_4 \cdot 3H_2O$) (Sigma Aldrich, St. Louis, MO, USA) for 24 h.

The fiber samples were reduced in 100 mM sodium borohydride for 24 h for noble metal nanoparticle growth [32]. The silk–metal composite gels were rinsed in deionized water for 24 h to remove excess reducing agent. To maintain the metal-coated nanofibrillar hydrogel network, samples were then dehydrated in a series of ethanol rinses at concentrations of 25, 50, 75, and 100% for 30 min each and then supercritically dried in CO_2 using a Leica EM CPD300 Automated Critical Point Dryer (Buffalo Grove, IL, USA) with a set point of 35 °C and 1200 psi.

To prepare sufficient sample material for X-ray diffraction, thermal gravimetric, and nitrogen gas adsorption analysis, silk–metal composites were prepared in a bulk monolith geometry. A HFIP–silk solution was cast in a 48-well cell culture dish (diameter of 10 mm) and crosslinked with 200 proof ethanol by casting on the top for 24 h. After crosslinking, silk hydrogels were rinsed in deionized water for 48 h to remove any remaining HFIP and ethanol. The hydrogels were equilibrated in 100 mM of sodium tetrachloropalladate (II) (Na_2PdCl_4), potassium tetrachloroplatinate (II) (K_2PtCl_4), or gold chloride trihydrate ($HAuCl_4 \cdot 3H_2O$) as in the case of the silk–metal fiber synthesis above but for 48 h. To ensure reduction throughout the volume of the silk gels, 2 M sodium borohydride ($NaBH_4$) and 2 M dimethylamine borane (DMAB) was used for the palladium-, and platinum- and gold-equilibrated silk gels, respectively. The high reducing agent concentration was used to drive diffusion into the depth of the gel. Electrochemical reduction proceeded for 24 h before rinsing for 48 h in deionized water. An ethanol solvent exchange was performed prior to supercritical drying in CO_2.

2.2. Scanning Electron Microscopy (SEM)

SEM was used to evaluate scaffold morphology. All the micrographs were taken with a TM-3000 Scanning Electron Microscope (Hitachi, Tokyo, Japan) or a FEI Helios 600 scanning electron microscope (ThermoFisher Scientific, Hillsboro, OR, USA). Samples were not coated in gold prior to imaging.

2.3. X-ray Diffractometry (XRD)

XRD measurements were performed using a PANalytical Empyrean (Malvern PANalytical, Almelo, The Netherlands) diffractometer with scans at 45 kV and 40 mA with Cu K_α radiation (1.54060 Å), a 2θ step size of 0.0130°, and 20 s per step for diffraction angles (2θ) performed from 5° to 90°. XRD spectra analysis was performed with High Score Plus software (Malvern PANalytical, Almelo, The Netherlands). Crystallite size (D) was determined with the Debeye–Scherrer formula $D = K\lambda(Bcos\theta)^{-1}$ with the shape factor (K), full width at half maxima (B), radiation wave length (λ), and Bragg angle (θ). A shape factor of K = 0.9 was used. High Score Plus software was used to analyze the XRD spectra (Version 4.6, Malvern PANalytical, Almelo, The Netherlands).

2.4. Thermal Gravimetric Analysis (TGA)

Thermal gravimetric analysis (TGA) was performed on a Thermal Instruments Q-500 (New Castle, DE, USA) in a ramp state with a temperature rate of 10 °C/min from ambient to 1000 °C. Samples were maintained under nitrogen gas flow at a rate of 60 ml/min.

2.5. Fourier Transform Infrared (FTIR) Spectroscopy

FTIR analysis of silk film samples was performed in a PerkinElmer Frontier Optica FIR spectrometer (PerkinElmer, Waltham, MA, USA) in attenuated total reflectance (ATR). Films were measured before and after the galvanic displacement. For each sample, 64 scans were collected with a resolution of 1 cm^{-1}, with a wave number range of 4000–650 cm^{-1}.

2.6. Porosity and Surface Area Analysis

Nitrogen gas adsorption–desorption measurements were performed according to International Union of Pure and Applied Chemistry (IUPAC) standards [33] using a Micromeritics ASAP 2020 Plus (Micromeritics, Norcross, GA, USA) to determine surface area and pore size. All the samples were vacuum degassed at 100 °C for 10 h prior to analysis. Brunauer–Emmett–Teller (BET) analysis [34] was used to determine the specific surface area from gas adsorption. Pore size distributions for each sample were calculated using the Barrett–Joyner–Halenda (BJH) model [35] applied to volumetric desorption isotherms. All the calculations were performed using Micromeritics' ASAP 2020 software (Micromeritics, Norcross, GA, USA).

3. Results and Discussion

3.1. Silk Fibroin Aerogel Synthesis

A fast and robust method was developed to form noble metal composite silk fibroin nanostructured fibers by crosslinking HFIP–silk solution with ethanol. The average molecular weight of the silk was 100 kDa and a concentration of 40 mg/mL. Figure 1 shows the synthesis scheme for the silk fibroin noble metal composite aerogels. In order to physically crosslink the silk, the HFIP–silk solution was injected into a silicon tubing mold with an inner diameter of 1.5 mm and 3 cm long and submerged in a bath of 100% ethanol (Figure 1c–d). Parameters such as silk fibroin concentration, noble metal concentration, and reducing agent concentration were the determining factors in the composite fibers. The type of noble metal dictated the noble metal particle size and allowed for particles less than 10 nm and noble metal deposition onto the silk template. Visible deposition of noble metal nanoparticles was observed by a visible color change compared with a control (Figure 1g–h). The formed fibers after reduction provided flexibility and were able to bend into geometric shapes.

To demonstrate the ability to deposit multiple noble metals, silk hydrogels were prepared as cylindrical gels and equilibrated in 100 mM noble metal ion solutions of palladium (Na_2PdCl_4) and platinum (K_2PtCl_6) (Figure S1). Reduction of the metal ions can occur from multiple reducing agents, such as DMAB or $NaBH_4$, as represented by palladium reduction and platinum reduction in Figure S1c,d, respectively. Furthermore, 100 mM metal ion solutions of gold ($HAuCl_4$) were also reduced in DMAB (Figure S2a–c). The noble metal type and reducing agent changes the morphology of the aerogel fibers as visible in SEM images (Figures S3 and S4). The silk–gold aerogels reduced in DMAB show thinner nanofibrils (30–50 nm) with smaller pore sizes compared with the silk–platinum aerogels reduced in $NaBH_4$, which have larger diameter nanofibrils (70–100 nm) and larger pores.

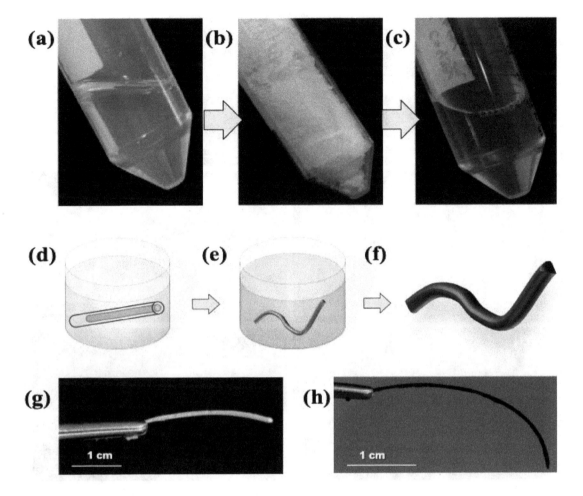

Figure 1. Silk fibroin aerogel fiber synthesis scheme. Different working solutions of silk fibroin starting with (**a**) regenerated silk fibroin solution, (**b**) lyophilized silk fibroin, and (**c**) hexafluoro-2-propanol (HFIP)–silk fibroin. Scheme depiction of the synthesis of noble metal silk fibroin aerogel fibers. (**d**) HFIP–silk fibroin in silicon tubing mold in an ethanol bath to induce physical crosslinking, (**e**) equilibrating in noble metal ionic solution, and (**f**) after reduction and supercritical drying. (**g**) The silk fibroin aerogel fiber without noble metal addition and (**h**) the silk–palladium aerogel fiber (scale bars are 1 cm).

3.2. Aerogel Morphology and Noble Metal Nanoparticles

The structure of the nanofibrils and growth of the noble metal nanoparticles indicates the effects of reduction on the silk fibroin. It has been previously demonstrated that the reduction of these noble metals can be completed using gelatin and cellulose materials with higher reducing agent concentrations [4,5]. Figure 2 shows the SEM images of the silk fibroin aerogel with platinum and palladium after reduction with 100 mM sodium borohydride. Figure 2a shows the fiber formation of the HFIP–silk aerogel fiber. At higher magnification, the presence of nanoparticles covers the silk fiber network for the silk–palladium composite aerogel fibers (Figure 2b–c). This is observed by the brighter regions indicative of nanoparticle growth. The silk–platinum composite aerogel fibers show a more even distribution of noble metal nanoparticle growth onto the silk fibroin (Figure 2d–e).

The morphology of the underlying HFIP–silk fibroin template shows an interconnected network of silk protein spherical particles with diameters of ~500 nm forming a string of pearl-like fibers with high porosity, which is caused by the gelation of the original HFIP–silk (Figure S3). The interconnected network is visible at higher magnification in the anchored silk–gold composite aerogel fibers (Figure S4a–c) when reduced with DMAB. The gold nanoparticles cluster around the larger silk nanofibrils in discrete nanoparticles. Chemical reduction with sodium borohydride of the silk–platinum composite aerogels shows an interconnected silk nanofibril network with nanoparticles dispersed on the surface with a diameter in the range of 5–20 nm (Figure S4d–f).

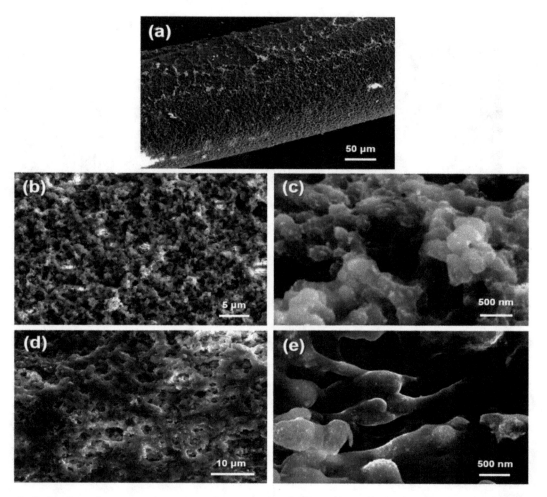

Figure 2. Scanning electron microscopy images. (**a–c**) The silk–palladium composite aerogel fibers. (**d–e**) The silk–platinum composite aerogel fibers.

The change in fiber morphology is possibly caused by the changes in pH of the noble metal ion solution where solutions of $HAuCl_4$ have lower pH values compared with K_2PtCl_6 or Na_2PdCl_4. Due to the lower pH, gold metal ions are more attracted to the silk molecular structure caused by electrostatic interactions, which is visible as a higher density of decorated surface nanoparticles compared with the silk–platinum and silk–palladium aerogels (Figure S3). This has been discussed in other biopolymer-related work [4,5]. Silk fibroin has a repeating pattern of glycine, serine, and alanine forming polymer blocks with shorter block regions of non-repeating sequences [36]. It is these amino acids that allow silk to grow noble metal nanoparticles.

3.3. XRD Characterization

Figure 3a shows the X-ray diffraction (XRD) spectra for the HFIP–silk composite aerogel composites synthesized with palladium and platinum, respectively. XRD spectra for aerogel composites prepared with palladium were indexed to the Joint Committee on Powder Diffraction Standards (JCPDS) reference number 01-087-0637 for palladium and 01-073-0004 for palladium hydride. For aerogel composites prepared with platinum, XRD peaks were indexed to JCPDS reference number 00-004-0802 for platinum. Both palladium and platinum phases are cubic crystal systems with Fm-3 m space groups. The shape evolution of the palladium and platinum nanocrystals of different morphologies can be directed by agents, such as silk and other secondary chemicals, during reduction [37]. The minor presence of palladium hydride in the Pd–silk aerogel composites is likely due to hydrogen gas evolution during electrochemical reduction and the tendency of palladium to store hydrogen gas within its crystal lattice [38]. The palladium hydride peaks shift the position of the fitted palladium phase peaks slightly from their indexed positions. For instance, the (111) palladium fitted peak at 39.5° is shifted right relative to the indexed reference peak position at 39.0° likely due to peak convolution with the (101) palladium hydride peak. The change in nanocrystal structure can be associated with the shape-directing capabilities of the hydrogen evolution during reduction, which is observed as the visible peak shift in the XRD spectra [37]. The crystallite sizes, determined by using the Debeye–Scherrer formula, and the (111) peaks were 3.6 nm and 2.6 nm for silk–palladium and silk–platinum, respectively, and suggest that the nanoparticles observed in the SEM images in Figure 2 are polycrystalline. The XRD spectra for the HFIP–silk composite aerogel fiber with gold is shown in Figure S5a. The peaks observed for both the silk–palladium and silk–platinum composite aerogels at approximately 20.8° are attributed to the silk protein templates [39–41].

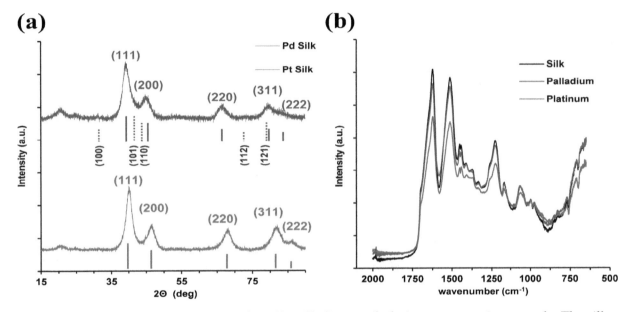

Figure 3. (a) X-ray diffraction spectra for silk palladium and platinum composite aerogels. The silk–palladium aerogel peaks are indexed to the Joint Committee on Powder Diffraction Standards (JCPDS) reference 01-087-0637 for palladium (blue lines), 01-073-0004 for palladium hydride (blue dashed lines; Miller indices labeled in gray). The silk–platinum aerogels are indexed to 00-004-0802 (red lines) for platinum. (b) Fourier transform infrared (FTIR) spectra for the silk, silk–palladium, and silk–platinum fiber aerogels.

3.4. FTIR Characterization

The secondary structure of the silk fibroin was completed by FTIR, examining the secondary structure in the Amide II and III band. The FTIR spectra for silk, silk–palladium, and silk–platinum are shown in Figure 3b [42]. The Amide I band is associated with 1600–1700 cm^{-1}, which shows the characteristic peak of beta-sheeted silk fibroin at 1625 cm^{-1} [20]. The alpha-helix structure (1658–1662 cm^{-1}) additionally shows a sharp peak associated with the shoulder near the Amide I band [20]. The FTIR spectra are characteristic of silk and are unchanged for the silk–palladium and silk–platinum aerogel fibers. After supercritical drying there is a high percentage of alpha-helix and beta-sheet content through the interaction at the molecular level. The same was found in the silk–gold aerogel fibers (Figure S5b). The high beta-sheet content observed is typical of HFIP–silk and has been shown in other studies where silk gels dried in ethanol or methanol are used for fracture fixation devices [27].

3.5. TGA Characterization

To characterize the mass composition of noble metals in the silk aerogels, thermogravimetric analysis (TGA) was performed with the results shown in Figure 4. When examined, the control silk, silk–palladium, and silk–platinum aerogel fibers showed an increase in the metal-to-silk weight ratio. Silk begins decomposition above 200 °C, but to ensure the complete degradation of the silk fibroin from the sample, analysis was conducted up to 1000 °C. The final mass percentage indicates the amount of noble metal to silk fibroin. The silk-only control sample shows 0% metal at 1000 °C [20,43,44], silk–palladium shows 10%, and silk–platinum indicates a mass percentage of 15%. The silk–gold aerogel fiber composites showed a higher metal mass percentage of 30%, which was observed in the morphological changes in the SEM images described above (Figure S6). The change in mass percentage of noble metal varied between each condition as stated above due to the electrostatic forces between the metal ions and silk fibroin. As stated above, the different pH values for 100 mM Na_2PdCl_4, K_2PtCl_4, or $HAuCl_4 \cdot 3H_2O$ cause different ionic interactions with the silk. Lower pH will deposit more metal ions, which, after reduction, shows a higher density of nanoparticles on the silk surface and is consistent with the mass percent differences observed with TGA.

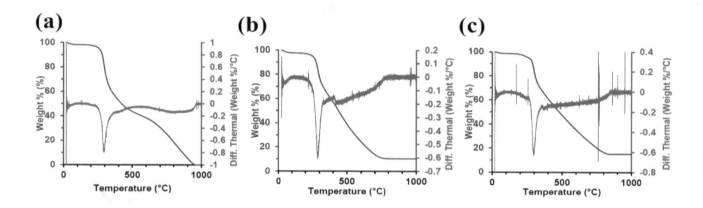

Figure 4. Thermogravimetric analysis (TGA) with differential thermal analysis (DTA) for the **(a)** silk aerogel, **(b)** silk–palladium aerogel, and **(c)** silk–platinum aerogel.

3.6. Porosity and Surface Area Characterization

Nitrogen gas adsorption isotherms were generated for the silk–noble metal aerogels prepared with 100 mM noble metals. Given the small mass of the silk aerogel composite fibers, bulk silk and silk composite aerogels were used to achieve nitrogen gas adsorption–desorption isotherms. The physisorption data shown in Figure 5 indicate type IV adsorption–desorption isotherms according to IUPAC classification standards, suggesting the presence of both mesoporous (2–50 nm) and macroporous (>50 nm) structures in all of the aerogel samples. The macropore features generally correlate with the pores observed in the SEM images in Figure 2 for the silk–palladium and silk–platinum composites. The mesoporous content suggests porosity within the silk phase of the composite aerogels not directly observed with SEM. After a sharp rise in adsorbed gas, no limiting adsorption plateau at high relative pressures (P/P_o) is observed. The maximum volume adsorbed at the highest relative pressure of P/P_o = 0.995 is 498, 482, and 254 cm^3/g for the silk, silk–palladium, and silk–platinum samples, respectively. All the sample isotherms exhibit type H3 hysteresis typical of mesoporous capillary condensation. The hysteresis loops close at higher relative pressures for the silk–palladium and silk–platinum composite aerogels compared with the silk only aerogels. This may be due to metal nanoparticles filling or blocking the smaller pores. This is supported by the pore frequency distributions seen in Figure 5b,d,e for silk, silk–palladium, and silk–platinum, respectively. The silk aerogels exhibit a high frequency of 3–4 nm pores, with a broad presence of mesopores up to 40 nm. The frequency of 3–4 nm pores decreases for the silk–palladium and silk–platinum aerogels with a commensurate increase in mesopores of diameters of 20–30 nm. The BET specific surface areas determined from desorption isotherms of the silk control, silk–palladium, and silk–platinum samples were 268, 170, and 72 m^2/g, respectively. The decrease in specific surface area from pure silk to silk–metal composite aerogels is consistent with nanoparticle pore blockage, which is consistent with SEM image analysis.

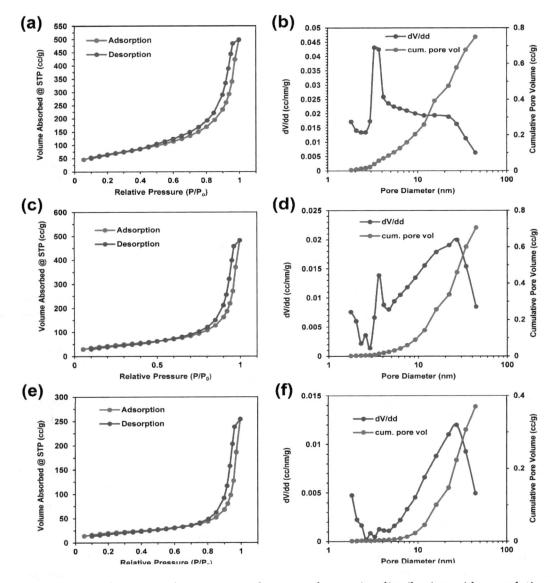

Figure 5. Nitrogen adsorption–desorption isotherms and pore size distribution with cumulative pore volume for the (**a**,**b**) silk aerogels, (**c**,**d**) palladium–silk aerogels, and (**e**,**f**) platinum–silk aerogels.

4. Conclusions

Here we have shown that HFIP–silk fibroin aerogel fibers can be utilized as a platform to anchor noble metal nanoparticles onto the surface of silk nanofibrils. The silk aerogels have a high surface area and are composed of different percentages of noble metals, which are based primarily on the nobility, not the concentration. This changes the mass percentage, porosity, surface area, and pore diameter. Compared with other biopolymers, which have acted as templates with noble metals, silk is much simpler to process and does not require any chemical crosslinking. Palladium, platinum, and gold nanoparticle growth have been shown on silk fibroin aerogel fibers. The molecular structure of silk is unchanged when anchored with different noble metals. These silk fibroin noble metal aerogels have potential utility in energy storage and conversion that can be used where degradable, flexible materials are required.

5. Patents

This work has been submitted with a preliminary patent and invention disclosure.

Supplementary Materials: The following are available online at Figure S1: Silk hydrogels equilibrated in 100 mM Na_2PdCl_4 and 100 mM K_2PtCl_6 reduced in 2M $NaBH_4$; Figure S2: Silk hydrogels equilibrated in 100 mM $HAuCl_4$ reduced with 2 M DMAB; Figure S3: SEM images of the HFIP–silk noble metal fibers; Figure S4: SEM images of the HFIP–silk aerogels for silk–palladium, silk–platinum, and silk–gold; Figure S5: X-ray diffraction (XRD) and Fourier transform infrared (FTIR) absorption spectra for the silk–gold composite aerogels, Figure S6: TGA curves of the silk–gold aerogels; Figure S7: Nitrogen gas adsorption–desorption curves for the silk–gold aerogels.

Author Contributions: Conceptualization, A.N.M. and F.J.B.; Data curation, A.N.M., F.J.B. and C.K.N.; Formal analysis, A.N.M. and F.J.B.; Funding acquisition, A.N.M. and F.J.B.; Investigation, A.N.M., F.J.B., C.K.N., M.Y.R., J.W. and R.K.S.; Methodology, A.N.M. and F.J.B.; Project administration, A.N.M. and F.J.B.; Resources, A.N.M., F.J.B. and J.K.W.; Supervision, A.N.M. and F.J.B.; Validation, F.J.B.; Original draft preparation, A.N.M. and F.J.B.; Review and editing of manuscript, A.N.M., F.J.B., C.K.N., E.A.N., M.Y.R., J.W., R.K.S., K.W. and J.K.W.

Acknowledgments: The authors acknowledge Stephen Bartolucci at Watervliet Arsenal and Benet Laboratories for providing use of their SEM. The authors also acknowledge David Kaplan from Tufts University for the raw silk cocoons.

References

1. Ulery, B.D.; Nair, L.S.; Laurencin, C.T. Biomedical Applications of Biodegradable Polymers. *J. Polym. Sci. B Polym. Phys.* **2011**, *49*, 832–864. [CrossRef] [PubMed]
2. Tan, J.; Saltzman, W.M. Biomaterials with Hierarchically Defined Micro- and Nanoscale Structure. *Biomaterials* **2004**, *25*, 3593–3601. [CrossRef] [PubMed]
3. Zhu, Z.H. Preparation and Characterization of Regenerated Bombyx Mori Silk Fibroin Fiber with High Strength. *eXPRESS Polym. Lett.* **2008**, *2*, 885–889. [CrossRef]
4. Burpo, F.J.; Mitropoulos, A.N.; Nagelli, E.A.; Rye, M.Y.; Palmer, J.L. Gelatin Biotemplated Platinum Aerogels. *MRS Adv.* **2018**, *3*, 2875–2880. [CrossRef]
5. Burpo, F.J.; Mitropoulos, A.N.; Nagelli, E.A.; Palmer, J.L.; Morris, L.A.; Ryu, M.Y.; Kenneth Wickiser, J. Cellulose Nanofiber Biotemplated Palladium Composite Aerogels. *Molecules* **2018**, *23*, 1405. [CrossRef] [PubMed]
6. Altman, G.H.; Diaz, F.; Jakuba, C.; Calabro, T.; Horan, R.L.; Chen, J.; Lu, H.; Richmond, J.; Kaplan, D.L. Silk-Based Biomaterials. *Biomaterials* **2003**, *24*, 401–416. [CrossRef]
7. Vepari, C.; Kaplan, D.L. Silk as a Biomaterial. *Prog. Polym. Sci.* **2007**, *32*, 991–1007. [CrossRef]
8. Lawrence, B.D.; Cronin-Golomb, M.; Georgakoudi, I.; Kaplan, D.L.; Omenetto, F.G. Bioactive Silk Protein Biomaterial Systems for Optical Devices. *Biomacromolecules* **2008**, *9*, 1214–1220. [CrossRef]
9. Lu, Q.; Hu, X.; Wang, X.; Kluge, J.A.; Lu, S.; Cebe, P.; Kaplan, D.L. Water-Insoluble Silk Films with Silk I Structure. *Acta Biomater.* **2010**, *6*, 1380–1387. [CrossRef]
10. Perry, H.; Gopinath, A.; Kaplan, D.L.; Dal Negro, L.; Omenetto, F.G. Nano- and Micropatterning of Optically Transparent, Mechanically Robust, Biocompatible Silk Fibroin Films. *Adv. Mater.* **2008**, *20*, 3070–3072. [CrossRef]
11. Jin, H.J.; Park, J.; Karageorgiou, V.; Kim, U.J.; Valluzzi, R.; Cebe, P.; Kaplan, D.L. Water-Stable Silk Films with Reduced β-Sheet Content. *Adv. Funct. Mater.* **2005**, *15*, 1241–1247. [CrossRef]
12. Wang, X.; Yucel, T.; Lu, Q.; Hu, X.; Kaplan, D.L. Silk Nanospheres and Microspheres from Silk/pva Blend Films for Drug Delivery. *Biomaterials* **2010**, *31*, 1025–1035. [CrossRef] [PubMed]
13. Kim, U.-J.; Park, J.; Li, C.; Jin, H.-J.; Valluzzi, R.; Kaplan, D.L. Structure and Properties of Silk Hydrogels. *Biomacromolecules* **2004**, *5*, 786–792. [CrossRef] [PubMed]
14. Floren, M.L.; Spilimbergo, S.; Motta, A.; Migliaresi, C. Carbon Dioxide Induced Silk Protein Gelation for Biomedical Applications. *Biomacromolecules* **2012**, *13*, 2060–2072. [CrossRef] [PubMed]
15. Mallepally, R.R.; Marin, M.A.; McHugh, M.A. CO2-Assisted Synthesis of Silk Fibroin Hydrogels and Aerogels. *Acta Biomater.* **2014**, *10*, 4419–4424. [CrossRef] [PubMed]
16. Marin, M.A.; Mallepally, R.R.; McHugh, M.A. Silk Fibroin Aerogels for Drug Delivery Applications. *J. Supercrit. Fluids* **2014**, *91*, 84–89. [CrossRef]
17. Tseng, P.; Napier, B.; Zhao, S.; Mitropoulos, A.N.; Applegate, M.B.; Marelli, B.; Kaplan, D.L.; Omenetto, F.G. Directed Assembly of Bio-Inspired Hierarchical Materials with Controlled Nanofibrillar Architectures. *Nat. Nanotechnol.* **2017**, *12*, 474–480. [CrossRef] [PubMed]
18. Zhao, C.; Yao, J.; Masuda, H.; Kishore, R.; Asakura, T. Structural Characterization and Artificial Fiber Formation of Bombyx Mori Silk Fibroin in Hexafluoro-Iso-Propanol Solvent System. *Biopolymers* **2003**, *69*, 253–259. [CrossRef] [PubMed]
19. Chen, X.; Shao, Z.; Knight, D.P.; Vollrath, F. Conformation Transition Kinetics of Bombyx Mori Silk Protein. *Proteins* **2007**, *68*, 223–231. [CrossRef]
20. Hu, X.; Kaplan, D.; Cebe, P. Determining Beta-Sheet Crystallinity in Fibrous Proteins by Thermal Analysis and Infrared Spectroscopy. *Macromolecules* **2006**, *39*, 6161–6170. [CrossRef]

21. Hirota, N.; Mizuno, K.; Goto, Y. Cooperative Alpha-Helix Formation of Beta-Lactoglobulin and Melittin Induced by Hexafluoroisopropanol. *Protein Sci.* **1997**, *6*, 416–421. [CrossRef] [PubMed]

22. Roccatano, D.; Fioroni, M.; Zacharias, M.; Colombo, G. Effect of Hexafluoroisopropanol Alcohol on the Structure of Melittin: A Molecular Dynamics Simulation Study. *Protein Sci.* **2005**, *14*, 2582–2589. [CrossRef] [PubMed]

23. Mandal, B.B.; Grinberg, A.; Seok Gil, E.; Panilaitis, B.; Kaplan, D.L. High-Strength Silk Protein Scaffolds for Bone Repair. *Proc. Natl. Acad. Sci. USA* **2012**, *109*, 7699–7704. [CrossRef] [PubMed]

24. Ling, S.; Qin, Z.; Li, C.; Huang, W.; Kaplan, D.L.; Buehler, M.J. Polymorphic Regenerated Silk Fibers Assembled through Bioinspired Spinning. *Nat. Commun.* **2017**, *8*, 1387. [CrossRef] [PubMed]

25. Zhang, W.; Ahluwalia, I.P.; Literman, R.; Kaplan, D.L.; Yelick, P.C. Human Dental Pulp Progenitor Cell Behavior on Aqueous and Hexafluoroisopropanol Based Silk Scaffolds. *J. Biomed. Mater. Res. Part A* **2011**, *97*, 414–422. [CrossRef] [PubMed]

26. Gil, E.S.; Kluge, J.A.; Rockwood, D.N.; Rajkhowa, R.; Wang, L.; Wang, X.; Kaplan, D.L. Mechanical Improvements to Reinforced Porous Silk Scaffolds. *J. Biomed. Mater. Res. Part A* **2011**, *99*, 16–28. [CrossRef] [PubMed]

27. Perrone, G.S.; Leisk, G.G.; Lo, T.J.; Moreau, J.E.; Haas, D.S.; Papenburg, B.J.; Golden, E.B.; Partlow, B.P.; Fox, S.E.; Ibrahim, A.M.S.; et al. The Use of Silk-Based Devices for Fracture Fixation. *Nat. Commun.* **2014**, *5*, 3385. [CrossRef] [PubMed]

28. Tsotsas, E.; Mujumdar, A.S. (Eds.) *Modern Drying Technology*; Wiley-VCH Verlag GmbH & Co. KGaA: Weinheim, Germany, 2011.

29. Williams, J.R.; Clifford, A.A.; Al-Saidi, S.H.R. Supercritical Fluids and Their Applications in Biotechnology and Related Areas. *Mol. Biotechnol.* **2002**, *22*, 263–286. [CrossRef]

30. Brunner, G. Applications of Supercritical Fluids. *Annu. Rev. Chem. Biomol. Eng.* **2010**, *1*, 321–342. [CrossRef] [PubMed]

31. Rockwood, D.N.; Preda, R.C.; Yücel, T.; Wang, X.; Lovett, M.L.; Kaplan, D.L. Materials Fabrication from Bombyx Mori Silk Fibroin. *Nat. Protoc.* **2011**, *6*, 1612–1631. [CrossRef]

32. Burpo, F.J.; Nagelli, E.A.; Morris, L.A.; McClure, J.P.; Ryu, M.Y.; Palmer, J.L. Direct Solution-Based Reduction Synthesis of Au, Pd, and Pt Aerogels. *J. Mater. Res.* **2017**, *32*, 4153–4165. [CrossRef]

33. Sing, K.S.W. Reporting Physisorption Data for Gas/solid Systems with Special Reference to the Determination of Surface Area and Porosity. *Pure Appl. Chem.* **1985**, *57*, 603. [CrossRef]

34. Brunauer, S.; Emmett, P.H.; Teller, E. Adsorption of Gases in Multimolecular Layers. *J. Am. Chem. Soc.* **1938**, *60*, 309–319. [CrossRef]

35. Barrett, E.P.; Joyner, L.G.; Halenda, P.P. The Determination of Pore Volume and Area Distributions in Porous Substances. I. Computations from Nitrogen Isotherms. *J. Am. Chem. Soc.* **1951**, *73*, 373–380. [CrossRef]

36. Jin, H.-J.; Kaplan, D.L. Mechanism of Silk Processing in Insects and Spiders. *Nature* **2003**, *424*, 1057–1061. [CrossRef] [PubMed]

37. Polavarapu, L.; Mourdikoudis, S.; Pastoriza-Santos, I.; Perez-Juste, J. Nanocrystal Engineering of Noble Metals and Metal Chalcogenides: Controlling the Morphology, Composition and Crystallinity. *CrystEngComm* **2015**, *17*, 2727–2762. [CrossRef]

38. Jewell, L.L.; Davis, B.H. Review of Absorption and Adsorption in the Hydrogen-Palladium System. *Appl. Catal. A Gen.* **2006**, *310*, 1–15. [CrossRef]

39. Zhang, H.; Li, L.L.; Dai, F.Y.; Zhang, H.H.; Ni, B.; Zhou, W.; Yang, X.; Wu, Y.Z. Preparation and Characterization of Silk Fibroin as a Biomaterial with Potential for Drug Delivery. *J. Transl. Med.* **2012**, *10*, 117. [CrossRef]

40. Lu, S.; Li, J.; Zhang, S.; Yin, Z.; Xing, T.; Kaplan, D.L. The Influence of the Hydrophilic-Lipophilic Environment on the Structure of Silk Fibroin Protein. *J. Mater. Chem. B* **2015**, *3*, 2599–2606. [CrossRef]

41. Wang, H.Y.; Zhang, Y.Q. Effect of Regeneration of Liquid Silk Fibroin on Its Structure and Characterization. *Soft Matter* **2013**, *9*, 138–145. [CrossRef]

42. Hu, X.; Kaplan, D.; Cebe, P. Dynamic Protein-Water Relationships during Beta-Sheet Formation. *Macromolecules* **2008**, *41*, 3939–3948. [CrossRef]

43. Lamoolphak, W.; De-Eknamkul, W.; Shotipruk, A. Hydrothermal Production and Characterization of Protein and Amino Acids from Silk Waste. *Bioresour. Technol.* **2008**, *99*, 7678–7685. [CrossRef] [PubMed]

44. Kang, K.Y.; Chun, B.S. Behavior of Hydrothermal Decomposition of Silk Fibroin to Amino Acids in near-Critical Water. *Korean J. Chem. Eng.* **2004**, *21*, 654–659. [CrossRef]

Permissions

The contributors of this book come from diverse backgrounds, making this book a truly international effort. This book will bring forth new frontiers with its revolutionizing research information and detailed analysis of the nascent developments around the world.

We would like to thank all the contributing authors for lending their expertise to make the book truly unique. They have played a crucial role in the development of this book. Without their invaluable contributions this book wouldn't have been possible. They have made vital efforts to compile up to date information on the varied aspects of this subject to make this book a valuable addition to the collection of many professionals and students.

This book was conceptualized with the vision of imparting up-to-date information and advanced data in this field. To ensure the same, a matchless editorial board was set up. Every individual on the board went through rigorous rounds of assessment to prove their worth. After which they invested a large part of their time researching and compiling the most relevant data for our readers.

The editorial board has been involved in producing this book since its inception. They have spent rigorous hours researching and exploring the diverse topics which have resulted in the successful publishing of this book. They have passed on their knowledge of decades through this book. To expedite this challenging task, the publisher supported the team at every step. A small team of assistant editors was also appointed to further simplify the editing procedure and attain best results for the readers.

Apart from the editorial board, the designing team has also invested a significant amount of their time in understanding the subject and creating the most relevant covers. They scrutinized every image to scout for the most suitable representation of the subject and create an appropriate cover for the book.

The publishing team has been an ardent support to the editorial, designing and production team. Their endless efforts to recruit the best for this project, has resulted in the accomplishment of this book. They are a veteran in the field of academics and their pool of knowledge is as vast as their experience in printing. Their expertise and guidance has proved useful at every step. Their uncompromising quality standards have made this book an exceptional effort. Their encouragement from time to time has been an inspiration for everyone.

The publisher and the editorial board hope that this book will prove to be a valuable piece of knowledge for researchers, students, practitioners and scholars across the globe.

List of Contributors

Udeni Gunathilake T.M. Sampath and Yern Chee Ching
Department of Mechanical Engineering, Faculty of Engineering, University of Malaya, 50603 Kuala Lumpur, Malaysia

Cheng Hock Chuah and Johari J. Sabariah
Department of Chemistry, Faculty of Science, University of Malaya, 50603 Kuala Lumpur, Malaysia

Pai-Chen Lin
Department of Mechanical Engineering, National Chung Cheng University, 621 Chiayi Country, Taiwan

Bo-Wei Du, Shao-Ying Hu, Ranjodh Singh, Tsung-Tso Tsai, Ching-Chang Lin and Fu-Hsiang Ko
Department of Materials Science and Engineering, National Chiao Tung University, 1001 University Road, Hsinchu 30010, Taiwan

Cindu Annandarajah
Department of Agricultural and Biosystems Engineering, Iowa State University, Ames, IA 50011, USA

Amy Langhorst, Alper Kiziltas and Deborah Mielewski
Ford Motor Company, Research and Advanced Engineering, Dearborn, MI 48124, USA

David Grewell
Department of Industrial and Manufacturing Engineering, North Dakota State University, Fargo, ND 58102, USA

Reza Montazami
Department of Mechanical Engineering, Iowa State University, Ames, IA 50011, USA

Estefanía Lidón Sánchez-Safont, Luis Cabedo and Jose Gamez-Perez
Polymers and Advanced Materials Group (PIMA), Universitat Jaume I, 12071 Castellón, Spain

Alex Arrillaga and Jon Anakabe
Leartiker S. Coop., Xemein Etorbidea 12A, 48270 Markina-Xemein, Spain

Fang Wang
Center of Analysis and Testing, Nanjing Normal University, Nanjing 210023, China

Zhenggui Gu
School of Chemistry and Materials Science, Nanjing Normal University Jiangsu, Nanjing 210023, China

Qichun Liu
Center of Analysis and Testing, Nanjing Normal University, Nanjing 210023, China
School of Chemistry and Materials Science, Nanjing Normal University Jiangsu, Nanjing 210023, China

Qingyu Ma
School of Physics and Technology, Nanjing Normal University, Nanjing 210023, China

Xiao Hu
Department of Physics and Astronomy, Rowan University, Glassboro, NJ 08028, USA
Department of Biomedical Engineering, Rowan University, Glassboro, NJ 08028, USA
Department of Molecular and Cellular Biosciences, Rowan University, Glassboro, NJ 08028, USA

Shaoyun Chen
College of Chemical Engineering and Materials Science, Quanzhou Normal University, Quanzhou 362000, China

Min Xiao and Shuanjin Wang
The Key Laboratory of Low-carbon Chemistry & Energy Conservation of Guangdong Province/State Key Laboratory of Optoelectronic Materials and Technologies, Sun Yat-Sen University, Guangzhou 510275, China

Luyi Sun
Department of Chemical & Biomolecular Engineering and Polymer Program, Institute of Materials Science, University of Connecticut, Storrs, CT 06269, USA

Carla Vilela, Catarina Moreirinha, Armando J. D. Silvestre and Carmen S. R. Freire
Department of Chemistry, CICECO – Aveiro Institute of Materials, University of Aveiro, 3810-193 Aveiro, Portugal

Adelaide Almeida
Department of Biology and CESAM, University of Aveiro, 3810-193 Aveiro, Portugal

Elena Manaila and Gabriela Craciun
Electron Accelerators Laboratory, National Institute for Laser, Plasma and Radiation Physics, 409 Atomistilor Street, 077125 Magurele, Romania

Maria Daniela Stelescu
National R&D Institute for Textile and Leather — Leather and Footwear Research Institute, 93 Ion Minulescu Street, 031215 Bucharest, Romania

Dongmei Han and Yuezhong Meng
The Key Laboratory of Low-carbon Chemistry &
Energy Conservation of Guangdong Province/State Key
Laboratory of Optoelectronic Materials and Technologies,
Sun Yat-Sen University, Guangzhou 510275, China
School of Chemical Engineering and Technology, Sun
Yat-Sen University, Guangzhou 510275, China

Guiji Chen
The Key Laboratory of Low-carbon Chemistry &
Energy Conservation of Guangdong Province/State Key
Laboratory of Optoelectronic Materials and Technologies,
Sun Yat-Sen University, Guangzhou 510275, China
Shanghai Kingfa Science and Technology Development
Co., Ltd., Shanghai 201714, China

Shou Chen and Xiaohua Peng
Shenzhen Beauty Star Co., Ltd., Shenzhen 518112, China

Maria Beatrice Coltelli
Department of Civil and Industrial Engineering,
University of Pisa, Via Diotisalvi, 2, 56122 Pisa, Italy

Laura Aliotta
Department of Civil and Industrial Engineering,
University of Pisa, Via Diotisalvi, 2, 56122 Pisa, Italy
Interuniversity National Consortium of Materials
Science and Technology (INSTM), Via Giusti 9, 50121
Florence, Italy

Pei-Wen Tai, Hsin-Yu Lo and Ting-Ya Wu
Department of Biomedical Engineering, National Yang-
Ming University, Taipei 11221, Taiwan

Tze-Wen Chung
Department of Biomedical Engineering, National Yang-
Ming University, Taipei 11221, Taiwan
Center for Advanced Pharmaceutical Science and Drug
Delivery, National Yang-Ming University, Taipei 11221,
Taiwan

Weng-Pin Chen
Department of Mechanical Engineering, National
Taipei University of Technology, Taipei 10608, Taiwan
Additive Manufacturing Center for Mass Customization
Production, National Taipei University of Technology,
Taipei 10608, Taiwan

Moritz Koch, Sofía Doello and Karl Forchhammer
Interfaculty Institute of Microbiology and Infection
Medicine Tübingen, Eberhard-Karls-Universität
Tübingen, 72076 Tübingen, Germany

Kirstin Gutekunst
Department of Biology, Botanical Institute, Christian-
Albrechts-University, 24118 Kiel, Germany

**Liang Li, Danfeng Wang, Pengfei Yan and Shaoping
Sun**
Key Laboratory of Chemical Engineering Process &
Technology for High-efficiency Conversion, College of
Heilongjiang Province, School of Chemistry and Material
Science, Heilongjiang University, Harbin 150080, China

Na Liang
Key Laboratory of Photochemical Biomaterials and
Energy Storage Materials, Heilongjiang Province,
College of Chemistry & Chemical Engineering, Harbin
Normal University, Harbin 150025, China

Yoshiaki Kawashima
Department of Pharmaceutical Engineering, School of
Pharmacy, Aichi Gakuin University, Nagoya 464-8650,
Japan

Fude Cui
School of Pharmacy, Shenyang Pharmaceutical University,
Shenyang 110016, China

**Maurizia Seggiani, Norma Mallegni and Vito
Gigante**
Department of Civil and Industrial Engineering,
University of Pisa, Largo Lucio Lazzarino 2, 56122 Pisa,
Italy

Andrea Lazzeri and Patrizia Cinelli
Department of Civil and Industrial Engineering,
University of Pisa, Largo Lucio Lazzarino 2, 56122
Pisa, Italy
Interuniversity National Consortium of Materials
Science and Technology (INSTM), Via Giusti 9, 50121
Florence, Italy

**F. John Burpo, Chi K. Nguyen, Enoch A. Nagelli,
Madeline Y. Ryu, Jenny Wang, Kamil Woronowicz and
J. Kenneth Wickiser**
Department of Chemistry and Life Science, United
States Military Academy, West Point, NY 10996,
USA

Alexander N. Mitropoulos
Department of Chemistry and Life Science, United
States Military Academy, West Point, NY 10996, USA
Department of Mathematical Sciences, United States
Military Academy, West Point, NY 10996, USA

R. Kenneth Sims
Department of Civil and Mechanical Engineering,
United States Military Academy, West Point, NY
10996, USA

Index

Natural Fibers, 42-43, 75, 148-150, 152, 154, 161, 207-208, 211, 217, 219

Natural Rubber, 54, 118, 121, 127, 129-130, 132-135

Noble Metals, 221-222, 224-225, 227-229, 231

Non-toxicity, 33-34, 195

O
Organic Thin-film Transistors, 33, 40

P
Paclitaxel, 194, 199, 201, 204-206

Palladium, 131, 221-231

Plasticized Starch, 118-119, 122, 132

Polycarbonate, 89-90, 136, 161

Polymer Chains, 64-65, 79, 118, 123, 125, 164, 212

Porosity, 1-2, 4, 9, 13-17, 20-26, 31, 114, 221-222, 224-225, 228-229, 231

Porous Silk, 221-222, 231

Porous Structure, 2, 5-6, 26, 103

R
Reactive Extrusion, 54, 68

S
Silk Fibroin, 73-77, 79-80, 82-88, 162, 164, 177-179, 221-231

Silk Nanofibril, 222, 225

Stiffness, 42-43, 47, 49, 53-54, 74, 77, 82, 86, 119, 148-150, 152, 155-157, 159, 161, 163, 210, 213

Stress-strain, 44, 61-63, 73, 82-84, 87-88, 111, 140, 166, 214

Synechocystis, 180-182, 189, 191-193

Synthetic Biopolymers, 1, 7-8, 26

T
Tensile Strength, 15, 43-46, 50, 54, 62, 68, 111, 119-122, 127, 130, 140, 149, 154, 158, 217, 221

Thermal Decomposition Kinetics, 89-90, 99

Thermal Degradation, 86, 89-91, 94, 100, 102, 110, 118, 131, 135, 146, 150, 209

Thermal Resistance, 54, 64-65

Thermoplastic Polyurethane, 53-54, 67, 70-71, 136-137, 147

Tissue Engineering, 2, 4, 9, 14-17, 21-22, 25-31, 86, 162-163, 168, 175, 177-178, 204

Toughness, 53-54, 62-65, 68-70, 74, 86, 136-137, 140-141, 143-144, 146, 149, 164, 208

Tyrosine, 77, 162-165, 171, 173

U
Unzipping Reaction, 89-90, 95, 100

V
Viscosity, 2, 16, 55, 129, 209

W
Water Remediation, 103-104, 113, 115

Water Uptake, 15, 118, 128, 132

Printed in the USA
CPSIA information can be obtained
at www.ICGtesting.com
JSHW051404091023
49903JS00006B/273